SURVIVABILITY AND TRAFFIC GROOMING IN WDM OPTICAL NETWORKS

The advent of fiber optic transmission systems and wavelength-division multiplexing (WDM) have led to a dramatic increase in the usable bandwidth of single-fiber systems. This book provides detailed coverage of survivability (dealing with the risk of losing large volumes of traffic data due to a failure of a node or a single fiber span) and traffic grooming (managing the increased complexity of smaller user requests over high-capacity data pipes), both of which are key issues in modern optical networks.

A framework is developed to deal with these problems in wide area networks, where the topology used to service various high-bandwidth (but still small in relation to the capacity of the fiber) systems evolves toward making use of a general mesh. Effective solutions, exploiting complex optimization techniques and heuristic methods are presented to keep network problems tractable. Newer networking technologies and efficient design methodologies are also described.

This book is suitable for researchers in optical fiber networking and designers of survivable networks. It would also be ideal for a graduate course on optical networking.

ARUN K. SOMANI is the Jerry R. Junkins Endowed Chair Professor of Electrical and Computer Engineering at Iowa State University. His research interests are in the area of fault-tolerant computing, computer interconnection networks, WDM-based optical networking, and parallel computer system architecture. He has served as an IEEE distinguished visitor and an IEEE distinguished tutorial speaker. He has been elected a Fellow of the IEEE for his contributions to the theory and applications of computer networks.

T0171921

SURVIVABILITY AND TRAFFIC GROOMING IN WDM OPTICAL NETWORKS

ARUN K. SOMANI

CAMBRIDGE
UNIVERSITY PRESS

CAMBRIDGE UNIVERSITY PRESS
Cambridge, New York, Melbourne, Madrid, Cape Town,
Singapore, São Paulo, Delhi, Tokyo, Mexico City

Cambridge University Press
The Edinburgh Building, Cambridge CB2 8RU, UK

Published in the United States of America by Cambridge University Press, New York

www.cambridge.org
Information on this title: www.cambridge.org/9780521369961

First published 2006
First paperback edition 2011

A catalogue record for this publication is available from the British Library

ISBN 978-0-521-85388-0 Hardback
ISBN 978-0-521-36996-1 Paperback

This book is dedicated to my parents, who have been an inspiration towards my achievements in life, and to all my students who helped turn my dream into a reality and my family who supported me unconditionally.

Contents

Preface

Before the 1970s, networks were primarily used to carry voice or telephone calls over circuit-switched networks. Failures and service outages in such transport networks were handled mainly at the circuit layer, and many a time manually. Most of the remedial actions included routing of the call by manual configuration of switches by network operators. Over time the capacity of the transport networks increased, data overlay networks were created and a large number of end-users instituted private voice and packet networks.

With the advent of fiber optic transmission systems and eventually wavelength-division multiplexing (WDM) the bandwidth of a single fiber soared. With increasing deployment of fibers in networks, the risk of losing large volumes of traffic due to a span cut or a node failure has also increased tremendously. In the 1990s Bellcore developed the SONET (synchronous optical network) standard and standardized the concept of self-healing rings. It was soon followed by the equivalent standard named SDH (synchronous digital hierarchy) in Europe. This appeared to be the final solution. Many service providers could replace all of their cumbersome and expensive point-to-point transmission systems with a few multi-node, self-healing SONET rings. Many carriers joined the SONET ring bandwagon.

With further developments in technology, more and more mesh-network topologies started emerging. Failure management still remained a part of the solution and recovering from them remained a challenging issue. It soon started to fuel the everlasting question that still prevails to this day: which is a better option, ring-based or mesh-based restoration? Over the years, as the traffic increased a mesh-based approach seemed to be a more viable option for providing restoration, compared to a traditional ring network. There are different tradeoffs between ring and mesh restoration. In a ring network, once the network is installed the restoration is automatic. In contrast, in a mesh network, most of the control is in the hands of a centralized control system, and hence it makes the control very complicated. This motivated the need for intelligent network elements which are capable of distributed routing, detecting failures of the links and passing topology update information amongst adjoining nodes using variations of link state protocols.

Another important element of the architecture is when and how much to do to recover at the time of establishing a connection. The range of options varies from identifying and provisioning the resources for recovery to take care of failure at the time of establishing the connection, to identifying all resources to recover when a failure occurs. The former is called protection and the latter is an extreme in restoration. In between options include identifying an alternate path only for a connection being established, identifying an alternate path and resources to be used on the path but not activate the path, or even activating the resources but not using them. Each option presents a different cost and impact on functionality of the network.

In present-day networks, SONET rings are more prevalent in metropolitan networks, where there is less geographic diversity, whereas mesh-based networks are more common in ultra-long-haul networks covering areas of vast geographical variations. Also, as optical cross-connects and WDM switching technologies mature, mesh-based restoration for pure optically switched networks are of increased interest because of the reduced costs of optical–electrical conversion and the economics of scale for integration of WDM and electronics.

Another important debate that has not been settled is which layer in networking should be responsible for protection and/or restoration. At the present time the internet protocol (IP) is the dominant mode of internetworking. The IP layer has its own protection and restoration strategy. For each destination, a source may have many possible paths and may choose to have a preferred path for routing packets to that destination. At the optical or physical network level, a network designer may adopt a strategy to recover from a complete fiber failure. On the other hand, many IP routes may be routed through the same fiber as a fiber does support many channels, each being used for a different IP route. Thus, a single fiber failure may result in multiple IP route failures. Moreover, optical networks use the concept of a virtual topology, in which most commonly used routes may be created using a concatenation of channels on multiple fibers. The IP layer may not even be aware of how the physical network has been utilized to create a virtual topology. Hence IP routing and restoration have some pitfalls. It is not clear which strategy or a combination of the two is better. It is believed by many researchers that protection/restoration at the WDM layer is more advantageous, but this is disputed by those who prefer IP protection/restoration.

This brings us to one of the two most important contributions of this book. We attempt to provide a brief overview of different optical networking trends and technologies followed by different network design and restoration architectures in mesh-restorable optical networks. Several chapters are devoted to studying various protection and restoration architectures, methods to model the problem, and algorithms to design functionality and operational aspects, and to study the performance of various schemes.

Another important problem in optical fiber networks with wavelength-division multiplexing is that of traffic grooming. The capacity of a single wavelength channel has been increasing constantly, reaching the level of 10–40 GHz/channel. At the same time individual user requirements are not increasing at the same pace, although the overall number of users and applications are increasingly dramatically. Thus, it is important to accommodate all the applications while utilizing the resources efficiently. The traffic grooming problem presents a whole new set of challenging problems.

The second half of the book focuses on traffic grooming. Traffic grooming is a technique for multiplexing different subwavelength capacity traffic requirements onto a single wavelength so that the wavelength and hence the capacity requirements of the whole network are minimized.

This book is written with two main communities in mind.

One of them consists of my colleagues in industry, research scientists, technology planners, network designers, and also corporate research laboratories. They are incessantly striving to access the economics of different architectural decisions and standards devolvement. Network operators are in a fiercely competitive market, striving for more and more productivity. Mesh networking studied in this book provides them with great productivity enhancements through greater network efficiencies and flexibility. Developers of network modeling, simulation, and network planning would be interested in many of the ideas presented in this book. Incorporating capabilities to design all types of architecture alternatives, and accessing the merits and demerits of each chosen alternative, the book fuels the interest of different network researchers.

The second main community who would benefit from this book are graduate-level teachers and researchers who want a self-contained volume to derive insights into aspects of transport network design and also to use this to teach a graduate-level course on optical networking. I expect both communities to benefit equally. Any designer of a new graduate course will hopefully continuously upgrade the material with more advanced developments as the technology improves.

In the following, the flow of the contents in the book is explained.

Chapters 1 and 2 serve as introductory chapters to networking using fiber optics technology and the rest of the book. Since the book specializes on two most important topics, understanding the rest of the concept is important. For that reason, these chapters serve as an introduction to wavelength-division multiplexing, broadcast-and-select-network designs, and different optical networking trends and technology. They also introduce several interesting aspects and issues in the design of such networks, which include the routing and wavelength assignment problems, optical packet switching and optical burst switching, traffic grooming and survivability in mesh restorable optical networks, and survivable traffic grooming in optical networks.

The book then moves on to discuss different restoration approaches and also the upgradeable network design problem. The whole concept is treated within a survivability framework. Mesh restoration architectures have natural ties into the network design problem. It can be viewed as a network design and network operation problem. The problem can be formulated using different scenarios and that subject is dealt with in the next chapter. Different formulations and heuristic approaches to solving the network design problems are discussed in Chapter 3. Chapter 4 deals with an alternate approach, called the p-cycle or protection cycle. This strategy allows the use of similar protection algorithms to those designed for ring networks in mesh-like architectures.

Chapter 5 concentrates on network operation. There are two important goals in network operation. The network can be optimized to use minimal capacity out of that available to serve the offered demand, assuming that the demands are not greater than the capacity. The goal here is to keep as much capacity free as possible to accommodate future requests. Alternately, the demands offered may be more than the available capacity. In that case, the goal is to optimize the operation and accept those requests that would maximize the service providers' gain. Depending on the gain matrix chosen, the optimization can be tuned to serve a specific aspect. The details of such operational optimizations are discussed in the form of the capacity minimization and the revenue maximization subproblems.

The optimization problems formulated in Chapter 5 are complex and difficult to solve in real time. We continue to discuss different relaxation techniques that can be applied to solve the integer linear programming (ILP) problems in Chapter 6. The goal here is to adopt those techniques that will result in a near-optimal solution while minimizing the required computation time. Some insight into the formulation is developed and used to derive near-optimal or optimal solutions while keeping the commutation time under control for near real-time and real-time on-line applications.

In Chapter 7, we discuss how to tolerate multiple link failures in a network. The problem is reviewed in detail and some solutions are studied. We also present an ILP approach to solve the dual-link failure problem in a WDM network. Several other techniques for solving the multi-link failure protection and restoration problem are also presented.

In the following chapter, Chapter 8, we present another approach called the subgraph-based routing strategy in mesh-restorable WDM optical networks. In this scheme no resources are reserved for protection. However, provision is made to make sure that all resources are available when an actual failure indeed occurs. It is demonstrated how subgraph routing can be used to protect a network against multiple-link and node group failures.

Chapter 9 introduces the concept of traffic grooming in WDM optical networks. Both static and dynamic traffic grooming concepts are presented. The chapter also discusses techniques for static traffic grooming in rings and presents the advantages of and issues in traffic grooming in WDM rings.

Chapter 10 presents a model to study the advantages of traffic grooming and quantifies the gains of traffic grooming. An analytical model of a WDM grooming network is presented for grooming on a single wavelength on a single- and a multi-hop path. The model also discusses the type of network where all or some of the nodes in the network may be capable of grooming.

One of the important issues in any traffic grooming is that of fairness. Since users may request different capacities, it is important that all requests are handled fairly based on chosen criteria. Although fairness means different things to different people, we present a model of fairness in Chapter 11 that we use in this work and evaluate the fairness of various routing and wavelength assignment algorithms used in WDM grooming networks.

Chapters 12 and 13 bring the two topics, survivability and traffic grooming, together. In Chapter 12, we discuss survivable grooming network design. Both routing and wavelength selection issues for survivable traffic grooming are presented and solutions are developed to utilize resources effectively in such networks. Chapter 13 deals with the design of networks that support static traffic grooming with survivability.

In the second part of the second topic, a new framework for dealing with traffic grooming is presented. This framework, called a trunk-switched network, presents a methodology to represent and analyze traffic grooming. The framework is a powerful one, which can support modeling of various traffic grooming mechanisms and analyze them. Chapter 14 presents the framework and modeling methodology for a WDM traffic grooming network. The next chapter, Chapter 15, is devoted to use of the network for performance analysis. Chapter 16 is devoted to validating the model and includes several examples to demonstrate the applications of trunk-switched WDM grooming networks. Both sparse and dense networks (defined based on the number of nodes that support traffic grooming) as well as homogeneous and heterogeneous networks (defined based on the type of grooming nodes being identical or different) are managed.

Chapter 17 is devoted to traffic routing and wavelength assignment algorithms in a traffic grooming network. Several algorithms are presented and analyzed.

Chapter 18 presents traffic grooming in an IP-over-WDM network. This is an important area as both IP and WDM have to eventually work together. Algorithms to route IP traffic efficiently over a WDM network are presented and analyzed.

The final chapter, Chapter 19, presents an innovative technique called the light trail architecture, which is used for traffic grooming in WDM optical networks for local and metropolitan areas and has the potential to integrate with wide area networking. It also discusses how restoration can be achieved in the light trail architecture and presents an ILP formulation for the survivable light trail design problem.

Additional material in a few appendices would help researchers from various communities to develop interest in the topic of survivability and traffic grooming in optical networks. The goal is to create a highly useful and interesting book that is imbued with new options and insights for industry and academia to enjoy and benefit from.

Acknowledgments

A work of this magnitude is a result of encouragement from and the efforts of many people.

First, there has to be inspiration. My father, a visionary, had provided me with a gift that I only appreciated after he passed away. Right from the beginning, he kept me focused on achieving the best and continuing to strive until a goal is achieved. He pushed me into situations where the chances of failures were high and encouraged me to succeed. Part of the reason for him probably was that he had to give up his studies sooner than he wanted or needed to. He fulfilled that dream of his through his children and felt his success through us, his children. I have been a tutor right from my third/fourth grade. Not only encouraging us in academics, he also continuously strived to make us good citizens and to help others.

Whenever he felt weak, such as when I had to leave home for college, my mother took charge. The encouragement from the two of them has resulted in what I am today. I am indebted to them.

And then my students, who were always full partners in the work I produced. I was blessed with excellent graduate students who always challenged me and kept me on my toes. Although I had many students working with me in many different areas, this work is mainly representative of the work I performed with Suresh Subramaniam, Ling Li, Sashisekhran Thiagarajan, Srinivasan Ramasubramanian, Murari Sridharan, Tao Wu, Jing Fang, Pallab Datta, Mike Fredericks, Nathan Vanderhorn, Wensheng He, Srivatsan Balasubramaniam, Nitin Jose, and Yana Ong, who have contributed to this work in many ways. The contribution made by each of them is great. They deserve a rich appreciation.

Among my colleagues, I was blessed to have worked with many good people. James A. Meditch initially inspired me to start looking at problems in this area. Murat Azizoglu was a great colleague to work with. Later on Ahmed Kamal, Robert Weber, and Mani Mina have been great allies. I never hesitated in asking them for any help, academic or otherwise, which I always received. Thanks go to all of them.

Finally, my wife, Manju, who never questions me when I return from the office and always tolerates any amount of my work frustration. Religiously she packs my lunch every morning at 7:30 a.m. and reminds me every evening, including most weekends, that it is 7:00 p.m. and I might be hungry. She always took care of the kids and provided the support they needed from both of us. My kids love me unconditionally, irrespective of the amount of time I found for them.

My special thanks to Pallab who has really assisted me in organizing the material in the book into its current shape. Without him, I have no idea when I would have been able to finish it.

It has been a real blessing to be surrounded by friends, colleagues, students, and family as I have. I am indeed thankful to all of them.

1

Optical networking technology

Technological advances in semiconductor products have essentially been the primary driver for the growth of networking that led to improvements and simplification in the long-distance communication infrastructure in the twentieth century. Two major networks of networks, the public switched telephone network (PSTN) and the Internet and Internet II, exist today. The PSTN, a low-delay, fixed-bandwidth network of networks based on the circuit switching principle, provides a very high quality of service (QoS) for large-scale, advanced voice services. The Internet provides very flexible data services such as e-mail and access to the World Wide Web. Packet-switched internet protocol (IP) networks are replacing the electronic-switched, connection-oriented networks of the past century. For example, the Internet is primarily based on packet switching. It is a variable-delay, variable-bandwidth network that provides no guarantee on the quality of service in its initial phase. However, the Internet traffic volume has grown considerably over the last decade. Data traffic now exceeds voice traffic. Various methods have evolved to provide high levels of QoS on packet networks – particularly for voice and other real-time services. Further advances in the area of telecommunications over the last half a century have enabled the communication networks to see the *light*. Over the 1980s and 1990s, research into optical fibers and their applications in networking revolutionized the communications industry. Current telecommunication transmission lines employ light signals to carry data over guided channels, called *optical fibers*. The transmission of signals that travel at the speed of light is not new and has been in existence in the form of radio broadcasts for several decades. However, such a transmission technology over a guided medium, unlike air, with very low attenuation and bit-error rates makes optical fibers a natural choice for the medium of communication for next-generation high-speed networks. The first major change with the development of the fiber technology was to replace copper wires by fibers. This change brought high reliability in data transmission, improved the signal-to-noise ratio and reduced bit-error rates.

1

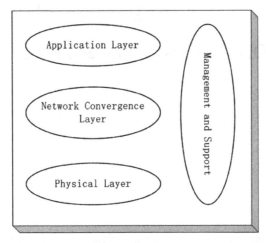

Fig. 1.1. A next-generation network architecture.

With the advent of optical transmission technology over optical fibers, the communication networks have attained an orders of magnitude increase in the network capacity. This development had a tremendous and dramatic impact on social and economic aspects of the lives of people around the world [1, 3, 4, 5]. With increasing data traffic and high QoS requirements on packet networks, it is becoming desirable to bring together various different networks around a single packet-based core network. The *next-generation network (NGI) architectures* are converging to share a common high-level architecture as shown in Fig. 1.1. It would integrate multiple different networks, e.g., PSTN, IP, ATM (asynchronous transmission mode), and SONET (synchronous optical networks)/SDH (synchronous digital hierarchy), etc. based networks into a single framework.

Initially, the migration from electronic to optical transmission technology was achieved by only replacing copper cables with optical fibers. Traditional time-division multiplexing (TDM) that allows multiple users to share the bandwidth of a link was employed. In TDM, the bandwidth sharing is in the time domain. Multiplexing techniques specific to optical transmission technology were not employed in the early networks. The synchronous optical network is the most popular network in this category. SONET is based on a ring-architecture, employing circuit-switched connections to carry voice and data traffic. The second-generation optical network uses wavelength-division multiplexing (WDM) which is similar to frequency-division multiplexing (FDM).

1.1 Wavelength-division multiplexing

The optical transport layer is capable of delivering multi-gigabit bandwidth with high reliability to the service platforms. The bandwidth available on a fiber is

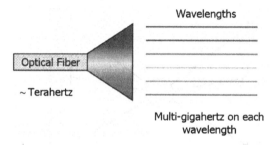

Fig. 1.2. A fiber divided among multiple wavelengths.

approximately 50 THz (terahertz). Increasing the transmission rates could not be adopted as the only means of increasing the network capacity. Transmission rates beyond a few tens of gigabits per second could not be sustained for longer distances for reasons of impairments due to amplifiers, dispersion, non-linear effects of fiber, and cross-talk. Hence, wavelength-division multiplexing was introduced, which divided the available fiber bandwidth into multiple smaller bandwidth units called *wavelengths*.

The WDM-based networking concept was derived from a vision of accessing a larger fraction of the approximately 50 THz theoretical information bandwidth of a single-mode fiber. A natural approach to utilizing the fiber bandwidth efficiently is to partition the usable bandwidth into non-overlapping wavelength bands. Each wavelength, operating at several gigabits per second, is used at the electronic speed of the end-users. The end-stations thus can communicate using wavelength-level network interfaces. Wavelength-division multiplexing turns out to be the most promising candidate for improving fiber bandwidth utilization in future optical networks. Figure 1.2 depicts the WDM view of a fiber link. The research, development, and deployment of the WDM technology evolved at a rapid pace to fulfill the increasing bandwidth requirement and deploy new network services.

The wavelength-division multiplexing mechanism divides the bandwidth *space* into smaller portions. Hence, the multiplexing is said to occur in the space domain. Different connections, each between a single source–destination pair, can share the available bandwidth on a link using different wavelength channels. Advanced features such as optical channel routing and switching supports flexible, scalable, and reliable transport of a wide variety of client signals at ultra-high speed. This next-generation network concept dramatically increases, and maximally shares, the backbone network infrastructure capacity and provides sophisticated service differentiation for emerging data applications. Transport networking enables the service layer to operate more effectively, freeing it from the constraints of physical topology to focus on the sufficiently large challenge of meeting the service requirements.

The application signals have widely varying characteristics, e.g., signal format, type of signal, and transmission speed. To transport the varied application signals

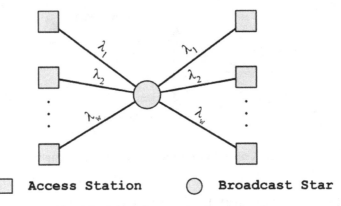

Fig. 1.3. A broadcast-and-select network.

over the optical transport network, a network service layer is needed to map the signals to optical channel signals along with an associated "overhead" to ensure proper networking functions. This layer, for example, captures today's IP and ATM capabilities with statistical multiplexing and a QoS guarantee. Protocols such as multi-protocol label switching (MPLS), resource reservation protocol (RSVP), and differentiated services (DiffServ) play a major role in supporting the required QoS across a wide set of applications. The network service layer relies entirely on the transport layer for the delivery of multi-gigabit bandwidth where and when it is needed to connect to its peers.

1.2 Broadcast-and-select networks

Early optical networks employed broadcast-and-select technology. In such networks, each node that needs to transmit data broadcasts it using a single wavelength and the receiving node selects the information it wants to receive by tuning its receiver to that wavelength. In a WDM network, many nodes may transmit simultaneously, each using a different wavelength on the same fiber or passing through the same node. A broadcast star architecture as shown in Fig. 1.3 is an example of such a network.

Nodes in the broadcast-and-select networks are connected by links and optical couplers. An optical coupler is a *passive* component that is used for either combining or splitting the signals into or from a fiber. The couplers can be configured to split/combine signals in specific ratios so as to achieve a proper energy mix/split. The central coupler in Fig. 1.3 couples all of the energy transmitted by all nodes and splits the combined energy equally among all receivers.

It should be evident that the data transmitted by a node is received by all other nodes. Every node only uses the data that is destined for it and discards the rest.

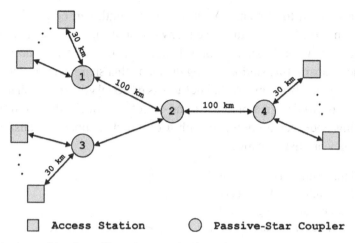

Fig. 1.4. A combination of broadcast-and-select elements used to form a network.

The major features of optical couplers include low insertion loss, excellent environmental stability, long-term reliability, and multiple performance levels. Such networks are typically found in local area networks and cable-TV or video distribution networks, networks that typically provide the end-user connectivity. Different receiving nodes can tune their receiver to different wavelength channels and receive the information pertinent to it from the appropriate source node.

Several broadcast star nodes can be used to design a broadcast-star network. A bigger network that employs several broadcast stars is shown in Fig. 1.4. In this network, the coupling nodes, marked as nodes 1 to 4, couple the energy received from all incoming fibers and redistribute it to all outgoing fibers. Only a single pair of two end nodes that wish to communicate can use one wavelength in a one-to-one communication setup. Thus if there are W wavelengths in the system, only W access node-pairs can communicate (W transmitter and W receivers) irrespective of the size of the network. Moreover, a centralized arbiter must control the wavelength usage. The last two items are the reason why such a network cannot be used in wide area networks or bigger network applications. In the case of a multicasting application, data transmitted from one node can be received by more than one node without any extra effort.

1.2.1 Broadcast-and-select network design

In order for a source node to communicate with a receiver, both nodes must be operating at the same wavelength. One way to organize nodes in the broadcast-and-select networks is to have each node transmit data on a specific wavelength. A network with N nodes would employ W wavelengths where every node is assigned a single wavelength to transmit. A unique wavelength can be assigned to every node to transmit if $N \leq W$. Otherwise many transmitting nodes will have to share a wavelength. If $W < N$ then a control mechanism would also be required to decide which transmitting station among the many that share the wavelength would use the channel at a given time. A channel access protocol governs the sharing. There are several ways to design a broadcast-and-select network with each transmitting and receiving node being equipped with a different number of transmitters and receivers. They are listed below.

(i) A fixed transmitter and single tunable receiver.
(ii) A fixed transmitter and W receivers.
(iii) W transmitters and a single fixed receiver.
(iv) $1 < k_1 < W$ tunable transmitters and $1 < k_2 < W$ tunable receivers.

Different schemes will require a different mechanism to control the network. For example, with a single transmitter and W receivers at a node, a node can transmit at any time as long as the wavelength is free and the central controller only assigns the transmission time to a node so that there is no collision. This is called a fixed transmitter scheme. A similar control is exercised in the third case, where the receive time needs to be allocated for a node along with the condition that the wavelength used by the receiver must be free for that duration. In the last case control is more complicated and various control mechanisms can be derived. Several schemes to control such a system have been developed in the literature.

Each receiver node may have one or more receivers that can be tuned to the wavelength of the transmitting station that is the current source of data to that particular node. This is called a tunable receiver scheme. In this environment, a control channel is used to establish a connection between source and destination nodes to allow the latter to tune their receivers to the transmitting wavelength of the source nodes. Quite a bit of time is lost in deciding who should transmit to whom and in tuning the receivers. The control channel may also be time slotted where data and control are separated. In a control round, each transmitting station requests the possible desired destination for which it has some data. The destination nodes arbitrate and decide on their respective current sources and correspondingly tune their receivers to the wavelengths of the transmitting nodes. The actual data is then transmitted. On the other hand, control could be asynchronous in which a transmitting station sends a request to the destination node and upon receiving an acknowledgement, transmits the data. An acknowledgement is sent in such a way that a tuning period is allowed at the destination node. The actual transmission only occurs after the tuning is complete.

In the case where the receiving node has W receivers, receiver arbitration is not required. Transmitter arbitration is still required irrespective of whether $N > W$

or $N < W$, to decide which node should transmit at a given time to a particular receiver since more than one node may have data to transmit.

In an alternate scenario, each receiver may have a fixed-wavelength receiver. This is called a fixed receiver scheme. In this case, a transmitting node has to tune to the wavelength of the receiver before transmitting. This is called a tunable transmitter scheme.

In the case where the sender node has W transmitters, the sender can transmit on the receiver's wavelength without arbitration. However, a control mechanism is required to decide which node should transmit to the destination at a given time. Again the control mechanism can be synchronous or asynchronous.

In another scenario the tuning of both transmitters and receivers to match the transmitter–receiver pairs using a control mechanism may be deployed.

The disadvantage of such a passive network is that its range is limited. Long-range networks, typically countrywide networks, cannot employ such a broadcast-and-select mechanism due to capacity inefficiency. The data is unnecessarily sent to all the nodes in the network, resulting in poor network utilization. Also, as the signals travel farther the quality of the signal degrades necessitating signal regeneration with reshaping and retiming.

1.3 Wavelength-routed WDM networks

In order to avoid unnecessary transmission of signals to nodes that do not require them, *wavelength routing* mechanisms were developed and deployed. A wide variety of optical components to build WDM networks were developed that included wide-band optical amplifiers (OAs), optical add/drop multiplexers (OADMs), and optical cross-connects (OXCs). Thus it became possible to route data to their respective destinations based on their wavelengths. All-optical networks employing wavelength-division multiplexing and wavelength routing are now viable solutions for wide area networks (WANs) and metropolitan area networks (MANs).

The use of wavelength to route data is referred to as wavelength routing, and networks that employ this technique are known as *wavelength-routed* networks [234, 236, 240]. In such networks, each connection between a pair of nodes is assigned a path and a unique wavelength through the network. A connection from one node to another node, established on a particular wavelength, is referred to as a *lightpath*. Connections with paths that share a common link in the network are assigned different wavelengths. The two end nodes may use any protocol and signal type such as analog or digital to modulate the optical signal. These wavelength-routed WDM networks thus offer the advantages of protocol transparency and simplified management and processing in comparison to routing in telecommunications systems using digital cross-connects.

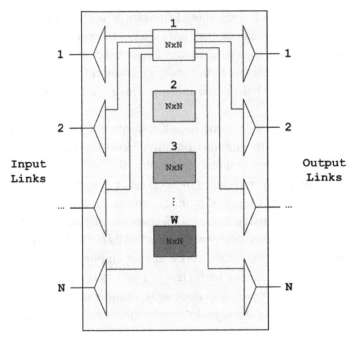

Fig. 1.5. Node architecture in a wavelength-routed WDM network without wave-length conversion.

In wavelength-routed networks the nodes employ *optical cross-connects* that can switch an individual wavelength from one link to another. In order to operate the network in a transparent manner, the switching of a wavelength is done in the optical domain. Figure 1.5 depicts an optical switching node. In this architecture, wavelengths on an incoming fiber are demultiplexed and separated. Then the same wavelengths from all fibers are switched together and routed to the outgoing fibers using a wavelength switching mechanism. There is a separate switch for each wavelength. At the output, all wavelengths being routed to one fiber are multiplexed and then sent out to the outgoing fiber.

A wavelength-routed WDM network is shown in Fig. 1.6. The figure shows connections established between nodes A and C, H to G, B to F, and D and E. The connections from nodes A to C and B to F share a link. Hence, they have to use different wavelengths on the fiber.

In a wavelength-routed WDM network, the path of a signal is determined by the location of the signal transmitter, the wavelength on which it is transmitted, and the state of the network devices. An example of such a network with two wavelengths on each link is shown in Fig. 1.7. There are two connections that are in progress, one from node 1 to node 2 using wavelength λ_1, and another from

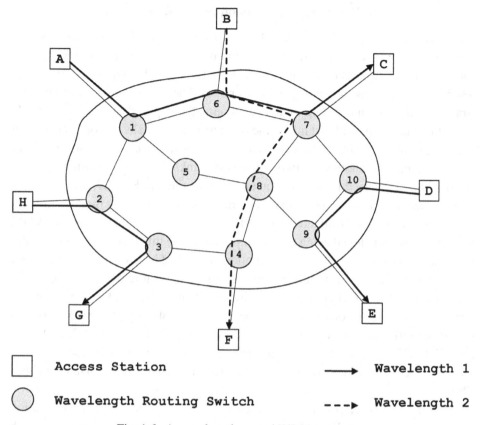

Fig. 1.6. A wavelength-routed WDM network.

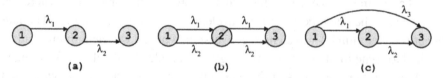

Fig. 1.7. A demonstration for the wavelength continuity constraint on a two-hop path.

node 2 to node 3 using wavelength λ_2. A connection request from node 1 to node 3 is blocked, although free wavelengths are available on both link 1 and link 2. This is because of the *wavelength continuity constraint*, that is, the same wavelength must be assigned to a connection on every link. Otherwise a wavelength converter is required at the switching node 2. Connection requests to set up lightpaths encounter a higher blocking probability than path setup requests do in electronic-switched networks because of the wavelength continuity constraint.

1.4 Wavelength conversion in WDM networks

As noted above in optical networks, the *wavelength continuity constraint* restricts a connection to occupy the same wavelength on every link of a chosen path from source to destination. The transmission of signals on the fiber has to follow certain restrictions due to the technological constraints. The wavelength continuity constraint could result in rejecting a call even though the required capacity is available on all the links of the path but not on the same wavelength. The reason for rejecting a request is due to the inability of intermediate nodes to switch the connection from one wavelength to another on two consecutive links. The wavelength continuity constraint and the need for conversion are demonstrated in Fig. 1.7.

The effect of wavelength conversion is analyzed using different models. Initially analytical models to evaluate the performance of wavelength-routed WDM networks were developed using the assumption that the link loads were statistically independent [10, 179, 180]. This assumption is justified based on the overall effect of routing various connections and the intermixing of traffic over different links. Thus the traffic on a link appeared to be independent of other links in the network. Analytical models for quantifying the benefits of employing wavelength conversion capabilities in a network were later developed [87, 164, 179, 197, 219, 267, 269, 270, 272, 279], continuing with the same assumption of statistical link load independence. The concept of link load correlation was put forward in later analysis [268].

Wavelength converters are also expensive devices. Thus another expectation in network design has been that either not every node can perform wavelength conversion or a node cannot convert any wavelength arbitrarily to any other wavelength. The former concept is named sparse-wavelength conversion, where only a few nodes in the network have full-wavelength conversion capability. The latter concept is called limited-wavelength conversion, where a wavelength may only be translated to some limited set of wavelengths at a node. A model for analyzing the network performance in terms of the blocking probability with sparse-wavelength conversion with link load correlation has been developed in [268]. Models for limited-range wavelength conversion can be found in [144, 235, 287, 294].

An alternative for wavelength conversion is a multiple-fiber network where each link consist of multiple fibers, say F. Thus every wavelength is available on every link F times. Analytical models for multi-fiber networks were developed by extending the models for a single-fiber wavelength-routed network [196].

Most of the analytical models assume fixed-path routing, i.e. the path that is chosen for establishing a connection from the source to the destination is known a priori. Analytical models that account for dynamic routing based on up-to-date network status are complex, and hence have received very little attention [24].

All-optical wavelength converters are being prototyped in research laboratories [153, 166, 235, 254]. However, the techniques have not matured yet. All-optical wavelength conversion retains the signal in the optical domain and is transparent to the clock-rate and frame format. However, these devices are prohibitively expensive to deploy widely in the networks. These benefits depend on the topology of the network, the traffic demand, the number of available wavelengths, and the routing and wavelength assignment algorithms [27, 206], among other factors. As the network becomes denser, one would expect the usefulness of converters to decrease, since the paths get shorter. In the limiting case with a link between every node-pair, wavelength converters have no effect on the blocking performance, since all connections are one-hop connections. This assumes that the direct link is always used in this case. If alternate-path routing is allowed, wavelength converters may still be of some benefit. On the other hand, a sparsely connected network tends not to mix the traffic well and thus causes a load correlation in successive links. This reduces the usefulness of wavelength converters [265].

1.4.1 Conversion technology

A WDM network without wavelength conversion is referred to as a *wavelength selective* (WS) network. As noted earlier, the performance of a network can be improved by using wavelength converters at the switching nodes. Networks with wavelength conversion are called *wavelength interchanging* (WI) networks [282].

Wavelength conversion mechanisms can be classified based on the range of wavelength conversion. *Fixed-wavelength conversion* allows the signal to be converted from a specific input wavelength to a fixed output wavelength. The choice of output wavelength is fixed for an input wavelength, hence the name. If the signal on a wavelength can be converted into any other wavelength, it is referred to as *full-wavelength conversion*. If the signal can be converted from one wavelength to a set of, but not all, wavelengths, it is referred to as *limited-wavelength conversion*. Figure 1.8 shows the different types of wavelength conversions at a node with four wavelengths denoted by $W1$ through $W4$.

The first solution for wavelength conversion is opto-electronic wavelength conversion, in which the optical signal is converted into the electronic domain first. The electronic signal is then used to drive the input of a tunable laser tuned to the desired wavelength of the output. Since this technique is not transparent to data bit rate and data format, which is one of the major advantages of using optical networking, opto-electronic wavelength conversion is not preferred for use in future networks. All-optical wavelength conversion, in which no opto-electronic conversion is involved, can be divided into the following categories.

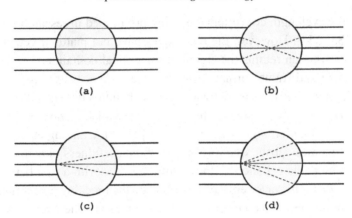

Fig. 1.8. Types of wavelength conversion: (a) no wavelength conversion; (b) fixed-wavelength conversion; (c) limited-wavelength conversion; (d) full-wavelength conversion.

- Wavelength conversion using wave mixing, including *four-wave mixing* [254] and *difference frequency generation (DFG)* [8].
- Wavelength conversion using cross-modulation, including *semiconductor optical amplifiers (SOAs) in XGM and XPM mode* [125, 282], and semiconductor lasers [196].

1.5 Optical packet switching

Thus far, the discussion has centered on circuit-switched lightpaths in WDM networks. One alternative to circuit switching is to use optical packet switching (OPS) [43, 181, 214, 314] technology in the backbone. The major advantages of OPS are the flexible and efficient bandwidth usage, which enables the support of diverse services. Pure OPS technology in which packet header recognition and control are performed in the all-optical domain is still many years away, and may not become a reality. OPS technology with electronic header processing and control is more realistic for medium-term network scenarios. A practical OPS experiment has been performed under the European ACT KEOPS (KEys to Optical Packet Switching) project [43]. In KEOPS, the header is sent with data (payload) but at a lower bit rate, and the header processing is still in the electrical domain. This potentially requires optical buffering at the input port to allow the header processing circuits to finish the job. At present, the buffering technology is not mature and has to overcome a number of technological constraints, such as the large and varying size of optical buffering. Header processing at high speeds is also an important issue.

1.6 Optical burst switching

Another viable alternative switching technology to transporting IP traffic directly over WDM networks is optical burst switching (OBS) [18, 49, 145, 281]. In a wavelength-switched network, once a lightpath is established, it remains in place for a relatively long time, perhaps months or even years. In OBS, the goal is to set up a wavelength channel for each single burst to be transmitted. OBS is an adaptation of an International Telecommunication Union–Telecommunication Standardization Sector (ITU-T) standard for burst switching in ATM networks, known as *ATM block transfer* (ABT) [162]. A burst that carries one or more IP packets is dynamically assigned to a wavelength channel upon its arrival. To establish a connection, an associated control packet is transmitted over a wavelength channel or a non-optical channel before the burst is transmitted. In the *tell-and-wait* (TAW) scheme [116], a burst is buffered while the control packet is being sent to set up switches and reserve bandwidth for establishing a connection. While in the *tell-and-go* (TAG) scheme [78, 116], a burst is sent immediately after its control packet without receiving a confirmation. If a switch along the path cannot carry the burst due to congestion, the burst is dropped. In this scheme, it may still be necessary to buffer the burst in the optical burst switch until its control packet has been processed [162]. Other schemes known as *just-enough-time* (JET) [49] and *just-in-time* (JIT) [141] have also been developed. An OBS architecture is described in [145]. Currently, no commercial OBS networks exist.

1.7 The rest of the book

With a brief description of development and the main concepts of WDM technology, the rest of the book concentrates on a few dominant issues. In Chapter 2, design issues are discussed and following that the book focuses on survivability and traffic grooming, in particular subwavelength level traffic grooming issues. Towards the end, new architectural concepts that are being developed will be discussed.

2

Design issues

Optical technology involves research into components, such as couplers, amplifiers, switches, etc., that form the building blocks of the networks. Some of the main components used in optical networking are described in Appendix A1. With the help of these components, one designs a network and operates it. Issues in network design include minimizing the total network cost, the ability of the network to tolerate failures, the scalability of the network to meet future demands based on projected traffic volumes, etc. The operational part of the network involves monitoring the network for proper functionality, routing traffic, handling dynamic traffic in the network, reconfiguring the network in the case of failure, etc. In this chapter, these issues are introduced in brief, followed by a discussion of the two main issues in network operation, namely survivability and how traffic grooming relates to managing smaller traffic streams.

2.1 Network design

Network design involves assigning sufficient resources in the network to meet the projected traffic demand. Typically, network design problems consider a static traffic matrix and aim to design a network that is optimized based on certain performance metrics. Network design problems employing a static traffic matrix are typically formulated as optimization problems. If the traffic pattern in the network is dynamic, i.e. the specific traffic is not known a priori, the design problem involves assigning resources based on certain projected traffic distributions. In the case of dynamic traffic the network designer attempts to quantify certain network performance metrics based on the distribution of the traffic. The most commonly used metric in evaluating a network under dynamic traffic patterns is the *blocking probability*. This is computed as the ratio of the number of requests that cannot be assigned a connection to the total number of requests. With this metric, one makes decisions on the amount of resources that need to be deployed in a network, the operational policies such as routing and wavelength assignment algorithms, and call acceptance criteria, etc.

To formulate a network design problem as an optimization problem, the inputs to the problem are the static traffic demand and some specific requirements, e.g., the required network reliability and fault tolerance requirements, the network performance in terms of blocking, the restoration time after a failure occurs, etc. The objective of the optimization problem is to find a topology that minimizes the resources, including the number of links and fibers, the number of wavelengths on each fiber, and the number of cross-connect ports, to meet the given requirements. The outputs include the network configuration, and the routes and wavelengths that are to be used for source–destination pairs. The network design problem can be formulated as an integer linear programming (ILP) or a mixed integer linear programming (MILP) problem. Since the number of variables and constraints can be very large in WDM networks, heuristics are usually used to find solutions faster.

In some design problems, even the network topology may be given. The design problem simply identifies resources to be deployed at (i) each node, in terms of the type of switch and the number of switching elements, cross-connects, etc. and (ii) each link in terms of the number of fibers in each link, the number of wavelengths in each fiber, etc.

Simulations can be used to measure the blocking performance of networks. However, simulation takes a long time to achieve meaningful results. In order to expedite this process, analytical models are also employed that serve as a coarse tool for evaluating network performance [30, 34, 88, 264]. The analytical models are employed in the network design phase to serve as an *elimination criterion*, where network designs with lower projected performance are rejected. Therefore, analytical models form a critical component of the network design phase.

After a network has been built, one critical problem is how to operate it such that its performance is optimized under dynamic traffic conditions. The intensity of the dynamic traffic is usually known while the individual demands arrive and depart randomly. Since network resources are typically not sufficient to guarantee that every dynamic demand can be accommodated in the network, the average blocking probability for a given utilization is one of the metrics used to measure network performance. Some other metrics include the control overhead and the algorithm complexity.

2.2 Network model

The optical layer model (shown in Fig. 2.1) consists of nodes interconnected by links that accommodate one or more fibers. In a single-fiber model, each fiber

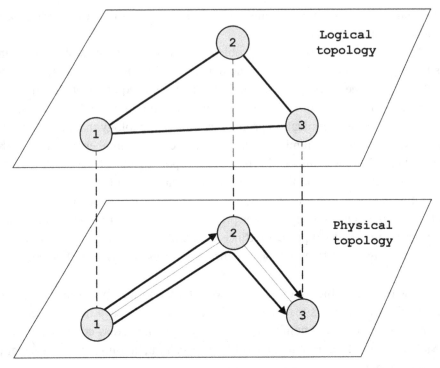

Fig. 2.1. Optical layer model.

carries multiple wavelengths, which are expected to increase from a few tens to a few hundred. A connection request between nodes is satisfied by establishing a lightpath from a source node to a destination node. A lightpath is an optical channel between two nodes that is assigned the same wavelength on all links along the route. A lightpath provides a circuit-switched connection between two nodes. A set of logical paths, constituting a logical topology G_v on a physical network topology G_p, is depicted in Fig. 2.1.

Each node in a physical topology consists of an optical cross-connect (OXC) and optical terminating equipment. This may not always be the case as some nodes may act as through nodes where optical channels are in transit. An optical channel passing through the optical cross-connect may be routed from an input fiber to an output fiber without undergoing optical to electrical to optical (O-E-O) conversions. The same wavelength is assumed to be assigned on all links along the route. Thus no wavelength conversion function is performed in the OXC, and all cross-connects are wavelength-selective. An optical channel is terminated by optical terminating equipment such as wavelength add/drop multiplexers (WADMs). WADMs are used to add or drop selected wavelengths to and from the fiber. So any node can be a source or destination for a connection.

2.3 Routing and wavelength assignment

Routing and wavelength assignment (RWA) algorithms are responsible for selecting a suitable route and wavelength among the many possible choices for establishing a connection. Good routing and wavelength assignment algorithms are critically important to improving the performance of WDM networks, in particular because routes here are circuit switched.

2.3.1 Routing algorithm

Routing algorithms have been extensively studied in telecommunications (circuit-switched) networks and computer (packet-switched) networks [38, 71, 75, 76, 100, 108, 127, 129, 147, 150, 170, 178, 239, 244, 263, 315]. The routing algorithms can be broadly classified into two classes, namely, *static routing* and *dynamic routing*. In static routing, the routes for node-pairs are fixed, i.e. the routes do not change with the network status. The static routing typically includes *fixed-path routing (FPR)* and *alternate-path routing (APR)*. In dynamic routing, the routes for node-pairs are dynamically selected according to the current network status. A typical example of dynamic routing is least-congestion routing (LCR). All of these routing algorithms have been previously proposed for circuit-switched networks and have been applied to optical WDM networks.

Fixed-path routing is the simplest among the three algorithms. Many researchers, who deal with wavelength assignment algorithms and develop analytical models, assume fixed-path routing because of its simplicity. However, this simplicity also results in performance degradation because only one path is provided for each node-pair. A connection request is blocked if no wavelength is available on that path. Alternate-path routing, in which more than one candidate path is provided for a connection request, improves the network performance significantly compared to fixed-path routing. However, the candidate paths and their orders are predetermined without considering the current network status. Performance cannot be further improved with these static routing algorithms. Least-congestion routing, which takes the current network status into account, selects the least congested path to establish a connection.

Figure 2.2 is used to demonstrate the routing algorithm. Suppose a path needs to be established from node 1 to node 6. The shortest-path algorithm will always choose the route 1–2–6 as it involves only two hops (assuming that the link costs

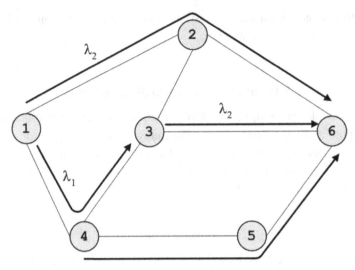

Fig. 2.2. Different wavelength assignment schemes.

are the same on all links). An alternate-path algorithm may have a list of three paths: path 1, 1–2–6; path 2, 1–4–3–6; and path 3, 1–4–5–6. The APR algorithm will explore these paths in order to pick a path that is available, i.e. for which a wavelength is free on all links. On the other hand, the LCR algorithm will evaluate all three paths, and choose the one on which the maximum number of wavelengths are free. Thus the shortest-path algorithm will reject the request if path 1–2–6 is not available whereas the other two algorithms may accept the request. Out of the APR and the LCR algorithms, the LCR algorithm is also likely to leave most resources free on the path used.

The performance results show that the blocking probability of using least-congestion routing is one to two orders of magnitude lower than alternate-path routing in mesh-torus networks as shown in Fig. 2.3.

2.3.2 Wavelength assignment

The wavelength assignment problem is unique to WDM networks. Unlike in circuit-switched networks, the same wavelength has to be free on all of the links of a path to establish a connection in all-optical WDM networks without wavelength conversion. If a wavelength converter is available at every node for every wavelength, assignment is a trivial problem. However, the technology of all-optical wavelength conversion is not mature as yet. Wavelength converters are expensive in time and space. Therefore, good wavelength assignment algorithms along with routing algorithms are critically important in improving network performance and reducing network cost.

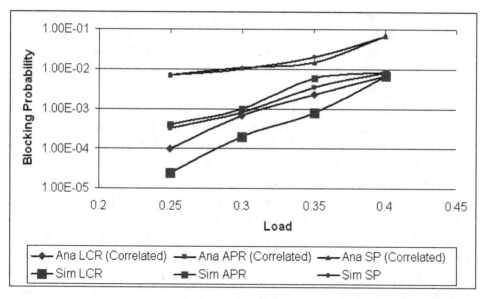

Fig. 2.3. Performance of different routing algorithms.

The wavelength assignment algorithms in the literature can be broadly classified into two categories. The following wavelength algorithms are commonly used.

(i) *Random (R)*: the random wavelength assignment algorithm chooses one of the available wavelengths free on all links randomly with a uniform distribution to establish a connection.

(ii) *First-fit (FF)*: this algorithm assumes that the wavelengths are arbitrarily ordered, e.g., $\lambda_1, \lambda_2, \ldots, \lambda_W$, where W is the maximum number of wavelengths per fiber. The first-fit algorithm checks the status of the wavelengths sequentially and chooses the first available wavelength to establish a connection.

(iii) *Most-used (MU)*: the free wavelength that is used on the greatest number of links in the network is chosen to establish a connection.

(iv) *Least-used (LU)*: the free wavelength that is used on the least number of links in the network is chosen to establish a connection.

Example. Figure 2.2 is used to demonstrate a wavelength assignment process. Suppose each link can carry three wavelengths, $\lambda_1, \lambda_2,$ and λ_3. In this figure, three paths have already been established. The path from node 1 to node 6 uses the route 1–2–6 and the wavelength λ_2. The path from node 1 to node 3 uses the route 1–4–3 and the wavelength λ_1. And finally the path from node 3 to node 6 uses the route 3–6 and the wavelength λ_2. Now suppose a new path from node 4 to node 6 needs to be established. A route 4–5–6 is the shortest hop-length path available and all wavelengths are available on both of the links. The random algorithm can use any of the three wavelengths. The first-fit algorithms will use wavelength λ_1 as that is the first wavelength available. The most-used algorithm will use wavelength λ_2 as that is used by most paths in the network. And finally the least-used algorithm will use wavelength λ_3 as it is not used by any connection. This example clearly demonstrates the differences between the different approaches taken by different wavelength assignment algorithms.

It is easier to analyze the behavior of random wavelength assignment, and therefore analytical models usually assume random wavelength assignment. However, the wavelengths used are randomly distributed and mixed with free wavelengths in the network. It may therefore be hard for a connection request to find a wavelength that is free on consecutive links from a source to a destination node, in particular when the connection involves multiple links in a heavily loaded network.

The least-used wavelength assignment algorithm attempts to route a connection on the least utilized wavelength in order to achieve a nearly uniform distribution of the load over the wavelength set. Both the random and the least-used algorithms distribute the load evenly over the wavelengths. The first-fit and the most-used methods attempt to pack the connections together to use fewer wavelengths, and leave more wavelengths consecutively free. The researchers [36, 52, 235, 237] have shown that the blocking probability of the random and least-used wavelength assignment algorithms is higher than that of the first-fit and most-used algorithms. The random assignment algorithm has a performance close to, but better than the least-used algorithm.

The first-fit algorithm with a fixed-path routing approach has a considerably lower blocking probability than an approach that uses the random wavelength assignment algorithm. However, the most-used assignment algorithm performs slightly better than the first-fit algorithm.

In the above discussions, the connection establishment procedure is separated into two steps: first select a route (if not fixed-path routing) and then select a wavelength from the available free wavelengths. The routing and wavelength assignment algorithm can also be solved jointly as proposed in [74, 163]. The route–wavelength pair that meets the specified criteria, i.e. maximizes the residual capacity, over all wavelengths and considered paths is selected jointly. These joint routing and wavelength assignment algorithms outperform the disjoint approaches.

2.3.3 Design of wavelength switching with conversion

As noted in Chapter 1, *wavelength conversion* is a mechanism by which an optical signal from one wavelength is converted into another. A device that performs such a conversion is referred to as a *wavelength converter*. Switch design [20, 55, 290] becomes quite complex with the availability of wavelength conversion. There are several options for deploying converters in optical switches.

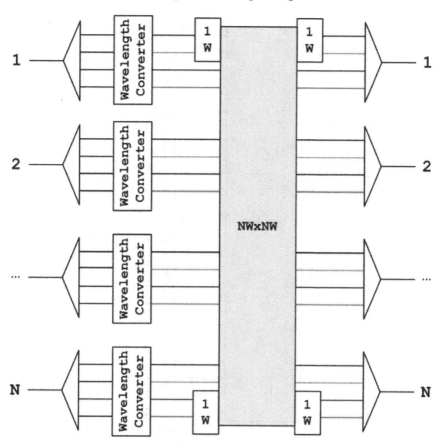

Fig. 2.4. Node architecture in a wavelength-routed WDM network with wavelength conversion at input ports.

In Chapter 1 Fig. 1.5 depicted the architecture of a node used in a wavelength-routed network without wavelength conversion. The architecture of a node employing full-wavelength conversion is shown in Fig. 2.4. The figure shows a node with four links with each link having four wavelengths. After demultiplexing the wavelengths from the links, the individual wavelengths are fed into tunable wavelength converters that convert the signals to the corresponding output wavelength. Every wavelength on every link is provided a dedicated wavelength converter. While such an architecture allows any wavelength to be converted to any other wavelength at the node, most of the connections could be routed without the need for wavelength converters. Hence, such an approach results in wastage of wavelength converter resources. Note that wavelength converters are expensive components, hence they cannot be employed at a node in large numbers unless there is a strong need for it.

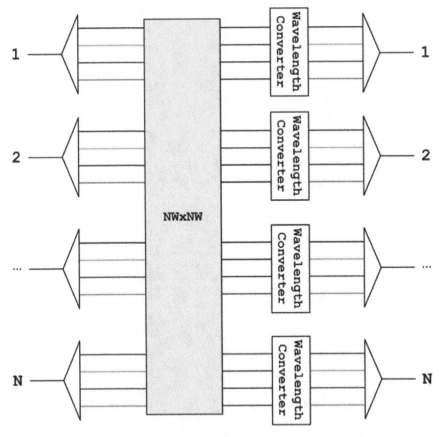

Fig. 2.5. Node architecture in a wavelength-routed WDM network with a wave-length conversion bank at the output ports.

Figure 2.5 shows the architecture of a node that employs a limited number of wavelength converters that can be shared by connections on different links. The set of wavelength converters are referred to as a *wavelength converter bank*. In this case, there are four converters available in the bank. The outputs of the wavelength converters are connected to a switch that could have more output ports than input ports. Such an architecture allows for more than one wavelength converter to be used to route connections on an output link. In the architecture shown in Fig. 2.5 up to four connections can avail the facility of wavelength conversion at the node with up to two connections on an output link employing wavelength converters. If the wavelength converters are not needed, then the connections are switched directly at the first switch to the respective output links.

The wavelength converter blocks that are shown in the node architectures can be implemented in several ways. A trivial yet practical solution is to receive the incoming signal in the electronic domain and to re-transmit in the desired output

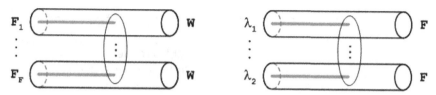

Fig. 2.6. Switching possibilities in a multi-fiber network employing fibers (F_1 to F_F) and W wavelengths per fiber.

wavelength. However, such an approach compromises on the transparency of the network. The intermediate nodes that employ wavelength conversion must know the clock-rate of the incoming signal and the frame format. Such a lack of transparency also limits the scalability of the network.

2.4 Multi-fiber networks

The expensive proposition of employing wavelength converters forced researchers to venture into alternatives for wavelength conversion. Multi-fiber networks [26] are an alternative to employing wavelength conversion. Multi-fiber networks employ more than one fiber on a link between two nodes. Hence, every link in the network has multiple fibers, each carrying multiple wavelengths. The advantage of having such a network is that a wavelength on a link from an input fiber can be switched to any of the fibers on the output link. Figure 2.6 shows the switching possibilities of different wavelengths at a node that does not employ wavelength conversion. Each link is assumed to have two fibers (F_1 and F_2) and two wavelengths (λ_1 and λ_2) per fiber. Such a multi-fiber approach is similar to that of a limited-wavelength conversion capability, refer to Fig. 1.8(d).

Employing multiple fibers in a network has a key advantage. Compared with a single-fiber network with the same link capacity, the number of wavelengths in a fiber can be reduced by a factor of f, where f is the number of fibers employed. A smaller number of wavelengths per fiber implies that the spacing between two wavelengths can be increased, resulting in the use of simpler and less expensive components. Such an approach allows one to increase the signal power on a particular wavelength, thereby improving the signal-to-noise ratio, and requires lesser amplification, resulting in an increased network span. Components requiring functioning in a narrow bandwidth are more expensive compared to those that work over a wider bandwidth. However, the down side of employing multiple fibers is that every fiber needs its own amplifier. Improvements in transmission and fiber technology have made it possible to transmit signals over longer distances without the need for amplification. With such improvements in technology, a multi-fiber approach seems to be an attractive alternative for wavelength conversion. The gain

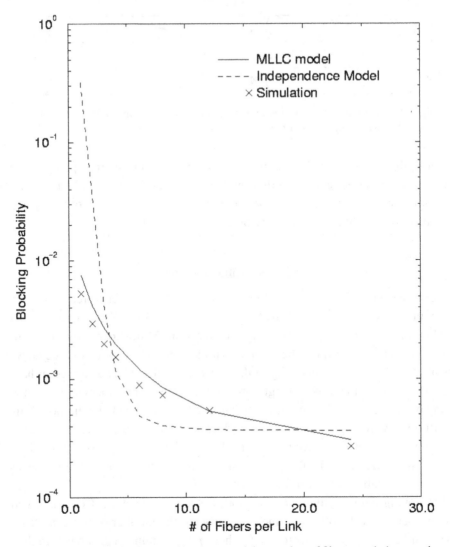

Fig. 2.7. Gain of multi-fiber architectures and the number of fibers needed to match the wavelength conversion gain, © IEEE. Source: L. Li and A. K. Somani, A new analytical model for multi-fiber WDM networks, in *Proc. Globecom'99* [156].

of a multi-fiber approach in different architectures that can match the performance of wavelength conversion is depicted in Fig. 2.7. From the figure, it is noticed that in most cases three to four fibers on each link with $W/3$ or $W/4$ wavelength on each fiber so that the fiber wavelength product on each link remains the same are sufficient to match the performance of wavelength conversion.

With the further development of networks and optical technology, more and more researchers have realized that a single-fiber network may not have enough capacity to support the dramatically increasing bandwidth requirement. In fact, most of the optical networks, if not all, consist of multiple fibers on each link. The wavelength continuity constraint is relaxed in multi-fiber networks because a wavelength that cannot continue on one fiber can be switched to another fiber using optical cross-connects as long as the same wavelength is available on the other fibers on the outgoing link. Several additional wavelength assignment algorithms have been developed for multi-fiber WDM networks in the literature as discussed below.

(i) *Least-loaded (LL)*: the LL algorithm proposed in [74] selects the wavelength that has the largest residual capacity on the most loaded link along a path.

(ii) *Minimum sum (MS)*: the MS algorithm proposed in [74] chooses the wavelength that has the minimum average utilization.

 Both MS and LL algorithms select the most used wavelength when multiple wavelengths are tied, hence they reduce to the most-used rule in the single-fiber case.

(iii) $M \sum$: the $M \sum$ algorithm in [273] chooses the wavelength that leaves the network in a "good" state for future requests. The goodness of a state is measured by a new concept called the *value* of the network. The value function $V(\alpha)$ of the resulting state α after a call is established is restricted to be functions of path capacities, i.e. $V(\alpha) = g([C(\alpha, p) : p : P])$, where P is the set of all possible paths, and $C(\alpha, p)$ is the path capacity of p in an arbitrary state ϕ.

(iv) *Relative capacity loss (RCL)*: the relative capacity loss algorithm in [311] chooses the wavelength that minimizes the relative capacity loss. The relative capacity loss of path p on wavelength λ^*, denoted by $R_c(p, \lambda^*)$, is defined as

$$R_c(p, \lambda^*) = \frac{P_c(p, \lambda^*) - P_c'(p, \lambda^*)}{\sum_\lambda P_c(p, \lambda)} \tag{2.1}$$

where $P_c(p, \lambda)$ is the wavelength path capacity of path p on wavelength λ [311].

Note that the random, first-fit, most-used, and least-used assignment algorithms proposed for a single-fiber network, can also be used in multi-fiber networks with or without modifications. The results in [74] show that the *least-loaded* and *minimum-sum* algorithms perform better than the random, first-fit, and most-used algorithms in multi-fiber networks. The $M \sum$ algorithm in [273] performs considerably better than other algorithms except the RCL algorithm at the cost of increased computational complexity.

2.5 Survivability

WDM networks are prone to catastrophic link failures that are caused due to a variety of reasons including optical fiber, transmitter, receiver, amplifier, router, and converter faults. For example, fiber-cut failures are as common as once every 4 days

[211]. Thus detection, location, and isolation of all of these fault scenarios are both very important and very possible. A link fault is detected easily as the receiver nodes detect a loss of light on the link. A network management algorithm is then invoked to first notify and then recover from the fault without causing a network failure.

A fiber-cut causes a link failure. When a link fails, all of its constituent fibers will fail. A node failure may be caused by the failure of the associated WXC. A fiber may also fail due to the failure of its end components. Since WDM networks carry high volumes of traffic, failures have very severe consequences. Thus the ability to reconfigure and re-establish communication upon failure must be provisioned at the time of establishing a connection. The actual process is managed by a combination of hardware and software methods.

2.6 Restoration methods

A lightpath that carries traffic during normal operation is known as the *primary lightpath*. When a primary lightpath fails, the traffic is rerouted over a new lightpath known as the *backup lightpath*. A multiple link failure is possible due to physical constraints such as common routing of fibers, longer maintenance times involved, etc. Thus both *single-link* as well as *multi-link failure* scenarios must be considered during the design and operation phases.

The restoration methods are broadly classified into reactive and pro-active methods. The reactive method is a simple way of recovering from failures. When an existing lightpath fails, a search is initiated to find a new lightpath that does not use the failed components. This has a low overhead as no resources are pre-reserved and remain unused. However, this method may not guarantee successful recovery as resources may not be available. In the case of a distributed implementation, contention among simultaneous recovery attempts for various failed connections may occur. In a pro-active method, backup lightpaths are planned and resources are reserved at the time the primary connection is established. This method obviously guarantees 100% restoration. The restoration time of a pro-active method is much smaller than the reactive method as the resources for restoration are already identified and reserved. In that sense, it is also a protection scheme. Thus the techniques can be classified into the following categories.

(i) *Protection*: a pro-active full redundancy-based scheme where both the primary and backup paths are established and provisioned at the time of setting up a connection.
(ii) *Restoration*: a partially pro-active approach wherein a backup connection is identified at the time of honoring the request and the primary lightpath is established, but the identified backup lightpath is only provisioned when a failure actually occurs.
(iii) *Recovery*: a reactive approach where the primary path is established at the time of setting up a connection and a restoration path is found when an actual failure occurs.

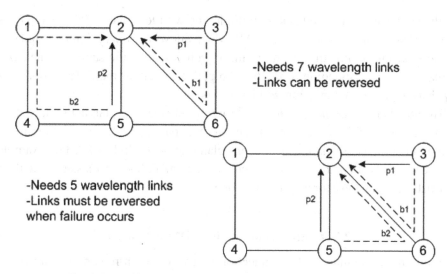

Fig. 2.8. Dedicated vs. backup path multiplexing reservation.

Two particular techniques, *D-connection*, in which explicit multiple paths set up for every connection based on a request and available resources are used, and *p-cycle* or *protection-cycle*, are employed for protection. To establish backup lightpaths, resource (wavelength channel) consumption becomes higher. Protection is efficient in time, but expensive in space as at least 50% or more capacity has to be reserved for protection alone.

2.6.1 Resource saving

Several techniques are used to save resources used for restoration. In particular, use of *backup multiplexing* and *subgraph-based L+1 routing* techniques to efficiently utilize the wavelength channels are very effective. They are demonstrated in the following example.

Example. Figure 2.8(a) shows two primary lightpaths p_1 and p_2 and their respective backup lightpaths b_1 and b_2 on a wavelength for dedicated protection. Figure 2.8(b) depicts the same two lightpaths, p_1 and p_2, that are link-disjoint. Hence they do not fail simultaneously upon a single-link failure. Therefore, the corresponding backup paths b_1 and b_2 can share the wavelength on link (6, 2) by simply selecting different backup paths.

The idea behind *backup multiplexing* is to allow two backup lightpaths to share a wavelength channel, if their primary lightpaths do not fail simultaneously. Upon a link failure, all the failed paths find their backup lightpaths readily available. Thus, these algorithms ensure a 100% *restoration guarantee*. The restoration guarantee is defined as the guarantee with which a failed lightpath finds its backup readily available. An alternative to backup multiplexing with less than 100% guarantee is to use *primary-backup multiplexing* [92], which allows a primary lightpath and one or more backup lightpaths for the primary paths to share the same channel. If a primary path corresponding to the backup path fails, either the failed primary or the sharing primary connection will have to be dropped.

The idea behind the *subgraph-based routing* scheme is to plan network routing such that for any failure, there exists an alternate path for every accepted request. This is in some sense failure-dependent planning at the link level. For example when a link fails, all paths that are affected by the link failure are reassigned to their new paths as they have already been planned.

2.7 Traffic grooming in WDM networks

The minimum granularity of a connection in a wavelength-routed network is the capacity of a wavelength. The transmission rate on a wavelength increases with advances in the transmission technology. However, the requirement of end-users such as internet service providers (ISPs), universities, and industries are still much lower than that of the capacity of a wavelength. The bandwidth requirement is projected to increase in the future; however, even doubling the current bandwidth would be more than sufficient to handle the projected demand for the near future. The current transmission rate on a wavelength is 10 Gbit/s (OC-192). At the time of writing, 40 Gbit/s (OC-768) technology is commercially available, however it is not widely deployed.

The large gap between the user requirement and the capacity of a wavelength has forced the need for wavelength sharing mechanisms that would allow more then one user to share the wavelength channel capacity. Wavelength sharing, similar to sharing a fiber using multiple wavelengths, can be done in several ways. One approach to sharing a wavelength is to divide the wavelength bandwidth into frames containing a certain number of time slots. A time slot on successive frames would then form a channel. Other approaches such as phase modulation and optical code division multiple access (OCDMA) can also be employed to share the capacity on a wavelength.

The merging of traffic from different source–destination pairs is called *traffic grooming*. Nodes that can groom traffic are capable of multiplexing or demultiplexing lower rate traffic onto a wavelength and switching them from one lightpath to another. The grooming of traffic can be either static or dynamic. In static traffic grooming, the source–destination pairs for which requirements are to be combined are predetermined. In dynamic traffic grooming, connection requests from different source–destination pairs are combined depending on the existing lightpaths at the time of the request.

Recent advances in optical switching technology, as in [41, 44, 131], have shown the possibility of realizing fast all-optical switches with switching times of less than a nanosecond. The use of such fast switches along with fiber delay lines as time-slot interchangers [56, 102] has opened up the possibility of realizing multi-fiber, multi-wavelength, optical time-switched networks. These networks are referred to as *WDM-TDM switched* or *WDM grooming* networks. WDM grooming networks employing WDM and TDM employ multiplexing mechanisms both in space and the time domain.

A device that translates the signal from one time slot on a wavelength to another time slot is called a *time slot interchanger* (TSI). As the signals are transmitted using photons, which do not have valency, it is not possible to store the energy. Thus there is no equivalent of a memory in the optical domain compared to that available in the electronic domain. The optical version of TSIs are designed using *fiber delay lines* that are loops of fibers through which the signal is passed from one end and received at the other. Depending on the length of the fiber, different delay times can be achieved. The length of a fiber needed to delay a signal by one time-slot interval is proportional to the product of the time unit and the speed of light. Delaying by more than one time slot can be achieved by cascading multiple delay units. Programmable delay lines can be implemented with switches and fiber delay lines by selectively activating specific stages of delay units.

We now define a WDM grooming network more formally here. A WDM grooming network consists of switching nodes interconnected by links. A link ℓ has F fibers with each fiber carrying W wavelengths. Every wavelength is divided into frames with T time slots per frame. A *channel* is defined as a specific time slot on successive frames. The link has a total of FWT channels in it. Every channel on a link is denoted by a four-tuple (l, f, w, t) that denotes the link, fiber, wavelength, and time-slot identifier on the link.

Requests or *calls* arrive in a network that requires a certain number of channels from a source to a destination node. The requests are accepted in the network along a certain path by assigning the requested number of channels on each link such that every intermediate node on the link can switch a specific input time slot to a specific output time slot. The switching capability at a node restricts the output channels to which an input channel can be switched. For example, if a node does not employ wavelength conversion and time slot interchange, then an input channel $(l, f, w, t)_i$ can be switched to $(l, f, w, t)_o$ if and only if $w_i = w_o$ and $t_i = t_o$. The restriction on being the same wavelength is removed if wavelength converters are available. Similarly, TSIs remove the constraint on the same time slot being maintained at the input and the output.

Different nodes in the network could implement different node architectures. For example, one node could employ wavelength conversion, but not TSIs while another node could implement TSIs but not wavelength converters.

WDM grooming networks can be classified into two categories [143]: dedicated-wavelength TDM (DW-TDM) networks and shared-wavelength TDM (SW-TDM) networks. In DW-TDM networks, each source and destination pair are connected by a *lightpath*, where a lightpath is defined as an all-optical connection between two nodes. Calls between the source and the destination are multiplexed on the lightpath. If the bandwidth required by a new call at a node is not available on any of the existing lightpaths to the destination, a new lightpath to the destination is established. On the other hand, in SW-TDM networks, if a call cannot be accommodated on an existing lightpath to the destination, it is allowed to be multiplexed onto an existing lightpath to an intermediate node. The call is then switched from the intermediate node to the final destination either directly or through other nodes. However, if none of the existing lightpaths from the node can accommodate the call, a new lightpath to the destination is established.

2.8 Optical packet switching

One alternative to circuit switching is to use optical packet switching (OPS) [25, 43, 181, 314] technology in backbone networks. The major advantage of OPS is the flexible and efficient bandwidth usage, which enables the support of diverse services. Pure OPS technology in which packet header recognition and control are performed in the all-optical domain is still many years away from becoming reality. OPS with electronic header processing and control is more realistic for medium-term network scenarios. A practical OPS experiment has been performed under the European ACT KEOPS (KEys to Optical Packet Switching) project [43]. In KEOPS, the header is sent with data (payload), but at a lower bit rate, and the header processing is still in the electrical domain. This potentially requires optical buffering at the input port to allow the header processing circuits to finish their job. However, there are still several critical technological challenges that need to be overcome before a practical OPS network becomes a reality. Firstly, the lack of an efficient way to store information in the optical domain is the major difficulty in the implementation of OPS nodes. At present, the buffering technology is not mature and has to overcome a number of technological constraints, such as the large and varying size of optical buffers. Secondly, in a highly dynamic traffic environment such as OPS, wavelength converters are required and play an important role in contention resolution. Wavelength converters can be integrated into the design of the optical buffer and switch architecture in OPS networks. An all-optical wavelength converter is desirable for OPS. However, the fabrication techniques for such wavelength converters are still not practical. The third issue is high-speed header processing in OPS. Currently, the processing of the header is performed in the electrical domain. All-optical header processing has received considerable attention [54, 98], but the technology is still in its early stages. A key enabling technology in OPS is the optical switch fabrics. To deal with packet-by-packet requests, an OPS node requires a switch fabric that is capable of rapid reconfiguration. When the data rate is at 40 Gbit/s and beyond, the switching times have to be of the order of a few nanoseconds. Finally, the other critical requirements include the reliability and scalability of the technology to high port counts, low loss and cross-talk, efficient energy usage, and so on. Unfortunately, none of today's available fabric technologies is eligible for building such reliable and cost-effective high-performance optical packet switches.

2.9 Optical burst switching

The concept of burst switching was proposed for conventional telephone networks in the early 1980s [250]. Fast circuit switching was originally developed to support statistical multiplexing of voice circuits, but it was also suitable for data communication at moderate rates. Starting from the middle 1980s, fast packet and cell switching took the place of circuit switching. At that time, fast circuit switching was implemented using time-division multiplexing in the electrical domain to provide distinct channels (time slots). This is essentially similar to ATM technology. The concept of burst switching has been extended for ATM networks. The International Telecommunication Union – Telecommunication Standardization Sector (ITU-T) standard for burst switching in ATM networks is known as *ATM block transfer* (ABT) [162]. Burst switching for optical networks, namely, optical burst switching, was proposed in the late 1990s [49, 190]. Optical burst switching (OBS) is maybe a viable alternative switching technology to transport IP traffic directly over WDM networks. In a wavelength-switched network, once a lightpath is established, it remains in place for a relatively long time, perhaps months or even years. In OBS, the goal is to set up a wavelength channel for each single burst to be transmitted. At the ingress node of an OBS network, various types of data are assembled as a data *burst*, which, for example, can carry one or more IP packets. In OBS, a burst is dynamically assigned to a wavelength channel upon its arrival and is later disassembled at the egress node. To establish a connection for an incoming burst, the ingress nodes send an associated control packet (request or set-up) over a dedicated wavelength channel or a non-optical channel before the burst is transmitted. The data burst is switched all-optical using the OBS fabric. Two primary techniques to transmit data are *tell-and-wait* (TAW) and *tell-and-go* (TAG). In the *tell-and-wait* scheme [116], a burst is buffered while the control packet is being sent to set up switches and reserve bandwidth for establishing a connection. In the *tell-and-go* scheme [78, 116] a burst is sent immediately after its control packet without receiving a confirmation. If a switch along the path cannot carry the burst due to congestion, the burst is dropped. In this scheme, it may still be necessary to buffer the burst in the optical burst switch until its control packet has been processed [162]. Other schemes, known as *just-enough-time* (JET) [49] and *just-in-time* (JIT) [141], have also been proposed in the literature. An OBS architecture is described in [145]. A number of research papers on OBS technology and its applications have been published by researchers around the world [160, 161]. Among them, the vast majority are based on JET. In the JET scheme, there is a delay between transmission of the control packet and transmission of the optical burst. This delay can be set to be long enough; for example, larger than the total processing time of the control packet along the path. Therefore, when a burst arrives at an intermediate node, the control packet has been processed and a channel on the output port has been reserved. Thus, there is no need to buffer the burst at the intermediate nodes. This is a very important feature of the JET scheme, since optical buffers are still difficult to implement. Improvements and variations of JET have also been studied extensively in the literature. Given limited success of burst switching in the 1980s, one may question why burst switching should be a promising approach to high-speed data communications now. As aforementioned, burst switching and ATM are essentially the same; however, the flexibility of ATM outperformed burst switching in the electrical domain. Some researchers believe that since optical fibers provide a virtually unlimited bandwidth resource, it makes sense to carry control information in a dedicated parallel channel so as to keep the data path simple. Besides, it is best to avoid queueing as much as possible, because both electrical and optical buffers are expensive at gigabit data rates. For this reason, many believe that OBS achieves good statistical multiplexing performance by transmitting many independent data channels in parallel.

2.9.1 Challenges

Just like OPS, OBS has to overcome several critical technological challenges before it really becomes practical. Some of these challenges might not seen very obvious at first glance. One important issue is synchronization at terminal nodes [145]. Consider an OBS network using passive optical components with no retiming of the data. A terminal that receives this burst must synchronize the received data at the bit level as well as the burst level. An alternate approach is to use retiming elements throughout the OBS network. This requires transmission components at every input to burst switches that recover timing information from the received data stream and to use the recovered timing information to regenerate the data and forward it using a local timing reference. Since the local timing reference may differ in frequency, it adds to the complexity of the transmission components at every input. This complexity makes the implementation of OBS more difficult. It is also worth mentioning that JET does not completely remove optical buffers from OBS networks. Notice that optical buffers are still required at the ingress nodes to generate the initial delay between a data burst and its control packet. The need for high-speed optical buffers remains as a notably intractable problem for OBS. Additionally, since the number of control channels is limited in optical networks, the control channels can become a bottleneck for the performance of the OBS networks. In OBS, guard bands are used in each burst to accommodate possible timing jitters along the path from source to destination. Owing to the relatively low speed of optical switching elements, a significant guard time has to be provided between control and data segments, which results in another significant overhead for OBS. Therefore, taking into account the large ratio between the switching delay and the IP burst duration, the network might be severely underutilized. Currently, commercial OBS networks do not exist. As yet it is not clear whether OBS will become an alternative technique for the core optical network or just an intermediate step towards all-optical packet switching.

3

Restoration approaches

The restoration schemes differ in their assumptions concerning the functionality of cross-connects, the traffic demand, the performance metric, and the network control. Survivability paradigms are classified based on their rerouting methodology as being path-/link-based, execution mechanisms as centralized/distributed, by their computation timing as precomputed/real time, and their capacity sharing as dedicated/shared [35, 72, 79, 82, 83, 101, 106, 107, 132, 135, 183, 184, 198, 199, 205, 210, 214, 251, 285, 288, 289, 291, 309]. This classification is shown in Fig. 3.1.

Pro-active vs. reactive restoration. A pro-active or reactive restoration method is either *link-based* or *path-based* [36, 208]. In a special case, a segment-based approach can also be used. In a segment-based detouring, a backup segment is assigned for more than one link. A link may be covered by more than one segment. The restoration path, as shown in Fig. 3.2, is computed for each path. In the case of a link failure, the backup segment is used.

Link-based restoration methods reroute disrupted traffic around the failed link, while path-based rerouting replaces the whole path between the source and the destination of a demand. Thus, a link-based method employs *local detouring* while the path-based method employs *end-to-end detouring*. The two detouring mechanisms are shown in Fig. 3.3. For a link-based method, all routes passing through a link are transferred to a local rerouting path that replaces that link. While this method is attractive for its local nature, it limits the choice of alternatives [36]. In the case of wavelength-selective networks, the backup path must use the same wavelengths for existing requests as that of their corresponding primary paths since the working segments are retained.

Computing disjoint paths. One mechanism for using the two link-disjoint paths is to first compute the shortest path between a given source–destination pair in a given network. Then, remove those links from the graph and recompute the shortest

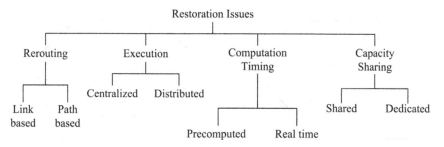

Fig. 3.1. Classification of restoration architectures for survivable networks.

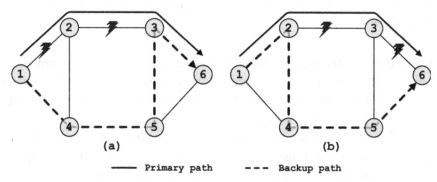

Fig. 3.2. Segmented path protection.

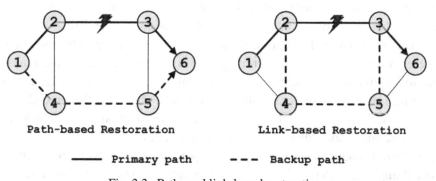

Fig. 3.3. Path- and link-based restoration.

path on the reduced graph. The problem with this approach is that the second path may be much larger, thus leading to a very high average path length. An alternate technique is to use the shortest-cycle approach, which will find a combination of two paths. Figure 3.4 demonstrates a network example with primary and backup path combinations when computed separately and also when computed using the shortest-cycle algorithm.

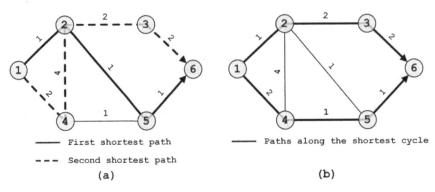

Fig. 3.4. The difference between the primary-backup path and shortest-cycle computation.

Centralized vs. distributed control. A restoration scheme may assume either centralized or distributed control. In large networks, distributed control is preferred over centralized control. A distributed control protocol requires several control messages to be exchanged between nodes.

The precomputed approach calculates restoration paths before a failure happens and the real-time approach does so after the failure occurs. The former approach allows fast restoration as the routes are precomputed while the latter approach is slow as the alternate route is computed after the failure has been detected. Centralized restoration methods compute primary and restoration paths for all demands at a central controller where current information is assumed to be available. The routes are then downloaded into the route tables for each node. These algorithms are usually path-based. The real-time computation method may also use a set of possible precomputed routes or identify routes in real time. The real-time approach needs to identify failure, ascertain the remaining topology and capacity, and then find the best alternate route for the affected demands. Therefore, this procedure can be very slow. Since restoration speed and fast failure isolation in optical networks is of utmost importance, this approach is not very attractive.

Centralized schemes that involve precomputed routes are more conducive for practical implementations. Maintaining up-to-date information, however, requires frequent communication between the nodes and the central controller. This overhead becomes a potential problem as the network size grows. Distributed methods may involve precomputed tables of discovered capacity and routes in real time. Real-time capacity discovery is slow and the capacity utilization may be inefficient. Distributed precomputation of the restoration route is an attractive approach.

Capacity sharing among the primary and restoration paths is dedicated or shared. The dedicated technique uses one-to-one protection where each primary path has a corresponding restoration path. In the shared case, several primaries can share the same backup path as long as the primaries are node- and link-disjoint. This scheme is called the backup multiplexing technique [91]. These paradigms serve as a good framework for analyzing the different design methodologies, as each methodology uses a restoration model that is a combination of the above paradigms.

P-cycle. Recently, a new link-based approach, called p-cycle (preconfigured protection cycles), has been introduced [301, 302]. In this approach, preconfigured protection cycles are embedded in a mesh network. Capacity is reserved on links belonging to p-cycles only to provide an alternate path for a failing link. This is equivalent to backup multiplexing. It achieves higher efficiency in densely connected networks and approaches the speed of a line-switched self-healing ring [302]. With p-cycle protection, when a link fails, only the nodes neighboring the failure need to perform real-time switching. This makes p-cycle comparable to SONET/SDH line-switched rings in terms of the speed of recovery from link failures. The approach is discussed in more detail in the next chapter.

Restoration in IP over WDM. The study in [158] considers an IP-over-WDM network in which each node employs optical cross-connects and IP routers and any two IP routers can be connected by a lightpath. The authors proposed two types of fault-management technique, under a single-link failure scenario for IP-over-WDM networks, namely WDM protection and IP restoration. In WDM protection, there is a link-disjoint backup path reserved for each lightpath at call setup time. The backup path is activated only when failure occurs and its corresponding primary path is disabled. The wavelength reservation for backup paths can be dedicated to a request or shared among different requests provided that their primary paths are not expected to fail simultaneously. The latter technique is also termed primary-backup multiplexing. In the IP restoration described in [158], the load from a source to a destination is distributed over multiple routes, and this load sharing is assumed to be performed by IP. That is, there is no backup reservation for the working paths at the IP layer. When failure occurs, the affected connection is rerouted through the rest of the routes. The stimulation and numerical results in [158] show that WDM protection outperforms IP restoration and the recovery times for WDM shared-path protection are much faster than the recovery times for IP restoration.

Shared mesh restoration. Shared mesh restoration is studied in [90] within the framework of the MPLS/GMPLS architecture. This work also focuses on single-link failure, although the proposed algorithm is extended to node and span failure protections, as special cases of multiple failures. To minimize the restoration delay after a failure, the restoration (backup) paths are preselected and are physically diverse from their service (primary) paths. The bandwidth resource is shared by multiple restoration paths, for which the service paths are not subject to simultaneous failures. In the GMPLS network, both OSPF (open shortest path first) and IS-IS (intermediate system–intermediate system) protocols have been extended to propagate additional link information, such as the available bandwidth, the bandwidth in service, etc. Since the proposed algorithm is based on full knowledge of bandwidth that is shared among restoration paths, it is called full information restoration (FIR). FIR is evaluated in comparison to two other well-known distributed restoration algorithms, namely, shortest-path restoration (SPR) [223] and partial information restoration (PIR) [177, 260]. The simulation results show that there is only a marginal difference between SPR and PIR in terms of their bandwidth utilization, while FIR reduces the amount of reserved restoration bandwidth dramatically. In terms of restoration overbuild, which is defined as the extra bandwidth required to meet the network restoration objective as a percentage of the bandwidth of the network with no restoration, FIR requires much less bandwidth overbuild (63–68%) compared with SPR and PIR (83–90%).

3.1 Restoration model

A dependable or protected connection request between a source–destination pair requires a primary route and a backup route. A non-dependable or unprotected connection is provided with a primary route only. Each path, primary or backup, always accommodates an operation, administration, and maintenance (OAM) channel terminated by the same source–destination pair as the path. The restoration model is shown in Fig. 3.1.

When a primary path fails, an alarm indication signal is generated by the node which detects the link failure and is transferred over this OAM channel by the end nodes. When the source receives the alarm signal on its OAM channel, it prepares to set up the precomputed backup path and sends messages to the controllers along the backup path to configure the ports accordingly. Since the backup is dedicated, the capacity is assumed to be reserved, so no run-time link capacity search needs to be performed. Once the backup path has been set up, the destination prepares itself to receive on that path. There is no restriction for the choice of wavelength on the backup path. It may or may not be the same as the primary path. The tuning time and the associated cost is assumed to be negligible.

3.2 Upgradeable network design

In mesh-restorable networks, fast restoration is provided using predetermined paths that are independent of the failure location and can use backup multiplexing techniques to improve wavelength utilization. Mesh networks provide better capacity efficiency than ring networks. In long-haul networks the greater distance-related cost makes capacity efficiency much more important.

The goal of the upgradeable network design problem is to select a topology to route the current set of connections given a network topology for a future traffic demand. For the current traffic demand, the problem is formulated using an integer programming approach to minimize the total facility cost. The output of the design problem is the number of fibers and wavelengths on each active link, the number of OXCs required for each node, and more interestingly, a subset of links in the final topology that need to be activated for the current traffic demand. Since the cost of provisioning and operating a link is significantly high, the current traffic demand, which is a subset of the future traffic demand, may avoid using some links in the final topology. The meaning in this case is that the cost-optimal routing and capacity planning for the current traffic demand may be fully realizable on a subgraph of the final topology.

An upgradeable network design approach may result in a significant cost reduction for the network service provider. As the traffic increases during the lifetime of the network, more spans are cost-effectively added to accommodate the increased traffic, thereby incrementally realizing the future topology for which the network was designed. The upgrade takes into account the technological advances that are input as design parameters into the problem formulation.

In the restoration design formulations, a 100% restoration guarantee is assumed for any single node or link failure for all protected connections. This guarantee means that primary and restoration paths of protected connections are allocated the same capacity, and are link-disjoint. The backup multiplexing technique is used to improve the wavelength utilization. This technique allows many restoration paths, belonging to demands of different node-pairs, to share a wavelength λ on link l if and only if their corresponding primary paths are link-disjoint. It should be noted that although every primary lightpath has a corresponding backup lightpath dedicated to it, wavelengths on a link are shared by restoration paths belonging to the demands of different node-pairs.

The following constraints govern the restoration model.

- The number of connections (lightpath) on each link is bounded.
- The levels of protection.
  - Full protection: every demand is assigned a primary and a backup path.
  - No protection: every demand is assigned only a primary path.
  - Best-effort protection: (i) every demand is assigned a primary path. A backup path is assigned if resources are available. (ii) Accept as many demands as possible with or without backup.
- No backups are admitted without a primary.
- Primary and backup paths for a given demand should be link-disjoint.
- Primary path wavelength restrictions: only one primary path can use a wavelength λ on link l; no restoration path can use the same wavelength λ on link l.

- Restoration path wavelength restrictions: many restoration paths can share a wavelength λ on link l if and only if their corresponding primary paths are link-disjoint.

3.3 Notation

The network design and static connection routing problem in a given topology are modeled using an integer linear programming approach. The topology is represented as a directed graph $G(N, L)$ with N nodes and L links with W wavelengths on each link. Two alternate paths, which are link-disjoint, are assumed to be given for each source–destination (s-d) pair, and are used to provide survivability. The following notation is used to develop the optimization formulations.

- $n = 1, 2 \ldots, N$: number assigned to each node in the network.
- $l = 1, 2 \ldots, L$: number assigned to each link in the network.
- $\lambda = 1, 2 \ldots, W$: number assigned to each wavelength.
- $i, j = 1, 2 \ldots, N(N - 1)$: number assigned to each s-d pair. For an N node system, there are $N(N - 1)$ s-d pairs.
- $K = 2$ alternate routes between every s-d pair.
- $p, r = 1, 2, \ldots, KW$: number assigned to a path for each s-d pair. A path has an associated wavelength (lightpath). Each route between every s-d pair has W wavelength continuous paths. The first $1 \leq p, r \leq W$ paths belong to route 1 and $W + 1 \leq p, r \leq 2W$ paths belong to route 2.
- $\bar{p}, \bar{r} = 1, 2, \ldots, KW$: if $1 \leq p, r \leq W$ (route 1), then $W + 1 \leq \bar{p}, \bar{r} \leq 2W$ (route 2) and vice versa.
- (i, p) : refers to the pth path for s-d pair i.
- d_i: demand for node-pair i, in terms of the number of lightpath requests. Each request is assigned a primary and a restoration route. There can be more than one request for each s-d pair.
- π_n^l: link termination indicator which takes a value one if one of the termination ends of link l is node n and is zero otherwise.

The following cost parameters are used as input to the problem formulation.

- C_l: cost of using a link l (data).
- C_w: cost of disrupting a currently working path (data). The restoration model assumes that if an existing connection is disturbed during rearranging of the connection to accommodate a new connection, there is a cost involved with this.
- C_{ND}: cost of a primary path (data).
- C_D: cost of a backup path (data).
- C_{lp}: cost of provisioning a link (data).
- C_f: fiber costs (data).
- C_λ: cost per wavelength channel (data).
- C_{oxc}: cost of a $\Omega \times \Omega$ cross-connect switch (data).

Information regarding whether two given paths are link and node-disjoint.

- $I_{(i,p),(j,r)}$ takes a value one if paths (i, p) and (j, r) have at least one link in common and zero otherwise. If two routes share a link, then all lightpaths using those routes have the corresponding I value set to 1, otherwise it is set to 0 (data).

The following design parameters and corresponding variables are used.

- Ψ: maximum number of wavelengths in each direction in a bidirectional fiber (technology-dependent data).
- w_l: number of wavelengths required on link l (integer variable).
- Ω: size of the minimum cross-point switching element (technology-dependent data).
- Θ_n: maximum allowable number of OXCs at node N (data).
- o_n: number of OXCs required at node N (variable).
- F_l: maximum number of fibers per link l (data).
- f_l: number of fibers required per link l (integer variable).
- m_l: it takes a one if the link l is being used, i.e. at least one fiber is being used ($f_l \geq 1$) in link l (binary variable).

The following notation is used for path-related information:

- $\delta^{i,p}$: a path indicator which takes a value one if (i, p) is chosen as a primary path and zero otherwise (binary variable).
- $v^{i,r}$: a path indicator which takes a value one if (i, r) is chosen as a restoration path and zero otherwise (binary variable).
- $\epsilon_l^{i,p}$: a link indicator, which takes a value one if link l is used in path (i, p) and zero otherwise (data).
- $\psi_\lambda^{i,p}$: a wavelength indicator, which takes a value of unity if wavelength λ is used by the path (i, p) and zero otherwise (data).
- $g_{l,\lambda}$: it takes a value of unity if wavelength λ is used by some restoration route that traverses link l and zero otherwise (binary variable).
- $\chi^{i,p}$: a path indicator which takes a value one if (i, p) is a currently working primary path and zero otherwise (data). However, the main interest would be in the primary paths of the current working connection as the restoration path is reassigned.

3.4 Cost model

The cost sources in a WDM network are mapped into the following four parameters: the link provisioning cost (C_{lp}), the fiber cost (C_f), the per channel cost (C_λ), and the cross-connect cost (C_{oxc}).

The link provisioning cost captures the investment required before any capacity on the link is used. This may include the digging cost, the cable cost, the leasing cost, the right-of-way cost, the maintenance cost, etc. Multiple fibers may be laid out as part of the initial investment, some of which may be lit and the dark fibers being used for future upgrades.

The fiber cost, C_f, is a combination of optical amplifier costs, the multiplexer and demultiplexer costs for fiber terminations, and the dispersion compensation components costs. The maximum number of wavelengths per fiber is a significant design parameter. Since the number of wavelengths per fiber decides the number of dispersion components and regenerators required, and the necessary laser power (hence the spacing between optical amplifiers), and the number of regenerators needed, the network provider needs to set this design parameter appropriately. In ultra-long-haul dense WDM (DWDM) backbone network design, the goal is to let the signal travel longer (thousands of kilometers) without regeneration. Since regenerators make up a significant part of the facility cost, reducing the number of regenerators results in a direct reduction in the total facility cost. Longer distances without regeneration typically mean that the signal-to-noise ratio is low, as each amplifier adds noise to the signal. The noise is alleviated by using forward error correction (FEC) and dispersion compensation. The total fiber cost is subdivided as follows: $C_f = A_f \times C_a + C_{mux} + C_{demux} + C_{dc}$, where C_a, C_{mux}, C_{demux}, and C_{dc} are the costs of optical amplifiers, multiplexers, demultiplexers, and dispersion compensation components, respectively, and A_f is the number of amplifiers along the fiber.

The per channel cost, C_λ, includes the receiver and transmitter card costs per wavelength and the power equalization required per channel. The power equalization is included as part of the transmitter cost. Since, depending on the current demand, the network provider may sub-equip fibers, this cost depends on the number of wavelengths currently used. $C_\lambda = C_r \times w_f + C_t \times w_f$, where C_t and C_r are transmitter and receiver card costs, respectively, and w_f is the number of wavelengths currently used in the fiber.

The number of cross-connects per node determines the switch size, and hence, the total facility cost. A typical wavelength routing switch architecture is shown in Fig. 3.5. The size of the switch includes both origin/destination traffic for the node, and the transit traffic. At each fiber port, the incoming wavelengths are demultiplexed and sent to a space switch where they are switched and sent to any output fiber port. The only constraint is that no two connections going to the same output fiber port can use the same wavelength. Connections on different wavelengths that are destined for the same output fiber port are multiplexed and sent out. The cost of the space switch for each wavelength depends on the size of the minimum cross-point switching element ($\Omega \times \Omega$) available in the market. Let the cost of a 2×2 ($\Omega = 2$) cross-point switching element be C_{oxc}. The number of such switching elements required for a $v \times v$ switch is $\frac{v}{2} \log_2 v$ (assuming the switches are

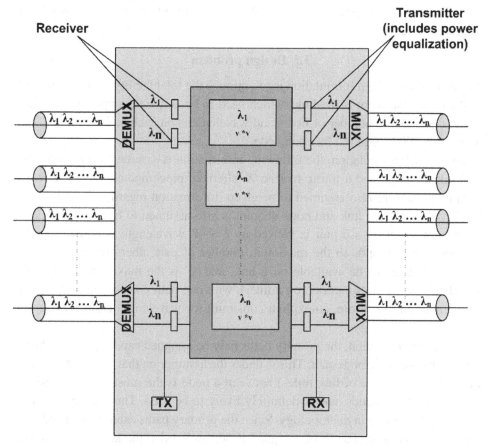

Fig. 3.5. Wavelength routing switch architecture.

implemented as a multi-stage interconnection network (MIN)). Hence, the cost of each MIN is $C_{\text{oxc}} \times \frac{\upsilon}{2} \log_2 \upsilon$.

The total facility cost (TFC) is given by the sum of all the link and node costs:

$$\text{TFC} = \sum_{l}^{L} (m_l C_{\text{lp}} + f_l C_f + w_l C_\lambda) + \sum_{n}^{N} \left[\Psi \times C_{\text{oxc}} \times \frac{(o_n)}{\Omega} \log_\Omega (o_n) \right] \quad (3.1)$$

where $m_l = 1, 0$ denotes whether a link l is used or not and f_l and w_l denote the number of fibers and wavelengths on a link l, respectively. o_n denotes the number of cross-connects needed in a node, and Ψ denotes the maximum number of wavelengths per fiber. The value of o_n is rounded off (ceil) to the nearest integral power of Ω. For the 2×2 case, the cost of a 6×6 switch ($2 \times (O_n = 3)$) is the same as that of an 8×8 switch ($2 \times (O_n = 4)$). In the second term in Eq. (3.1), the cost of a MIN in each node is multiplied by Ψ, since there is a MIN switch for each wavelength.

3.5 Design problem

The goal of the design formulation is to optimize the total facility cost. The output of the design problem is the links which need to be active to support the current traffic demand, the number of fibers and wavelengths on each active link, and the number of OXCs required for each node.

For the ILP formulation, the following information is assumed to be given: the network topology and a traffic matrix. K alternate precomputed routes between each node-pair are also assumed to be given. Information regarding whether any two given routes are link and node-disjoint is also assumed to be available. Each route between every s-d pair is viewed as $F \times W$ wavelength continuous paths (where F corresponds to the maximum number of dark fibers that are already laid, or expected to be available on a link, and W is the maximum number of wavelengths that are supported on a fiber), with one path corresponding to every wavelength and therefore, an explicit constraint for wavelength continuity is not needed.

As an initial solution, the primary paths may be assigned capacity on the shortest path between the node-pair. This is under the assumption that the shortest route (in this case in terms of link miles) between a node is the most cost effective in terms of route demands, and is definitely likely to be used. Thus, these links are automatically active in the topology. Since the primary paths cannot be shared and the working capacity has to be assigned if demands exist, the key is to optimize spare capacity allocation using techniques such as backup multiplexing to improve the wavelength utilization, thus optimizing the total network cost. This effective assumption greatly simplifies the problem formulation as explained here. Validating if backup multiplexing is performed for two given connections requires identifying where the primary paths for the connections are routed. For the case when there are exactly two disjoint alternate routes between any given node-pair, formulating the backup multiplexing constraint is simplified as shown in [186, 189]. If taken dynamically this decision makes the problem formulation very complex and it becomes computationally intractable. In this case, since there are K possible alternate routes, validation requires identifying the primary paths dynamically in the formulation. In order to make the problem more tractable, the formulation assumes that the shortest route between a node-pair is the most cost effective one in terms of route demands, and hence, the working capacity is pre-assigned.

The ILP solution determines the links that need to be active, the number of fibers and wavelengths required for each active link, and the number of OXCs required for each node.

3.5.1 Problem formulation

Objective: the objective is to minimize the total facility cost. This is the sum of all the link costs and the node costs as discussed in Section 3.4. Note that the actual cost of each node is computed using Eq. (3.1). The objective here is to minimize the number of cross-connects (o_n) required per node, thereby reducing the switching cost.

Minimize

$$\sum_{l=1}^{L} m_l C_{\text{lp}} + \sum_{l=1}^{L} f_l C_{\text{f}} + \sum_{l=1}^{L} w_l C_\lambda + \sum_{n=1}^{N} o_n \qquad (3.2)$$

Link provisioning constraint: Eqs. (3.3) and (3.4) set $m_l = 1$ if $f_l \geq 1$

$$m_l \leq f_l \qquad (3.3)$$

$$N(N-1)KFWm_l \geq f_l \qquad (3.4)$$

$$1 \leq l \leq L, \quad 1 \leq f_l \leq F_l$$

Notice that the second equation will require m_l to be at least 1, if $f_l = 0$, and practical values of f_l will never require m_l to be greater than 1.

Link capacity constraint: the total capacity required on a link depends on the total working and spare capacity assigned to the link. The following equations determine the number of wavelengths and fibers required on each link, given that the maximum number of wavelengths per link is Ψ,

$$\sum_{i=1}^{N(N-1)} \sum_{p=1}^{FW} \delta^{i,p} \epsilon_l^{i,p} + \sum_{\lambda=1}^{FW} g_{l,\lambda} \leq w_l \qquad (3.5)$$

$$w_l \leq \Psi f_l \qquad (3.6)$$

$$1 \leq l \leq L, \quad 1 \leq f_l \leq F_l$$

Cross-connect constraint: the total number of OXCs required at node n to carry traffic, which includes both connections originated/terminated at node n and those that are in transit, is governed by

$$\sum_{l=1}^{L} w_l \pi_n^l \leq \Psi o_n \qquad (3.7)$$

$$1 \leq n \leq N(N-1), \quad 1 \leq o_n \leq \Theta_n$$

Restoration path wavelength usage indicator constraint: $g_{l,\lambda}$ takes a value of unity if wavelength λ is used by some restoration route (i, r) that traverses link l.

Equations (3.9) and (3.10) set $g_{l,\lambda} = 1$, if $X_{l,\lambda} \geq 1$. $X_{l,\lambda}$ counts the number of paths using link l and wavelength λ for backup as shown in Eq. (3.8):

$$X_{l,\lambda} = \sum_{i=1}^{N(N-1)} \sum_{r=FW+1}^{KFW} v^{i,r} \epsilon_l^{i,r} \psi_\lambda^{i,r} \tag{3.8}$$

$$g_{l,\lambda} \leq X_{l,\lambda} \tag{3.9}$$

$$N(N-1)KFW g_{l,\lambda} \geq X_{l,\lambda} \tag{3.10}$$

$$1 \leq l \leq L, \quad 1 \leq \lambda \leq FW, \quad X_{l,\lambda} \geq 0$$

Demand constraints for each node-pair: the primaries are mapped on the shortest route (route 1) between the given node-pair. These constraints also implicitly satisfy topological diversity (i.e. the primary and the restoration path of a given demand should be node and link-disjoint):

$$\sum_{p=1}^{FW} \delta^{i,p} = d_i \tag{3.11}$$

$$1 \leq i \leq N(N-1) \tag{3.12}$$

$$\sum_{r=FW+1}^{KFW} v^{i,r} = d_i \tag{3.13}$$

$$1 \leq i \leq N(N-1) \tag{3.14}$$

Primary path wavelength usage constraint: only one primary path can use a wavelength λ on link l. No restoration path can use the same λ on link l:

$$\sum_{i=1}^{N(N-1)} \sum_{p=1}^{FW} \delta^{i,p} \epsilon_l^{i,p} \psi_\lambda^{i,p} + g_{l,\lambda} \leq 1 \tag{3.15}$$

$$1 \leq l \leq L, \quad 1 \leq \lambda \leq FW$$

Backup multiplexing constraint: if $I_{(i,1),(j,1)} = 1$, then the primary paths for node-pairs i and j share links on its backup routes. Note primary paths are already mapped on the shortest route. Thus, if $I_{(i,1),(j,1)} = 1$, only one of the restoration paths can use a wavelength λ on a link l as a backup, since their primary paths share link(s) on their route:

$$\left(v^{i,p} \epsilon_l^{i,p} \psi_\lambda^{i,p} + v^{j,r} \epsilon_l^{j,r} \psi_\lambda^{j,r} \right) I_{(i,1),(j,1)} \leq 1 \tag{3.16}$$

$$1 \leq i, \quad j \leq N(N-1), \quad FW+1 \leq p, \quad r \leq KFW$$

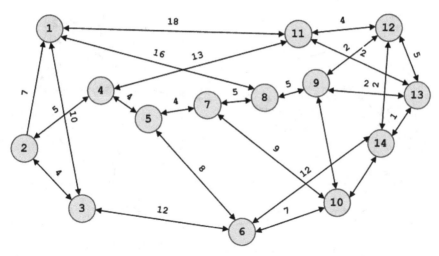

Fig. 3.6. A 14-node 24-link network topology (span distance in hundreds of miles).

3.5.2 Summary and example

The following example illustrates the topology sparsening effect and the applicability of the design formulations to study the effect of various costs and technology trends on network design and upgrades as the network evolves. The objective of the optimization problem is to minimize the total facility cost. Since the cost of provisioning and operating a link is significantly high, the current traffic demand, which is a subset of the future traffic demand, may avoid using some links in the final topology. The meaning in this case is that the cost-optimal routing and capacity planning for the current traffic demand may be fully realized on a subgraph of the final topology.

The eligible routes, between any given node-pair, for primary and backup routing are generated using the shortest-path routing algorithms. The primary paths for a given node-pair are assigned using the shortest-route mapping. Eligible restoration routes for a node-pair are generated to be link-disjoint from the shortest route where the primaries are mapped. To keep the problem sizes computationally manageable the number of eligible restoration routes is typically restricted to three or four per node-pair. The design formulation has a choice of restoration routes for each node-pair to better optimize the spare capacity using backup multiplexing techniques.

Topology sparsening. The 14-node, 24-link NSFnet network is used for experimental studies as shown in Fig. 3.6. This network is derived by adding links to the NSFnet T1 backbone shown in [313]. Given that this network topology is designed for some future traffic demand, the network is designed for the current traffic.

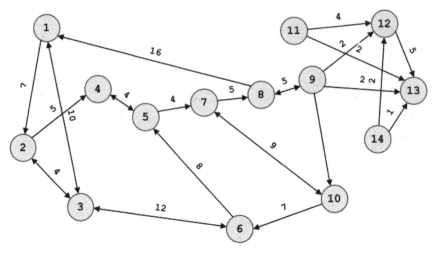

Fig. 3.7. Active links for current traffic demand (span distance in hundreds of miles).

The topology sparsening effect is demonstrated for a traffic demand matrix consisting of 44 requests distributed over 16 node-pairs. The cost values assumed for the design are $C_{lp} = 100$, $C_f = 10$, $C_\lambda = 1$. The resultant "sparse" topology is shown in Fig. 3.7. This sparsening effect is due to the high cost for provisioning and operating a new link. The links used by the primary paths of a node-pair are automatically active in the solution. The key then is to optimize spare capacity allocation among eligible restoration routes. The design formulation tries to maximize the use of a link which is provisioned by routing more demands through it. Identifying these links and backup multiplexing spare capacity to obtain a cost-optimal result is the goal of the design problem. It should be noted that resources such as fibers and wavelengths cannot be increased infinitely on a link. The increase is limited by restrictions such as the maximum number of fibers per link l (F_l), and the maximum number of cross-connects (Θ_n); hence the maximum size of the switching fabric in node n. These restrictions are captured in this formulation.

As evident in Fig. 3.7, some of the active links are unidirectional because the traffic matrix is a unidirectional matrix. This means that resources such as amplifiers are active only in one direction. For some designs, depending on the traffic, it may be cost effective to provision only one direction and when it is upgraded, the other direction may be activated. The installation costs can be avoided when considering the link for upgrade since only a fiber-related investment such as amplifiers, may be required.

The upgraded network for an increased traffic demand is shown in Fig. 3.8. Additional new traffic is introduced between two node-pairs which did not have any direct traffic between them. This may provision some new links as shown in

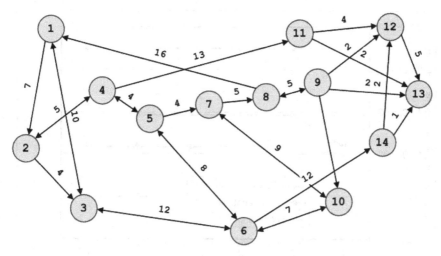

Fig. 3.8. Upgraded network for increased traffic demand (span distance in hundreds of miles).

the figure. When the design problem is solved for an increased traffic demand, the current topology and the resources already committed are input to the formulation. For example, resources such as active links, the number of fibers, and the wavelengths that are already committed, are input as lower bound values to the design problem when the network is considered for upgrade. These resources have already been budgeted and invested in the network.

3.5.3 Evolution with cost and technology trends

To study the effects of cost and technology trends in another experiment the following cost values are used: $C_{lp} = \$16\,000$ per mile, $C_f = \$10$ per meter, and $C_\lambda = \$5000$. The fiber laying costs or the leasing costs (in the case where dark fiber is available) can vary a lot depending on the location. It may be cheaper to lay fibers in rural areas compared to a business district in a metropolitan city. The cost values are conservative estimates and may vary. The number of wavelengths per fiber, which is a technological constraint, may range anywhere from four to more than a 1000 [258]. For the example, a set of $\Psi = \{4,8,12\}$ wavelengths are assumed per fiber.

A key factor is the economy of scale (i.e. how the cost scales with increases in capacity). The cost of fiber (C_f) may scale as $C \times kx$ (increasing the capacity C times results in k times the cost). A $3 \times 2x$ scaling is assumed in the above formulations. This means that tripling the capacity comes at twice the cost. The cost of a 2×2 ($\Omega = 2$) cross-point switching element is $C_{oxc} = \$1000$. The total facility cost is computed using the formula given in Eq. (3.1).

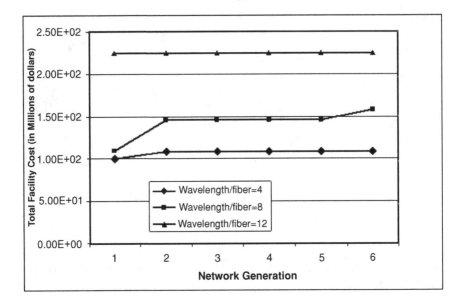

Fig. 3.9. Technology and cost effects on network evolution. (Note: the conclusions drawn here are not to be generalized, as the trends are highly dependent on the network traffic and the underlying topology assumed.)

The effects of cost and technology trends are studied on a nine-node 17-link network for a period of 6 years. In the first year the traffic demand consists of 70 demands distributed uniformly over 20 node-pairs. Every year the traffic is increased by a conservative growth estimate of 20–30%. Every 2 years, additional new traffic is introduced between four new node-pairs. As shown in Fig. 3.9, the total facility cost is plotted for different values of Ψ as the network evolves.

A network designer has to decide what technology to use for implementation, which links to provision and how many fibers and wavelength cards need to be put so that the network is cost effective. Overprovisioning and wrong technology choices can have an adverse effect on revenues. Looking at Fig. 3.9, the following information is obtained. It is evident that $\Psi = 8$ wavelengths per fiber is the most cost effective. The choice of technology (Ψ) is not obvious. It depends on the traffic patterns and the underlying topology. The network design also affects the overall budget in terms of how much to provision.

For the case of $\Psi = 8$, which is cost optimal for the given traffic and topology, the difference between the network cost in year 1 and year 4 is roughly $25 million. Depending on the budgeting constraint the designer may choose to provision the resources required to support year 1 and year 2 traffic and upgrade at the end of 4 years, or if the difference in cost is not significant, the designer may choose to deploy the resources required to support year 4 traffic in the beginning.

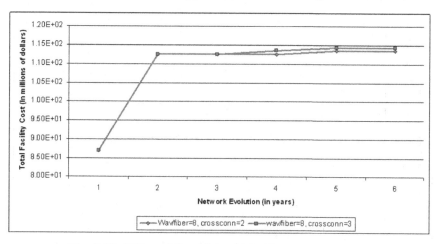

Fig. 3.10. Effect of Ω on the total facility cost for $\Psi = 8$.

On the other hand, as in the $\Psi = 4$ case, the cost differences between the years are significantly high. The difference in cost between year 3 and year 1 is about $32 million and that between year 5 and year 3 is $250 million. It may not be feasible to provision the full topology with all the required capacity at once, hence the network is upgraded as it evolves.

The designer can also study the effect of different switch sizes on the cost. The total facility cost for two different cross-point switching element sizes ($\Omega = 2, 3$) for the $\Psi = 8$ case is shown in Fig. 3.10. The cost of $\Omega = \{2, 3\}$ is assumed to be $C_{oxc} = \{\$1000, \$3000\}$ respectively. The cost difference in this case is fairly low. So the designer may use 3×3 cross-point switching elements to build the switching fabric.

Figure 3.11 shows the total fiber length in the network versus the number of wavelengths used per fiber Ψ for traffic in year 6. The fiber miles decreases as Ψ increases. This does not necessarily mean a reduction in network cost as the total cost depends on various other parameters as seen earlier.

3.6 Heuristic approach for network design

The ILP formulations of the last section for the design and upgrade problem of mesh-restorable optical networks is effective only for small networks with a moderate number of services or demand sets [188]. The complexity of the optimization problem grows exponentially as the size of the network grows or the demand sets increases. Under such scenarios some heuristic approach is needed. Simulated annealing (SA) is an elegant technique to solve the same problem in a reasonable amount of time.

Simulated annealing is a Monte Carlo approach for minimizing multivariate functions. The simulated annealing progresses by lowering the temperature slowly

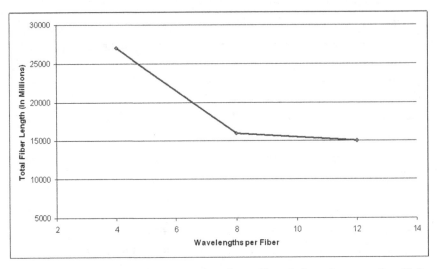

Fig. 3.11. Total fiber length vs. wavelength per fiber Ψ (based on year 6 traffic).

until the system "freezes" and no further changes occur. At each temperature the simulation should run long enough for the system to reach a steady state or equilibrium. This process is known as *thermalization*. The sequence of temperatures and the number of iterations applied to thermalize the system at each temperature comprise an annealing schedule.

Simulated annealing is used as a network design tool to optimize the total network facility cost used to route static connections. To achieve 100% survivability each request is assigned a primary path and a backup path computed using the shortest-cycle (SC) algorithm. The primary and backup paths for each arriving request are calculated dynamically based on the current link weights of the network.

After all the requests of the demand matrix are routed, the total network facility cost is computed. This total cost becomes the *initial solution* for the annealing process. In the thermalization stage (which comprises multiple subtransitions at the same temperature), the request set is shuffled based on different parameters. These steps are repeated until the system reaches a stable state.

To apply simulated annealing to the problem at hand, the following steps are used.

(i) First the system is initialized with a particular initial solution configuration.
(ii) A new configuration is constructed by imposing a displacement by shuffling.

3.6.1 Simulated annealing steps

Three different shuffling metrics were considered: random shuffling, one based on the descending hop length of connections, and another on the ascending hop length required to route the connection. The request set considered for shuffling could be the complete initial request set or a subset of the initial demand matrix.

If the energy of this new state is lower than that of the previous one, the change is accepted unconditionally and the system is updated. If the energy is greater, then the new configuration is accepted probabilistically. This ensures the system moves consistently towards lower energy states, yet may still *jump* out of local minima due to the probabilistic acceptance of some upward moves. It also allows the search to explore a larger search space without being trapped in local optima prematurely.

(iii) The shuffling of request demands is repeated for a fixed number (say X) of times at a particular temperature. At the end of each such subtransition, the objective function is recomputed. If the total network facility cost used to route the shuffled set of connections is less than the initial cost, the latest solution is used as the best objective function, otherwise no update is made.

(iv) After S subtransitions at a particular temperature, an equilibrium state is obtained wherein all the requests of the demand matrix are routed obeying a particular metric such that the overall network cost is minimized. At the end of each such S subtransitions the link utilization on all links are computed. The link utilization of the ith link, given by L_i is the total primary capacity used on the link divided by the total capacity available on a link.

(v) At the end of X subtransitions, the link weights used for the computation of the *SC* are mutated by a factor of $\gamma(1 - L_i)$ during the first transition boundary. This is done such that the shortest cycle explores different possible primary and backup path combinations for each request during each transition boundary at a fixed temperature. This process is repeated a fixed number of (say Y) times.

(vi) The temperature is scaled down linearly and the *thermalization* process restarts again. The objective function is recomputed after the routing of the predetermined set of static requests. If the objective function is reduced from the optimum value the new solution is used to update the objective function value. If the objective function value is greater than the optimal value it is accepted with a certain probability which depends on two parameters, the difference between the objective values δ and the control temperature T at that point of time.

(vii) The probability of acceptance is generally given by $p_a = \exp(-\delta/KT)$, where K is *Boltzmann's constant* and T is determined by the so-called annealing or cooling scheme described in the next section. If this calculated probability at any point given by say (X_i, Y_j) is greater than a particular random number R_k (varying between 0 and 1), then the inferior solution is accepted and is used to update the current solution, otherwise it is rejected. Initially p_a is very high, (i.e. close to 1 and hence greater than R_k), so all bad solutions are accepted. This allows SA to explore bad solution states in the beginning. T decreases as the search proceeds, thus gradually decreasing p_a, the probability of acceptance of a bad solution. As T approaches zero, the search reduces to a greedy search and is trapped in the nearest local optimum.

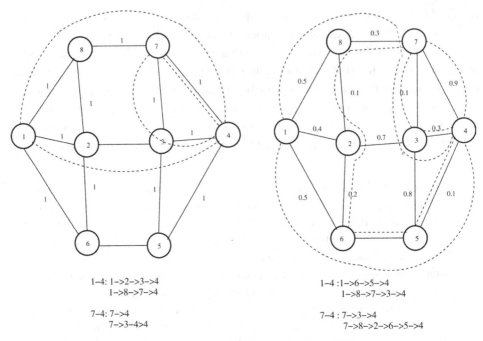

Fig. 3.12. Path mutation in simulated annealing.

Figure 3.12 depicts an example network. The initial primary and backup routes selected by the SC algorithm for requests $1 \to 4$ and $7 \to 4$ for uniform link weights are the shortest paths between the corresponding sources and destinations. But as the edge weights are mutated between two transitions based on the link utilization, the shortest cycle yields two new primary and backup paths for the same connection demands. Hence, the routing through successive edge mutations is equivalent to exploring a larger number of route sets as part of the shortest-cycle algorithm.

The fact that the simulated annealing technique tries different arrangements helps to obtain the best possible backup multiplexing and in turn improves the capacity efficiency, and also for a given node-pair, this in effect emulates the K generalized shortest cycles because of edge weight mutations. The above search is terminated by either repeating the *annealing* process for some predetermined number of iterations or when the search experiences no improvement in the best objective value for a fixed predefined ψ number of annealing steps.

There are different possible annealing schemes to update the temperature T. One may use an annealing scheme where the temperature varies as $T_n = \alpha T_{n-1}$, where T is the temperature at the nth temperature update and α is an arbitrary constant between 0 and 1. The parameter α decides how slowly T decreases. Typical values of α lie between 0.9 and 0.95. The parameters Y, α, and the initial value of T play a critical role for the performance of the simulated annealing. In the experimental studies, the temperature update for the annealing scheme is made as $T_n = T_0/(1 + \alpha T_{n-1})$. The typical value of α is of the order of 0.01–0.1 to have a graceful degradation of the temperature.

3.7 Network upgrade

The method of routing static connections using the simulated annealing technique can also be extended to the upgrading of networks. As a network evolves over time, the challenge lies in how to route the incremental traffic demands over the pre-existing network such that the total network facility cost is minimized at every instant in the process of network evolution [109, 283]. The goal is to reuse most of the resources from the previous generation of the network and to evolve out of it to accommodate the additional traffic set.

A link in the network may be used to realize the working and spare capacity requirements. The objective would be to incur a one-time *fixed charge* for the acquisition and to incur an incremental stepwise increase in cost with increasing capacity as additional transmission systems are turned on.

In the most generalized model a traffic model is represented as $\{R_1, R_a\}$, where R_1 is the number of connections that leave the system during the transition from one generation to the next and R_a defines the number of connections that arrive in the next generation.

In the experimental studies there are two scenarios: (1) all the demands from the first generation get carried over to the next generation (i.e. $\{|R_1| = 0, |R_a| \geq 0\}$) and (2) the number of demands increases (i.e. $|R_1| \geq |R_a|$), hence the network needs an upgrade. In the second scenario the traffic matrix might be entirely or partially different from the previous generation. The routes of the demands which do not carry on to the next generation are removed, but the installed capacity in the network (in terms of the number of fibers, cross-connects at nodes, etc.) are taken as lower bounds for the simulated annealing framework in the next generation.

At first the static connections of the first generation are optimally routed using the simulated annealing approach. The total optimal network state at the end of the first generation is taken as the lower bound of the SA while routing the connections during the second generation of network evolution. The same process is repeated for each successive generation of network evolution. Hence the final topology is the subset of links that need to be activated, the subset of fibers that need to be lit up, the number of wavelength cards that need to be installed and the cross-connects that need to be configured to realize the traffic.

3.8 Methodology validation

The 14-node, 23-link NSFnet network is used for experimental studies. Each of the links in the network is assumed to be bidirectional. The maximum number of fibers F in the network is assumed to be five; since fibers are expensive components, it is advisable to minimize the installation of fibers. The maximum number of wavelengths considered per fiber is 40. Each node in the network is assumed to be homogeneous (i.e. they employ the same switching architecture). A connection at a node is assumed to be switched between an incoming wavelength on one fiber to the same wavelength on the same or different outgoing fiber. The following cost values are applied for the experimental studies. The link provisioning cost $C_{lp} = \$160$ per mile, the fiber provisioning cost is kept at $C_f = \$1000$ per mile, the wavelength cost is kept at $C_\lambda = \$1$, and the cost of a $(2 \times 2)\,(\omega = 2)$ cross-connect is kept at $\$1000$. The cost values employed here are conservative estimates obtained from the literature.

The total facility cost is computed using the equation derived in the cost model. The value of Boltzmann's constant is chosen to be of the order of $5\,000$–$10\,000$ such that

$$0 \le \exp(-\delta/Kt_i) \le 1$$

where t_i is the temperature at the ith iteration. The temperature mutation parameter α is taken to be between 0.005 and 0.01 so that the temperature does not drop abruptly. Higher values of α lead to a fast convergence for the simulated annealing procedure. The edge weight mutation parameter γ is chosen to be between 0.5 and 1.0.

The network evolution is considered for a period of 6 years. The initial numbers of requests considered for the first generation were 75, 100, 150, and 200 distributed uniformly over the 22 node-pairs. Every year the traffic is increased by a conservative growth estimate of 15–20%. For the second model of traffic upgrade some connections from the first generation are probabilistically terminated so that they are absent in the next generation and new connections are added such that the overall connections increase across generations.

The total number of subtransitions at a given temperature considered for SA is limited to 10 and the number of transitions across different temperatures is considered to be 20. The numbers chosen are to show different possible solution sets.

3.8.1 Observations

Several observations are made from the study of the simulated annealing approach for routing static connections.

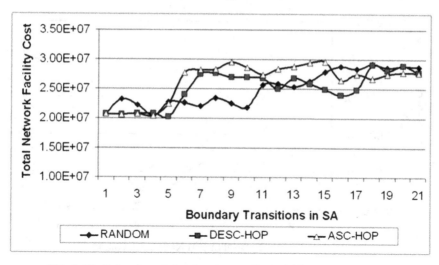

Fig. 3.13. Behavior of simulated annealing within one generation.

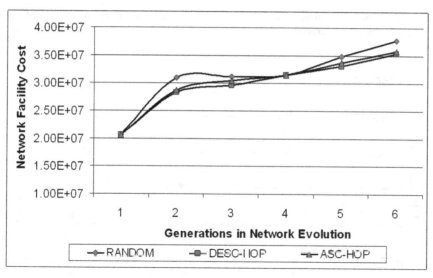

Fig. 3.14. Network facility cost across generations when the initial demand set remains intact.

- Figure 3.13 shows a simulated annealing progress curve for the three different shuffling schemes. As is observed from the figure, the higher values of the objective function were seen by each scheme because the SA was trying to explore a bad solution, which would otherwise not be explored. Simulated annealing hence actually tries to emulate the K shortest paths (K shortest cycles in this case) and does so in a much more unconstrained manner.
- The mutation scheme based on the descending hop length of connections gives the best solution for routing connections as depicted in Fig. 3.14.

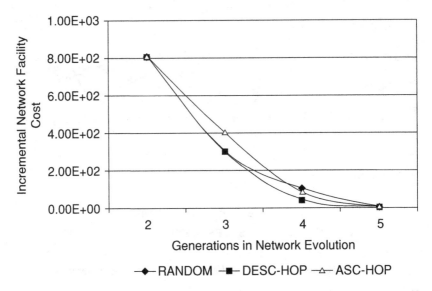

Fig. 3.15. Incremental network facility cost when we have an entirely new traffic pattern in future generations.

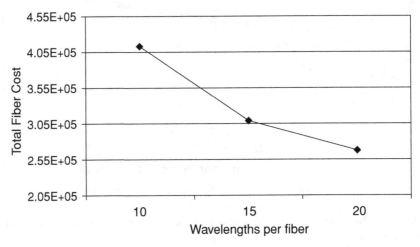

Fig. 3.16. Fiber cost vs. wavelengths per fiber.

This is an intuitive result as the higher hop count connections should be routed prior to shorter hop connections, such that all the requests are ideally packed in. The scheme based on ascending hop length and that based on random mutation performed significantly more poorly compared to the first scheme. In fact, as one progresses to higher and higher generations, a few longer hop connections start getting blocked due to unavailability of a route.

Table 3.1. *Comparison of different SA mutation schemes*

Metric	DESC hop length			ASC hop length			Random		
No. req.	Object	CPU	*E, F*	Object	CPU	*E, F*	Object	CPU	*E, F*
75	3.23E7	1	14, 1	3.58E7	1	14, 1	3.77E7	1	14, 1
100	3.53E7	1.85	18, 2	3.63E7	1.86	19, 2	3.81E7	1.93	19, 3
150	3.63E7	3.39	22, 3	4.16E7	3.34	22, 4	3.97E7	3.54	22, 3
200	4.50E7	5.33	22, 4	4.61E7	5.25	22, 4	4.73E7	5.48	22, 4

Table 3.2. *Comparison of simulated annealing with ILP*

No. req.	Simulated annealing (objective value)	ILP (objective value)
72	1.1687×10^8	1.14×10^8
92	1.684×10^8	1.46×10^8
112	1.8256×10^8	1.72×10^8

- Figure 3.15 shows the incremental costs that are incurred for the second upgrade model. This figure shows that the simulated annealing is successful in routing the connections in successive generations over the facility installed in the initial design itself. The descending hop length scheme performs the best. In future generations, the ascending hop length scheme performs almost similar to the random mutation scheme.

The point to be noted here is that the initial design solution for the first generation itself is an inexpensive one and tries to route traffic using a subset of links out of the complete topology. The successive incremental costs that come in future generations try to reuse the resources from the previous generation and sometimes add new links to accommodate new connections. This is a significant point as every network designer would like to build a network which not only accommodates current traffic, but also can carry a significant amount of future workload.

- Figure 3.16 shows the decrease in fiber cost across six generations as the number of wavelengths per fiber is increased while handling the same traffic demand.
- Table 3.1 represents the performance of the different mutation techniques for different initial traffic demands. No. req. represents the initial number of requests to start with at the beginning of the first generation. The traffic demands are assumed to grow by a conservative estimate of 20% in between two successive generations. Object represents the optimal value of the objective function (i.e. the total network facility cost found at the

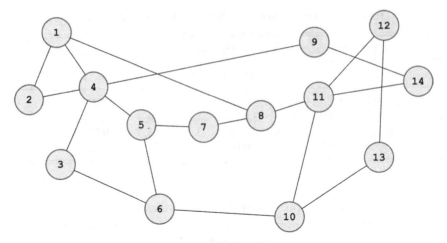

Fig. 3.17. Original topology.

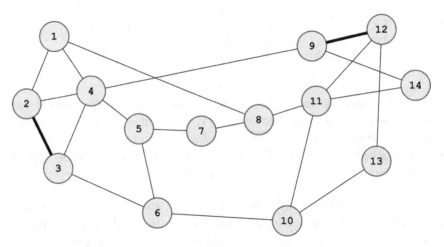

Fig. 3.18. Sparsening effects in simulated annealing.

end of the final generation). CPU represents the total CPU time needed to arrive at the solution for the complete design problem for six generations. E represents the number of edges that were activated after finding a solution and F represents the total number of fibers needed. As is observed from the table, the descending hop length metric scheme performs significantly better than the other two schemes as lower traffic demands are considered. But as the traffic demand increases, the solutions given by all the schemes tend to merge.

• Table 3.2 represents the performance of simulated annealing as compared to an ILP solution. It is noted that the simulated annealing objective value is within 10% of the ILP optimal value. For the above chart a demand set of 72 requests in the first generation, followed by an incremental demand of the order of 20 for the next two generations, is considered. The traffic demand used for this comparison is different from the traffic demand used for the previous experiments with SA.

Another interesting observation is the effect of topology sparsening, which happens in the design study as shown in Fig. 3.17 which illustrates some missing links during a certain generation. Figure 3.18 indicates the links that get added on to accommodate the additional traffic. Simulated annealing illustrates that the topology needed to route a set of connections for any given generation was actually the proper subset of the topology obtained for routing the traffic for a future generation.

4

p-cycle protection

The p-cycle (preconfigured protection cycles) is a cycle-based protection method introduced in [298, 299]. It can be characterized as embedding of multiple rings to act as protection cycles in a mesh network. The p-cycles are configured with spare network capacity to provide protection to connections. The design goal of p-cycle protection is to retain the capacity efficiency of a mesh-restorable network, while approaching the speed of a line-switched self-healing ring [298]. In p-cycle protection, when a link fails, only the end nodes of the failed link need to perform real-time switching. This makes p-cycle similar to SONET/SDH line-switched rings in terms of the speed of recovery from link failures. The key difference between p-cycle and ring protection is that p-cycle protection not only protects the links on the cycle, as is the case for ring protection, it also protects straddling links. A straddling link is an off-cycle link for which the two end nodes are both on the cycle. This important property effectively improves the capacity efficiency of p-cycles. Figure 4.1 depicts an example that illustrates p-cycle protection. In Fig. 4.1(a), A–B–C–D–E–A is a p-cycle formed using reserved capacity on the links for protection. When an on-cycle link A–B fails, the p-cycle can provide protection as shown in Fig. 4.1(b). When a straddling link B–D fails, each p-cycle protects two working paths on the link by providing two alternate paths as shown in Figs. 4.1(c) and (d), for the entire traffic on the link in both directions.

4.1 Design of p-cycle restorable networks

The design of a p-cycle restorable network is usually formulated as an integer linear programming problem, as in [298, 299]. The set of all simple distinct cycles up to some limiting size is generated using cycle generation algorithms. To select the set of p-cycles, an ILP solution identifies the optimal set of p-cycles in spare network capacity by choosing the number of copies of each elemental cycle to be configured as a p-cycle. Two design objectives can be set to design the protection cycles.

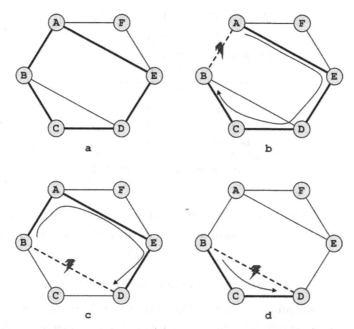

Fig. 4.1. (a) An example of a p-cycle. (b) Protecting on-cycle links. (c), (d) Protecting straddling links.

(i) Maximizing the restorability with a given amount of placement of spare capacity.
(ii) Minimizing the total amount of capacity for a full restorable p-cycle network.

Additional constraints such as the length of a p-cycle or other physical constraints may also apply, as depicted in the example below.

Example 1. Consider the example of Fig. 4.1 again. In this example, all cycles of length up to five are A–F–E–A, A–F–E–D–B–A, A–E–D–B–A, A–E–D–C–B–A, and B–D–C–B. Although among these two cycles, A–F–E–A and B–D–C–B, cover all nodes. However, it is not an acceptable set as the cycles in this set do not cover all links. A set of two cycles, A–F–E–D–B–A and B–D–C–B, cover all links. So does the set consisting of A–F–E–A and A–E–D–C–B–A. Note that the cycle A–F–E–D–C–B–A covers all links and nodes (it is a hamiltonian cycle), but it is not acceptable in this case as it consists of six links and a restriction of five links has been placed on the cycle length in this example.

4.2 Cycle selection algorithms

An algorithm to compute all cycles in a network is described in Section 4.5. For a given network, there may be a large number of cycles. Therefore the ILP approach

Table 4.1. *TS and AE values of all*
cycles in Example 1

Cycle	TS	AE
A–F–E–A	3	1
A–F–E–D–B–A	7	1.4
A–E–D–B–A	4	1
A–E–D–C–B–A	7	1.4
B–D–C–B	3	1

is only suitable in practice for small or medium size networks, because the number of cycles in a graph grows exponentially with the number of nodes and edges.

4.2.1 Use of preselecting a p-cycle candidate set

To reduce the computation complexity, the idea of preselecting a subset of a cycle has been proposed in [109, 300]. In [300], a subset of cycles that have "high merit" are preselected and then provided to an otherwise unchanged optimal solution model. Two preselection metrics have been proposed.

• Topological score (TS):

$$TS(j) = \sum_{i \in S} x_{ij} \tag{4.1}$$

• A priori efficiency (AE):

$$AE(j) = TS(j) \left/ \sum_{(i \in S | x_{ij} = 1)} C_i \right. \tag{4.2}$$

where S is the set of links. C_i is the cost or distance of link i. $x_{ij} = 1$ if link i is part of cycle j, $x_{ij} = 2$ if link i straddles cycle j, and $x_{ij} = 0$ otherwise.

For example, Table 4.1 lists the TS and AE values of all cycles in Example 1. C_i is assumed to be 1 for all of the links in the network. The cycles are sorted first by either TS or AE measures. Instead of using all distinct cycles in the ILP formulation, only a limited number of the top-ranked candidates are used in the optimization solution model.

Another algorithm has been developed in [47] to compute a set of candidate p-cycles. The algorithm generates a combination of high-efficiency cycles and short cycles so that the densely distributed and sparsely distributed working capacity can be efficiently protected. High-efficiency cycles are generated in the early stage by using a weight that considers the efficiency of cycles to control the order of exploring cycles in a depth-first search.

4.2.2 Straddling link algorithm

The straddling link algorithm (SLA) [109] aims to generate a single p-cycle that straddles each network link if possible. The algorithm first computes the shortest path between the end nodes of a link. It then computes the second shortest path between the end nodes of this link. The second shortest path is node-disjoint from the first one. The link itself is excluded from both the paths. The intent of the process is to generate a set of cycles with exactly one straddling link each. For a link that cannot be covered as a straddling link of any cycle, the algorithm generates a cycle formed by the link itself and the shortest path between the end nodes of the link, hence the link is covered as an on-cycle link.

Consider the example network of Fig. 4.1 again. Link A–E is covered by the cycle A–F–E–D–B–A, which is formed by two node-disjoint paths A–F–E and A–B–D–E. Similarly, the link B–D is covered by the cycle A–E–D–C–B–A. Links A–F and E–F in algorithm SLA are covered by cycle A–F–E. Similarly, the links B–C and C–D are covered by the cycle B–C–D, and the links A–B and D–E are covered by the cycle A–B–D–E. SLA can be implemented by two calls of Dijkstra's algorithm. It is simple and very fast. However, the cycles produced are generally inefficient for an overall p-cycle network design because most of the primary cycles generated only have one straddling link.

4.2.3 High-efficiency p-cycle optimization heuristic algorithm

An algorithmic approach for the p-cycle optimization problem has been proposed in [122]. The approach has two steps.

In the first step, a set of candidate p-cycles is computed. Three p-cycle operations are proposed to transform the set of cycles from the SLA into more efficient p-cycles. For each primary cycle generated by the SLA, the operations replace an on-cycle link by a path between the end nodes of the link, thus converting the on-cycle link to a straddling one. The p-cycles generated by these operations have a higher average AE. However, it generates more candidate p-cycles than SLA does.

In the second step, one p-cycle is chosen iteratively from the candidate p-cycle set and used to reduce the unprotected working capacity until all working capacities are protected. In each iteration, the p-cycle with the highest *actual efficiency* is selected, as defined by

$$AE(j) = w_i \times x_{ij} \left/ \sum_{(i \in S | x_{ij} = 1)} C_i \right. \tag{4.3}$$

where w_i is the amount of unprotected working capacity on link i. Unlike the AE measure, the *actual efficiency* depends not only on the number of on-cycle and straddling links, but also on the unprotected working capacity on these links.

4.3 Joint optimization of p-cycle design

The p-cycle design is generally formulated as a non-joint optimization problem, i.e. the working demands are routed via shortest paths first, and then a corresponding minimum spare capacity allocation problem using p-cycles is formulated. In an alternate approach, joint optimization design, one attempts to optimize the choice of routing working connections in conjunction with the p-cycle selection so that the total capacity is minimized. Solving the joint optimization problem is more difficult because of the additional computation complexity. In [300], a preselection heuristic is used to provide a set of "high-merit" p-cycle candidates in order to avoid enumerating all distinct cycles. The joint optimal design may reduce redundancy by up to 25% over non-joint optimal design, as shown in [300].

In another study [67] using the Pan-European Cost 239 network as a test network, it has been shown that in a WDM network without wavelength conversion, the spare capacity used by p-cycles of practical lengths (maximum length 4000–6000 km) consumes about half of the working capacity, while in the presence of wavelength conversion, the ratio of spare to working capacity is between 70% and 90%. The p-cycle design issues have also been studied in [22, 23, 64, 65, 66, 69, 80, 93, 95, 316].

4.4 A p-cycle-based design for dynamic traffic

In a dynamic environment, the demands arrive at a network one by one in a random manner. Therefore, not all information concerning the demands is known in advance. In [94], the performance of *protected working capacity envelopes* (PWCE) based on p-cycle protection was studied. The concept of *protected working capacity envelopes* was proposed in [296] to deal with dynamic traffic. The idea of PWCE is to provision over inherently protected capacity, as opposed to explicit provision protection for every dynamically arriving connection. In PWCE, an envelope of working capacity and a separate spare capacity part are created by an offline planning process in such a way that the envelope of the working capacity is protected by the spare capacity part. Therefore, provisioning a protected service for a dynamically arriving connection is done simply by routing a connection in the protected working envelope. The PWCE concept has the potential to offer implication and operation advantages, and to be more scalable and robust than shared-path-based protection.

In p-cycle protection, for on-cycle links, up to half of the capacity has to be set aside for protection, whereas for straddling links, there is no need to reserve spare capacity at all. In the case of dynamic traffic, to use the p-cycle concept for protection, the following two aspects have to be considered.

- *Selecting a set of p-cycles.* Since not all information about the demands is known in advance, every link in the network has to be protected, as every link may carry traffic in the future. In order to do this, a set of cycles needs to be selected in such a way that every link is either on at least one cycle or is a straddling link of a cycle, i.e. the end nodes of every link are on at least one duplicate cycle. This set of cycles serves as p-cycles.
- *Capacity allocation.* Again, due to a lack of full information in advance, to provide 100% protection against any single link failure, up to half of the capacity on the links of selected p-cycles needs to be reserved for protection.

To minimize the total reserved capacity, it is desirable to minimize the total length of the selected set of p-cycles. Thus, the p-cycle design for managing dynamic traffic is based on the following steps.

(i) Compute a set of cycles so that the two end nodes of each link in the network are at least on the same cycle, and the total length of all cycles is minimum. These cycles serve as p-cycles. The cycle length may be restricted.
(ii) For each link on a cycle, reserve half of the capacity for protection purposes.
(iii) For each arrived connection request, route the request using the reserved capacity in the network.

4.4.1 Performance matrix

Two redundancy metrics are defined to characterize the capacity utilization: network redundancy and instant redundancy.

Network redundancy (NR) is defined as the ratio of the total reserved capacity for protection to the total available capacity for working traffic. It is determined by the network topology and the total available capacity when the network is built.

Instant redundancy (IR) is defined as the ratio of the total used capacity for protection to the total used capacity for working traffic at an instant in time. IR characterizes the capacity utilization for the traffic that has arrived.

NR can be computed using

$$NR = \frac{\sum_{k=1}^{L'} 0.5 \times C \times \text{length}(k)}{\sum_{j=1}^{L} C \times \text{length}(j) - \sum_{k=1}^{L'} 0.5 \times C \times \text{length}(k)} \tag{4.4}$$

where j and k are link IDs, L is the total number of unidirectional links in the network, and L' is the total number of links that are on any of the cycles that are selected as p-cycles. Assume that every link in the network has equal length, and initial available capacity on every link is the same, denoted as C, then

$$NR = \frac{0.5CL'}{LC - 0.5CL'} = \frac{L'}{2L - L'} \tag{4.5}$$

If the network is represented as an undirected graph, the set of selected cycles needs to cover every node in the network at least once. Therefore $L' \geq N$, where N is the number of nodes in the network. Then

$$NR \geq \frac{N}{2L - N} \tag{4.6}$$

If the network is represented as a directed graph, and we assume that the connection between any two nodes in the network is bidirectional, it is necessary that the set of selected cycles covers every node in the network at least twice, one in each direction. In this case, $L' \geq 2N$. Therefore,

$$NR \geq \frac{N}{L - N} \tag{4.7}$$

Equations (4.6) and (4.7) provide lower bounds for NR under the assumptions that every link in the network has equal length and the initial available capacity on every link is the same.

4.4.2 Determining an optimal set of p-cycles

The problem of finding an optimal set of p-cycles is the first step in the design and is defined as follows. *Given a network topology, represented as an undirected graph $G(V, E)$, where $|V| = N$ and $|E| = L$, identify a set of cycles with minimum total length so that for $\forall j \in E$, such that at least one cycle contains both the end nodes of j in the set of cycles.* This problem can be formulated as an integer linear programming (ILP) problem. Suppose P denotes the set of all simple distinct cycles in the graph; see [57] to compute all simple cycles. The ILP solution then selects a set of p-cycles.

The following notation is used in the ILP formulation. The problem is first formulated for an undirected graph, and then modified for a directed graph.

4.4.2.1 Notation

- $j = 1, 2 \ldots, P$: number assigned to a cycle.
- L_l: length of link l.
- ω_j^l: link indicator, which takes a value of unity if link l is on cycle j and zero otherwise (data).
- σ_j^l: protection indicator. In an undirected graph, it takes a value of unity if link l is on cycle j or is a straddling link of cycle j. It takes a value of zero otherwise (data).
- δ_j: takes a value of unity if cycle j is chosen as a p-cycle in the design and zero otherwise (a binary variable).

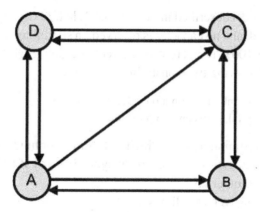

Fig. 4.2. An illustrative example of p-cycle protection in a directed graph.

4.4.2.2 Problem formulation for an undirected graph

(i) *Objective*: minimize the total length of all p-cycles:

$$\min \sum_{l=1}^{L} \sum_{j=1}^{P} \delta_j \, \omega_j^l L_l \tag{4.8}$$

(ii) *Subject to constraint*: both end nodes of each link are at least on the same cycle:

$$\sum_{j=1}^{P} \delta_j \sigma_j^l \geq 1 \qquad \forall l \in L \tag{4.9}$$

4.4.2.3 Directed-link networks and p-cycle protection

Although the idea of p-cycle protection in a directed graph is the same as in an undirected graph, there are some necessary modifications.

- In a directed graph, the links and cycles are unidirectional, therefore a p-cycle provides only one restoration path for its straddling links. Thus a p-cycle can only protect the working traffic on a straddling link up to the capacity reserved on each link of the p-cycle.
- A p-cycle can provide protection for a link $x \rightarrow y$ if the counter-direction link $y \rightarrow x$ is on the cycle.

To illustrate these, consider the example depicted in Fig. 4.2. Suppose cycle $A \rightarrow D \rightarrow C \rightarrow B \rightarrow A$ is a p-cycle. It can provide protection for links $A \rightarrow B$, $B \rightarrow C$, $C \rightarrow D$ and $D \rightarrow A$, for which counter-direction links are on the cycle. For the straddling link $A \rightarrow C$, the p-cycle $A \rightarrow D \rightarrow C \rightarrow B \rightarrow A$ provides only one restoration path $A \rightarrow D \rightarrow C$. There is no restoration path in the other direction on the cycle.

In the network model described in Section 4.4.1, half of the capacity on each on-cycle link is reserved, and no capacity is reserved on straddling links for protection. In order to provide 100% protection in a directed graph, each link l in the graph must meet one of following two constraints.

- The counter-direction link of l is on a p-cycle.
- Link l is a straddling link between two p-cycles.

To formulate the constraint, σ_j^l is modified as follows: for unidirectional networks, it is 2 if the counter-direction link to l is on cycle j. It is 1 if link l is a straddling link of cycle j. It is 0 otherwise.

The formulation is given as follows.

(i) *Objective*: same as Eq. (4.8).
(ii) *Subject to constraint*: for each link l, the counter-direction link to l is on a cycle or link l is a straddling link of two p-cycles.

$$\sum_{j=1}^{P} \delta_j \sigma_j^l \geq 2 \quad \forall l \in L \qquad (4.10)$$

4.4.3 Accommodating connections: routing strategies

Three different routing strategies can be used to route connections:

(i) Shortest-path routing (SPR).
(ii) Least-loaded routing (LLR).
(iii) Most-free routing (MFR).

Three routing strategies are defined as follows.

- *Shortest-path routing*: a connection is always routed on its first shortest path of the node-pair of this connection.
- *Least-load routing*: the maximum load link on a path is defined as the link that has maximum load among all the links on the path. Here the load is given in terms of the number of wavelengths used. The load on a path is defined as the load on the maximum load link on the path. When a call arrives, the load on the alternate paths of the node-pair are compared. The path that has the least load is chosen for routing.
- *Most-free routing*: a least-free link on a path is defined as the link that has least free capacity among all the links on the path. The free capacity on a path is defined as the free capacity on the least-free link on the path. When a call arrives, the free capacity on the alternate routes of the node-pair are compared. The path that has the most free capacity is chosen for routing. Note that although it is assumed that the initial capacity on every link in the network is the same, after half of the capacity on every link of selected p-cycles is reserved for protection, the available capacity for routing working traffic on every link is not the same. Therefore, LLR and MFR are not the same.

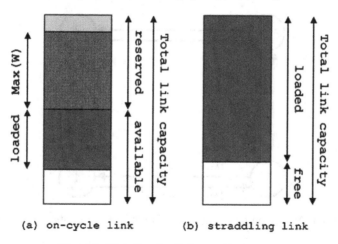

Fig. 4.3. Notation for link capacity usage.

For each of the routing strategies, a set of k edge-disjoint shortest paths are pre-computed for each node-pair. Figure 4.3 describes the notation for link capacity usage. The loaded capacity on a link at an instant of time is the capacity allocated to carry working traffic on the link. For every link on the p-cycles, half of the total capacity has been reserved for protection, as shown in Fig. 4.3(a). At an instant time t, the actual used spare capacity on every link of p-cycles is the same, denoted as Max(W), as shown in Fig. 4.3(a). It can be calculated as follows. At time t, the maximum number of wavelengths used for carrying working traffic on any of the links on p-cycles is denoted as Max(W_{cycle}). The maximum number of wavelengths used for carrying working traffic on any of the straddling links of p-cycles is denoted as Max($W_{\text{straddling}}$). Max(W) is the number of wavelengths required on every link of p-cycles to provide 100% protection at time t. Therefore Max(W) = Max(Max(W_{cycle}), $\frac{1}{2}$Max($W_{\text{straddling}}$)). There is no reserved capacity on straddling links. The rest of the capacity is called free capacity on a link.

4.4.3.1 Wavelength continuity constraint

It is assumed that no wavelength conversion is available in the network. Therefore the wavelength continuity constraint needs to hold for a primary path. Since p-cycle protection is a link-based protection method, in the absence of wavelength conversion, the p-cycle has to use the same wavelength as the working traffic for protection. Fiber-lever protection is employed to meet the wavelength continuity constraint. In this case, every link in the network is assumed to have two fibers, one in each direction. Half of the capacity on every fiber is reserved in such a way that working traffic using a wavelength on one fiber in one direction can always be backed up by the same wavelength on another fiber in another direction.

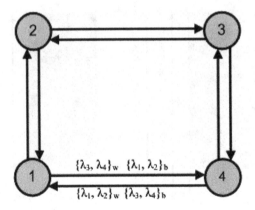

Fig. 4.4. Wavelength continuity constraint in a p-cycle-based design.

Figure 4.4 gives an example. Suppose two cycles $1 \to 4 \to 3 \to 2 \to 1$ and $1 \to 2 \to 3 \to 4 \to 1$ are p-cycles. Assume that every link has four wavelengths λ_1 to λ_4. Wavelengths λ_1 and λ_2 are reserved on every link on the cycle $1 \to 4 \to 3 \to 2 \to 1$ for protection. Wavelengths λ_3 and λ_4 are reserved on every link on the cycle $1 \to 2 \to 3 \to 4 \to 1$ for protection. Wavelengths λ_3 and λ_4 on the links of the cycle $1 \to 4 \to 3 \to 2 \to 1$ can be used for carrying working traffic, which can be backed up by the same wavelengths on the cycle $1 \to 2 \to 3 \to 4 \to 1$. Similarly, λ_1 and λ_2 on the links of cycle $1 \to 2 \to 3 \to 4 \to 1$ can be used for carrying working traffic, which can be backed up using the same wavelengths on the cycle $1 \to 4 \to 3 \to 2 \to 1$. In Fig. 4.4, the set of wavelengths reserved on a link for protection is denoted as $\{\lambda_i\}_b$, the set of wavelengths that can be used for carrying working traffic on a link is denoted as $\{\lambda_i\}_w$.

4.4.4 Effectiveness of the p-cycle-based design

To study the effectiveness of the p-cycle-based design, simulation experiments are carried out using four topologies. Figure 4.5 shows the Pan-European COST 239 network [207] with 11 nodes and 26 links. To study the effect of the average nodal degree of a network on the performance of a p-cycle-based design, three topologies are created by deleting edges from the COST 239 network. Instead of arbitrarily selecting edges to be deleted, the following process is used to select the edges to be deleted.

The requests in the traffic matrix shown in Table 4.2 are routed one by one using shortest-path routing. The traffic matrix is obtained by dividing every entry of the traffic matrix in [207] by 2.5 Gbit/s. The demand unit in the matrix is one wavelength. Note that the final outcome is not dependent on the order of routing requests, since every request is always routed on the shortest path between the

Table 4.2. *Request matrix for the COST 239 network*

Node	1	2	3	4	5	6	7	8	9	10	11
1	0	1	1	3	1	1	1	1	1	1	1
2	1	0	5	8	4	1	1	10	3	2	3
3	1	5	0	8	4	1	1	5	3	1	2
4	3	8	8	0	6	2	2	11	11	9	9
5	1	4	4	6	0	1	1	6	6	1	2
6	1	1	1	2	1	0	1	1	1	1	1
7	1	1	1	2	1	1	0	1	1	1	1
8	1	10	5	11	6	1	1	0	6	2	5
9	1	3	3	11	6	1	1	6	0	3	6
10	1	2	1	9	1	1	1	2	3	0	3
11	1	3	2	9	2	1	1	5	6	3	0

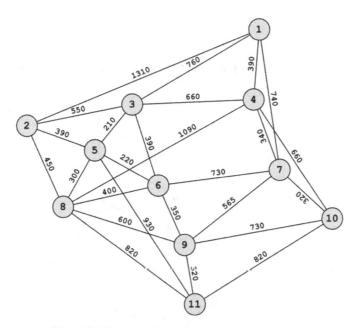

Fig. 4.5. The Pan-European COST 239 network.

source–destination node-pair. After all the requests are routed, the load on every link is calculated, and the links are sorted in increasing order of their link loads. The networks in Figs. 4.6–4.8 are created by deleting the first 5, 9, and 12 links in the sorted list, respectively. The above link selection process is based on the belief that removing the link that carries the least load would have the least impact on other links in the network. The length-based cost is considered, i.e. the length of a link is the real distance between the two cities in kilometers.

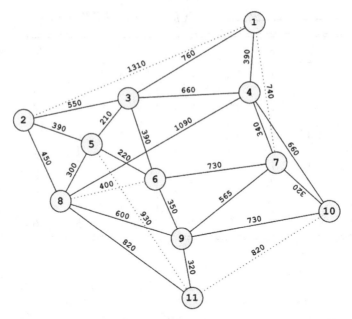

Fig. 4.6. Modified Pan-European COST 239 network with 21 links.

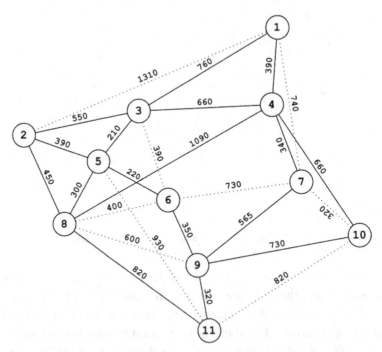

Fig. 4.7. Modified Pan-European COST 239 network with 17 links.

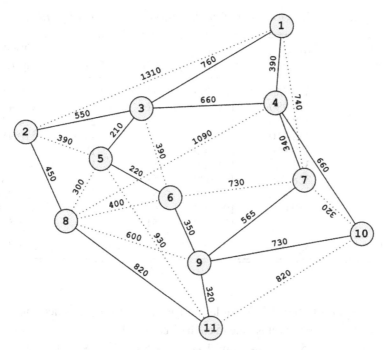

Fig. 4.8. Modified Pan-European COST 239 network with 14 links.

The dynamic traffic is generated based on the traffic matrix shown in Table 4.2. The requests are randomly generated one at a time using the distribution in this table. The service time of every request is assumed to be infinite, i.e. the connections are not released once they are established. Here dynamic traffic means that the network has no information on what requests are due to arrive. That is, upon the arrival of a request, the information on existing lightpaths for previous requests is available. No request is allowed to be rerouted to accommodate the most recently arrived request. The network selects the best route for the current request at the time of arrival, depending on the current network status.

A request is routed using the routing algorithms defined in Section 4.4.3. Two scenarios are considered for the performance evaluation. First, it is assumed that the initial capacity of the network is enough to accept all connection requests. Next, the blocking performance is studied, assuming that the network has limited resources.

4.4.5 Results and discussion

4.4.5.1 COST 239 network

Finding an optimal set of p-cycles. Assuming the network is bidirectional, a total of 7062 simple distinct cycles are computed. This entire cycle set is provided

Table 4.3. *Performance of the least-loaded routing algorithm*

Simulation index	Working capacity	Spare capacity	IR (%)	Max(W)	Total capacity
1	316 750	161 500	51	17	478 250
2	316 980	161 500	51	17	478 480
3	328 930	190 000	58	20	518 930
4	330 410	180 500	55	19	510 910
5	350 900	171 000	49	18	521 900
6	323 470	171 000	53	18	494 470
7	328 785	171 000	52	18	499 785
8	339 430	199 500	59	21	538 930
9	310 620	199 500	64	21	510 120
10	334 585	171 000	51	18	505 585
Average	328 086	177 650	54	19	505 736

as candidate cycles to the ILP formulation for optimal p-cycle selection. The ILP solution identifies two p-cycles, one in each direction: $1 \to 4 \to 7 \to 10 \to 11 \to 9 \to 6 \to 8 \to 2 \to 5 \to 3 \to 1$ and $1 \to 3 \to 5 \to 2 \to 8 \to 6 \to 9 \to 11 \to 10 \to 7 \to 4 \to 1$. The total length of the two cycles is 9500 km.

Capacity utilization. Assume that every link in the network has 60 wavelengths. The total length of all links in the network (considered to be bidirectional) is 30 090 km. The total capacity in the network is therefore 1 805 400 wavelength × km. A total of 30 wavelengths on the links of two cycles are reserved for protection. The total reserved capacity is thus 285 000 wavelength × km. The network redundancy is 18.8%, computed using Eq. (4.4) in Section 4.4.1.

Ten dynamic request sequences are generated using the method described in Section 4.4.4. That is the same distribution as in Table 4.2. However, in each scenario, the request arrives in a different order. The requests are routed one by one using the three algorithms defined in Section 4.4.4. Tables 4.3 and 4.4 list the capacity utilization using the LLR and MFR algorithms, respectively. The working capacity on a link is the capacity used to carry the working traffic on a link, computed by multiplying the number of wavelengths by the length of the link. The working capacity in column II is the total of the working capacity required on all links. Column III gives the total spare capacity used on the p-cycles after the last request has been routed. It is calculated using the following equation:

$$\text{Total spare capacity} = \sum_{k=1}^{L'} \text{length}(k) \times \text{Max}(W) \qquad (4.11)$$

Table 4.4. *Performance of the most-free routing algorithm*

Simulation index	Working capacity	Spare capacity	IR (%)	Max(W)	Total capacity
1	332 530	152 000	45	16	484 530
2	334 625	180 500	53	19	515 125
3	334 970	171 000	51	18	505 970
4	333 600	142 500	43	15	476 100
5	341 320	171 000	50	18	512 320
6	326 925	133 000	41	14	459 925
7	348 500	171 000	49	18	519 500
8	342 980	161 500	47	17	504 480
9	341 955	171 000	50	18	512 955
10	344 150	171 000	50	18	515 150
Average	338 155	162 450	48	17	500 605

where $\text{Max}(W)$ is defined in Section 4.4.3. Column IV in Tables 4.3 and 4.4 gives the instant redundancy, as defined in Section 4.4.1. Column V gives $\text{Max}(W)$. Column VI gives the total capacity used after the last request has been routed. The unit for capacity is (wavelength $\times$ km).

Since it is assumed that all calls stay in the network, the total working capacity and the total spare capacity for SPR are the same for each of the ten request sequences. The total working capacity is 270 230 and the total spare capacity is 247 000. The instant redundancy is 91%.

SPR uses the least working capacity, as each request is always routed on the shortest path, and requires the most spare capacity among the three routing algorithms. On the other hand, MFR uses the most working capacity and requires the least spare capacity. The reason is that MFR distributes the load most evenly. Hence $\text{Max}(W)$ is the least among all three. On average, MFR uses the least total capacity after the last request has been routed and SPR uses the most total capacity.

Comparison with backup multiplexing protection. For the purposes of a comparison, simulations for backup multiplexing protection (BMP) are conducted for the same ten request sequences. For each node-pair, three edge-disjoint shortest paths are precomputed. As a call arrives, a pair of paths from three alternate paths of the node-pair are selected as primary and backup paths in such a way that the total increase in capacity due to routing this connection is minimum. The backup capacity is always shared when this is possible and does not violate 100% protection. The wavelength continuity constraint applies to both primary and backup paths.

Table 4.5. *Path protection with backup multiplexing*

Simulation index	Working capacity	Spare capacity	IR (%)	Total capacity
1	300 720	225 905	75	526 625
2	306 080	208 660	68	514 740
3	295 200	237 635	80	532 835
4	304 170	244 135	80	548 305
5	305 505	217 170	71	522 675
6	296 375	265 120	89	561 495
7	309 995	238 010	76	548 005
8	298 090	261 385	87	559 475
9	301 135	242 635	80	543 770
10	295 520	261 570	88	557 090
Average	301 279	240 222	79	541 501

The first-fit policy is used for wavelength assignment. The results are depicted in Table 4.5. On average, BMP uses less working capacity than LLR and MFR, and more working capacity than SPR. BMP uses almost the same spare capacity as SPR. Among all schemes, BMP uses the most total capacity. Therefore, in terms of total used capacity, the p-cycle-based design using the above routing algorithms performs better than BMP for this network.

Blocking performance. To study the performance under the condition of insufficient resources in the network, every link is assumed to have 30 wavelengths. The total capacity in the network is then 902 700 wavelength × km. 15 wavelengths on the links of two cycles are reserved for protection. The total reserved capacity is 142 500 wavelength × km. As the network load increases, some of the requests are blocked. Table 4.6 presents the results for the same ten request sequences.

The p-cycle-based design with SPR has the largest number of blocked requests. This is expected, as there is only one route allowed for each request in SPR. The p-cycle-based design with LLR or MFR have an average of one blocked request, which is less than the average of four blocked requests in the SBPP scheme. This is consistent with the result in Section 4.4.5.1 where sufficient capacity is assumed. Under the assumption that there is enough capacity in the network, it is observed that after all 110 requests are routed, BMP uses slightly more total capacity than the p-cycle design. Therefore, when the network capacity is limited, BMP leads to a greater number of blocked requests than LLR or MFR.

Table 4.6. *Blocking (assume 30 wavelengths per link)*

Simulation index	Number of blocked requests			
	SPR	LLR	MFR	BMP
1	9	2	0	4
2	10	0	1	2
3	9	1	0	3
4	9	3	3	4
5	8	1	4	3
6	8	2	0	3
7	10	1	1	3
8	10	1	1	3
9	9	0	1	7
10	8	1	2	4
Average	9	1	1	4

Table 4.7. *Average nodal degree of four topologies*

Topology	Average nodal degree
COST 239	4.7
Modified COST 239 with 21 links	3.8
Modified COST 239 with 17 links	3.1
Modified COST 239 with 14 links	2.5

4.4.5.2 *Effect of the network connectivity*

The network connectivity may affect the performance of both the p-cycle-based design and backup multiplexing. The reason for this is that in a high-connectivity network, there are several choices for finding backup paths as well as p-cycles to provide protection. The average nodal degree is used to measure the connectivity of a network. To study this effect various topologies as shown in Figs. 4.6–4.8 with different numbers of links and average nodal degree, as shown in Table 4.7. They are variations of the COST 239 topology with links progressively removed to reduce the network connectivity and the nodal degree. Network simulations are performed using the same set of ten request sequences as earlier.

For the modified 21-link network in Fig. 4.6, the ILP solution identifies two hamiltonian cycles as p-cycles, one in each direction:

(i) $1 \rightarrow 3 \rightarrow 6 \rightarrow 5 \rightarrow 2 \rightarrow 8 \rightarrow 11 \rightarrow 9 \rightarrow 10 \rightarrow 7 \rightarrow 4 \rightarrow 1$.
(ii) $1 \rightarrow 4 \rightarrow 7 \rightarrow 10 \rightarrow 9 \rightarrow 11 \rightarrow 8 \rightarrow 2 \rightarrow 5 \rightarrow 6 \rightarrow 3 \rightarrow 1$.

Table 4.8. *Performance of original COST 239 network*

Routing schemes	Average working capacity	Average spare capacity	Average IR (%)	Average total capacity
SPR	270 230	247 000	91	517 230
LLR	328 086	177 650	54	505 736
MFR	338 155	162 450	48	500 605
SBPP	301 279	240 222	79	541 501

Table 4.9. *Performance of modified COST 239 network with 21 links*

Routing schemes	Average working capacity	Average spare capacity	Average IR (%)	Average total capacity
SPR	275 430	266 760	97	542 190
LLR	318 619	218 539	69	537 037
MFR	315 511	194 940	62	510 450
SBPP	313 665	258 003	82	569 668

For the modified 17-link network in Fig. 4.7, five cycles are preselected for p-cycle-based protection by the ILP as the best option:

(i) $1 \rightarrow 3 \rightarrow 2 \rightarrow 8 \rightarrow 5 \rightarrow 6 \rightarrow 9 \rightarrow 10 \rightarrow 4 \rightarrow 1$.
(ii) $1 \rightarrow 4 \rightarrow 7 \rightarrow 9 \rightarrow 11 \rightarrow 8 \rightarrow 5 \rightarrow 3 \rightarrow 1$.
(iii) $4 \rightarrow 10 \rightarrow 9 \rightarrow 7 \rightarrow 4$.
(iv) $2 \rightarrow 3 \rightarrow 5 \rightarrow 8 \rightarrow 2$.
(v) $5 \rightarrow 8 \rightarrow 11 \rightarrow 9 \rightarrow 6 \rightarrow 5$.

For the modified 14-link network in Fig. 4.8, four cycles are preselected for p-cycle-based protection by the ILP as the best option:

(i) $1 \rightarrow 3 \rightarrow 2 \rightarrow 8 \rightarrow 11 \rightarrow 9 \rightarrow 10 \rightarrow 4 \rightarrow 1$.
(ii) $1 \rightarrow 4 \rightarrow 7 \rightarrow 9 \rightarrow 6 \rightarrow 5 \rightarrow 3 \rightarrow 1$.
(iii) $2 \rightarrow 3 \rightarrow 5 \rightarrow 6 \rightarrow 9 \rightarrow 11 \rightarrow 8 \rightarrow 2$.
(iv) $4 \rightarrow 10 \rightarrow 9 \rightarrow 7 \rightarrow 4$.

Assuming the initial capacity of the network is sufficient to accept all connection requests, Tables 4.8–4.11 list the average working capacity, the average spare capacity, the average IR and the average total capacity of a p-cycle-based design and shared backup path protection (SBPP) on four networks.

Figure 4.9 shows the relationship between the total average capacity utilization and the average nodal connectivity for four topologies. Figure 4.10 shows the

Table 4.10. *Performance of modified COST 239 network with 17 links*

Routing schemes	Average working capacity	Average spare capacity	Average IR (%)	Average total capacity
SPR	290 970	412 200	142	703 170
LLR	349 546	374 573	107	724 118
MFR	347 925	373 240	107	721 165
SBPP	324 391	271 765	84	595 616

Table 4.11. *Performance of modified COST 239 network with 14 links*

Routing schemes	Average working capacity	Average spare capacity	Average IR (%)	Average total capacity
SPR	386 430	623 770	161	950 200
LLR	408 304	603 357	148	1011 702
MFR	406 026	602 039	148	1008 155
SBPP	363 269	412 931	114	776 199

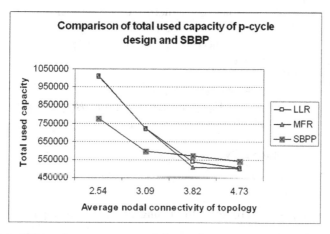

Fig. 4.9. Effect of average connectivity on the average total used capacity.

relationship between the average redundancy and the average nodal connectivity for four topologies.

As the average connectivity of the network increases, the average total used capacity decreases for both p-cycle-based designs and BMP, but the decreasing for the p-cycle-based design is faster than that for BMP. The same trend is observed for the comparison of IR between a p-cycle-based design and BMP. Therefore, the p-cycle-based design outperforms BMP in terms of capacity utilization for dense

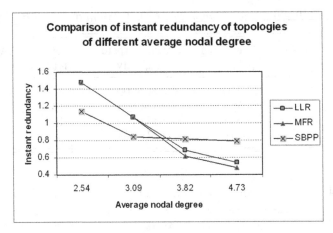

Fig. 4.10. Effect of average connectivity on the average instant redundancy.

$Findallcycles(G)$

```
{
    s = 1;
    while(s < N) do
    {
        A_K = adjacency structure of strong component
              K with least vertex in subgraph
              induced by {s, s + 1, ..., N};
        if(A_k ≠ ∅)
        {
            s = least vertex in V_k;
            for(i ∈ V_K){blocked(i) = false; B(i) = ∅;}
            Findcycles(s); s = s + 1;
        }
        else s = N;
    }
}
```

Fig. 4.11. Procedure for finding all cycles in a graph.

networks, since in dense networks it is likely that more links can be protected as straddling links; whereas BMP performs better than a p-cycle-based design when the network is sparse.

4.5 Algorithm for finding all cycles

The first step in p-cycle design is to find all the simple and distinct cycles in a graph. Finding all cycles in a graph is also used in other network designs. This section

Procedure *Findcycles(s)*
```
{
    for v ∈ A(s) Cycle(s, v);
}
```

Fig. 4.12. Procedure for finding cycle.

Procedure *Cycle(s,u)*
```
{
    flag = false;
    blocked(u) = true;
    for v ∈ A(u)
    {
        if (v = s)
        {
            output "cycle is found";
            found cycle = path + v;
            flag = true;
        }
        else if
        {
            path = path + v;
            flag = flag || Cycle(s,v)
        }
    }
    if (flag = true)
    Unblock (v);
    else
    {
        for v ∈ A(U)
        B(v) = B(v) + u;
        path = path − u;
    }
    return flag;
}
```

Fig. 4.13. Procedure for a cycle.

describes an algorithm developed by D. B. Johnson for directed graphs [57]. A modified version for bidirectional graphs can be found in [297].

The algorithm accepts a directed graph $G = (V, E)$. F is a subgraph of G induced by $W \subseteq V$ and $F = (W, \{(u, v) \mid u, v \in W \text{ and } (u, v) \in E\})$. An induced subgraph F is a strong component of G if for all $u, v \in W$ there exist paths $p_{u,v}$ and $p_{v,u}$

Procedure *Unblock(u)*
{
    blocked(*u*) = false;
    for *w* ∈ *B*(*u*)
        Unblock(*w*);
}

Fig. 4.14. Procedure to unblock bandwidth.

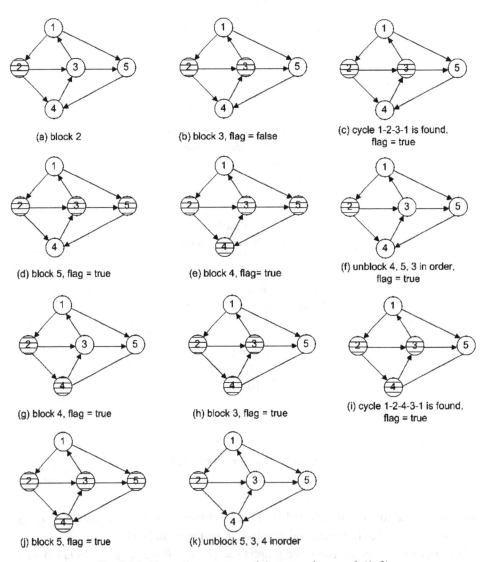

(a) block 2

(b) block 3, flag = false

(c) cycle 1-2-3-1 is found,
flag = true

(d) block 5, flag = true

(e) block 4, flag= true

(f) unblock 4, 5, 3 in order,
flag = true

(g) block 4, flag = true

(h) block 3, flag = true

(i) cycle 1-2-4-3-1 is found,
flag = true

(j) block 5, flag = true

(k) unblock 5, 3, 4 inorder

Fig. 4.15. Execution process of the procedure *cycle*(1, 2).

and this property holds for no subgraph of G induced by a vertex set W' such that $W \subset W' \subseteq V$. Adjacency structure A_G is composed of an adjacency list $A_G(v)$ for each $v \in V$.

The algorithm assumes that vertices are represented by integers from 1 to N. It finds all cycles in the graph by finding cycles from each root vertex s in the subgraph induced by s and vertices "larger than s" in some ordering of the vertices. Therefore the output is grouped according to least vertices of cycles. The pseudocode of the algorithm is shown in Figs. 4.11–4.14.

The main program *Findallcycles*(G) starts from $s = 1$ and increases by 1 in every loop. The *do* loop first finds the strong component K with least vertex in the subgraph induced by $\{s, s + 1, \ldots, s\}$ and A_K, i.e. the adjacency structure of K. If A_K is empty, the program ends, since there are no more cycles in the subgraph. If A_K is not empty, the procedure *Findcycles*(s) is called. *Findcycles*(s) finds all cycles rooted in s in the subgraph induced by s and vertices "larger than s." It is done by calling procedure *cycle*(s, v) for each $v \in A_K$.

The procedure *cycle*(s, v) uses a depth-first search to enumerate all cycles starting from (s, v), i.e. $(s, v, \ldots, s)$ as shown in Fig. 4.15. Figure 4.15 depicts a process of procedure *cycle*$(1, 2)$ in the example network. The search continues by adding edges to the path. When a vertex is appended to a path, it is blocked by setting *blocked* $(v) =$ true, so that v cannot be used twice on the same path. The search continues until a cycle is found or the path is blocked by a blocked node. Unblocking is always delayed sufficiently so that any two unblockings of v are separated by either an output of a new cycle or a return to the main procedure.

5

Network operation

The two most important objectives for network operation are:

(i) capacity minimization
(ii) revenue maximization.

For capacity minimization, there are three operational phases in survivable WDM network operation: (i) initial call setup, (ii) short-/medium-term reconfiguration, and (iii) long-term reconfiguration. All three optimization problems may be modeled using an ILP formulation separately. A single ILP formulation that can incorporate all three phases of the network operation is presented in this chapter. This common framework also takes service disruption into consideration. Typically, most of the design problems in optical networks have considered a static traffic demand and have tried to optimize the network cost assuming various cost models and survivability paradigms. Fast restoration is a key feature addressed in the designs. Once the network is provisioned, the critical issue is how to operate the network in such a way that the network performance is optimized under dynamic traffic.

The framework for revenue maximization is modified to include a service differentiation model based on lightpath protection. A multi-stage solution methodology is developed to solve individual service classes sequentially and to combine them to obtain a feasible solution. Different cost comparisons in terms of the increase in revenue obtained for various service classes with the base case of accepting demands without any protection show the gains of planning and operation efficiency.

5.1 Capacity minimization

Among the three phases of capacity minimization the initial call setup phase is a static optimization problem where the network capacity is optimized for the given topology and the traffic matrix to be provisioned on the network. After the network has been provisioned, demands are admitted based on a routing and wavelength assignment algorithm. The network cannot afford to run optimization procedures to route every call that arrives dynamically. As a result, the utilization of the network capacity slowly degrades to a point where calls may start getting blocked. This triggers various reconfiguration stages that try to better utilize the network capacity. In short-/medium-term reconfiguration, the goal is to optimize resource consumption for backup paths while not disturbing the primary paths of the currently working connections. Since backup paths are used only when the primary path fails, reconfiguring backups does not affect the service. If further optimization is required, a long-term reconfiguration is triggered that requires whole network setup.

The long-term reconfiguration problem can again be treated as a static formulation by allowing rerouting of all working connections and optimizing the network capacity for the expected demand set, comprising both the current working demands and the new demands. The current working demands may not be disrupted (by removing the capacity used by the current working demands) and the network capacity can be optimized for the backup requests and the new demands. The former treatment provides the best capacity optimization. It is possible, however, that all the current connections may be disrupted, which may not be acceptable. The latter case avoids disruption to the current working paths, which may result in poor capacity utilization. To address this tradeoff in the long-term reconfiguration problem, service disruption is captured by adding a penalty term for disrupting existing connections as explained in the optimization section. The need for triggering different stages in network operation are of primary research importance. The network control and management monitors the network dynamics and triggers different reconfiguration stages.

5.2 Revenue maximization

Network service providers offer varying classes of services based on the choice of protection, which can vary from full protection to no protection [91, 203, 286]. Based on the service classes, the traffic in the network is divided into one of the three classes, namely, full protection, no protection, and best-effort.

(i) The first class comprises high-priority traffic that requires full protection in the optical layer. Many carriers may have already greatly invested in their networks, and their equipment may not support protection, and such applications have to rely on the optical layer for legacy network protection.

(ii) The second class comprises high-priority traffic that requires no protection in the optical layer, as they may already be protected by higher layers such as SONET.

(iii) The best-effort class tries to provide protection for the connections based on the resources available. These connections may include IP traffic that have their own slower
protection mechanisms, and hence optical layer protection may be beneficial.

Additionally, traffic which does not have any stringent protection requirements
can pay for protection if the network has enough resources available. The network
typically relies on the best-effort traffic for maximizing revenue. Two variations of
the best-effort class are:

(i) every demand is assigned a primary path – a backup path is assigned if resources are
available;
(ii) the goal is to accept the maximum number of demands with or without a backup.

Managing online reconfiguration. One of the difficulties in adapting a formulation for online reconfiguration in larger and more practical networks arises due
to the combinatorial nature of the optimization problem. These problems typically
take hours to solve for a few hundred demands in small networks with a few tens
of wavelengths. This is still acceptable, as it takes a few weeks to provision a new
connection for a WDM optical network.

The following ILP formulations for network capacity minimization are adapted
to include service differentiation, based on lightpath protection, for revenue maximization in wavelength-routed optical networks.

The following information is assumed to be given: the network topology, a
demand matrix consisting of the new connections to be established for each class, a
set of current working connections and two alternate precomputed routes between
each node-pair. Since each route between every source–destination pair is viewed
as W wavelength continuous paths (lightpaths), one for each wavelength, there
is no need for an explicit wavelength continuity constraint. Information regarding
whether any two given routes are link- and node-disjoint is also assumed to be given.
The ILP solution determines the primary and backup lightpaths for the demand set,
and hence, determines the routing and wavelength assignment.

5.3 Capacity minimization: problem formulation

The objective of this problem is to minimize the network capacity. The first term in
the objective function (Eq. 5.1) denotes the capacity consumed by primary paths,
the second term denotes the capacity consumed by backup paths, and the last term
is a penalty term. If a currently working connection ($\chi^{i,p} = 1$) is reassigned in the
final solution ($\delta^{i,p} = 0$), then the objective value is penalized by adding a cost C_w

to it. Some of the constraints are similar to those in the last chapter and are repeated here for completeness.

Minimize

$$\sum_{i=1}^{N(N-1)} \sum_{p=1}^{KW} \delta^{i,p} \sum_{l=1}^{L} \epsilon_l^{i,p} C_l + \sum_{l=1}^{L} \sum_{\lambda=1}^{W} g_{l,\lambda} C_l$$

$$+ \sum_{i=1}^{N(N-1)} \sum_{p=1}^{KW} \chi^{i,p}(1 - \delta^{i,p}) C_w \tag{5.1}$$

Restoration path wavelength usage indicator constraint: $g_{l,\lambda}$ takes a value of one if wavelength λ is used by some restoration route (i, r) that traverses link l. Constraints (5.3) and (5.4) set $g_{l,\lambda} = 1$, if $X_{l,\lambda} \geq 1$

$$X_{l,\lambda} = \sum_{i=1}^{N(N-1)} \sum_{r=1}^{KW} v^{i,r} \epsilon_l^{i,r} \psi_\lambda^{i,r} \tag{5.2}$$

$$g_{l,\lambda} \leq X_{l,\lambda} \tag{5.3}$$

$$N(N - 1)WK g_{l,\lambda} \geq X_{l,\lambda} \tag{5.4}$$

$$1 \leq l \leq L, \quad 1 \leq \lambda \leq W, \quad X_{l,\lambda} \geq 0$$

Link capacity constraint:

$$\sum_{i=1}^{N(N-1)} \sum_{p=1}^{KW} \delta^{i,p} \epsilon_l^{i,p} + \sum_{\lambda=1}^{W} g_{l,\lambda} \leq W \qquad 1 \leq l \leq L \tag{5.5}$$

Primary path wavelength usage constraint: only one primary path can use a wavelength λ on link l, no restoration path can use the same λ on link l,

$$\sum_{i=1}^{N(N-1)} \sum_{p=1}^{KW} \delta^{i,p} \epsilon_l^{i,p} \psi_\lambda^{i,p} + g_{l,\lambda} \leq 1 \tag{5.6}$$

$$1 \leq l \leq L, \quad 1 \leq \lambda \leq W$$

Backup multiplexing constraint: if $I_{(i,\bar{p}),(j,\bar{r})}$ is one, then only one of the restoration paths can use a wavelength λ on a link l as a backup, since the primary paths share link(s) on their route

$$\left(v^{i,p} \epsilon_l^{i,p} \psi_\lambda^{i,p} + v^{j,r} \epsilon_l^{j,r} \psi_\lambda^{j,r} \right) I_{(i,\bar{p}),(j,\bar{r})} \leq 1 \tag{5.7}$$

$$1 \leq i, \quad j \leq N(N - 1), \quad 1 \leq p, \bar{p}, r, \bar{r} \leq KW$$

Constraint for topological diversity of primary and backup paths: primary and restoration paths of a given demand should be node- and link-disjoint

$$\sum_{p=1}^{W} \delta^{i,p} = \sum_{r=W+1}^{KW} v^{i,r} \tag{5.8}$$

$$\sum_{p=W+1}^{KW} \delta^{i,p} = \sum_{r=1}^{W} v^{i,r} \tag{5.9}$$

Demand constraints for each node-pair:

$$\sum_{p=1}^{KW} \delta^{i,p} = d_i \qquad 1 \le i \le N(N-1) \tag{5.10}$$

$$\sum_{r=1}^{KW} v^{i,r} = d_i \qquad 1 \le i \le N(N-1) \tag{5.11}$$

The ILP can be used in different phases of network operation by appropriately setting the C_w value. For example, in the initial call setup phase, all $\chi^{i,p}$ values are zero as there are no working connections. Hence the third term in Eq. (5.1) is zero. The higher the value of C_w is, the better the guarantee that primary paths of the working connections will remain unaffected. In the short-/medium-term reconfiguration phase, the cost of C_w is typically set very high for the primary paths of the working connections. A high value of C_w does not guarantee that the primary path will not be rerouted in the final solution. In order to avoid disruption to primary paths of working connections, the capacity consumed should be removed and the backup capacity consumption should be optimized. In the long-term reconfiguration phase, an intermediate value of C_w is chosen to capture the tradeoff between possibly disrupting all connections and avoid disrupting any connection.

5.4 Revenue maximization: problem formulation

Objective: the objective function in a revenue maximization problem is to maximize the revenue. Each demand translates into a primary path and a backup path for full protection classes; only a primary path for no protection classes and either only the primary or both the primary and the backup path for the best-effort class, depending on the capacity available. The first term in Eq. (5.12) denotes the revenue generated by primary paths, and the second term denotes the revenue from backup paths. The last term indicates that if a currently working connection ($\chi^{i,p} = 1$) is reassigned in the final solution ($\delta^{i,p} = 0$), then the objective value is penalized by subtracting a cost C_w from it.

Maximize

$$\sum_{i=1}^{N(N-1)} \sum_{p=1}^{KW} \delta^{i,p} C_{ND} + \sum_{i=1}^{N(N-1)} \sum_{p=1}^{KW} v^{i,p} C_D - \sum_{i=1}^{N(N-1)} \sum_{p=1}^{KW} \chi^{i,p}(1 - \delta^{i,p}) C_w \quad (5.12)$$

The constraints (5.2)–(5.7) in Section 5.3 all apply for Eq. (5.12). The new constraints for this revenue maximization problem are as follows.

Constraint for topological diversity of primary and backup paths:

$$\sum_{p=1}^{W} \delta^{i,p} = \sum_{r=W+1}^{KW} v^{i,r} \qquad 1 \le i \le N(N-1) \qquad (5.13)$$

$$\sum_{p=W+1}^{KW} \delta^{i,p} = \sum_{r=1}^{W} v^{i,r} \qquad 1 \le i \le N(N-1) \qquad (5.14)$$

Demand constraints for each node-pair: only one of the service classes described below is active in the formulation. To solve the combined problem for all classes, a different procedure can be adopted, as explained in Section 5.5.

- *Full protection:* every demand is assigned a primary and a backup path. The number of full protection demands for node, pair i is denoted by d_{i1},

$$\sum_{p=1}^{KW} \delta^{i,p} = d_{i1} \qquad 1 \le i \le N(N-1) \qquad (5.15)$$

$$\sum_{r=1}^{KW} v^{i,r} = d_{i1} \qquad 1 \le i \le N(N-1) \qquad (5.16)$$

- *No protection:* every demand is assigned only a primary path. The number of no protection demands for node-pair i is denoted by d_{i2},

$$\sum_{p=1}^{KW} \delta^{i,p} = d_{i2} \qquad 1 \le i \le N(N-1) \qquad (5.17)$$

$$\sum_{r=1}^{KW} v^{i,r} = 0 \qquad 1 \le i \le N(N-1) \qquad (5.18)$$

- *Best-effort protection* (i): only one variation of the best-effort service class can be used in the formulation. This assumption holds when the problem is solved for all classes. Every demand is assigned a primary path. A backup path is assigned if resources are available.

The number of best-effort demands for node-pair i is denoted by d_{i3},

$$\sum_{p=1}^{KW} \delta^{i,p} = d_{i3} \qquad 1 \le i \le N(N-1) \qquad (5.19)$$

$$\sum_{r=1}^{KW} v^{i,r} \le d_{i3} \qquad 1 \le i \le N(N-1) \qquad (5.20)$$

• *Best-effort protection* (ii): accept as many demands as possible with or without backup. The number of best-effort demands for node-pair i is denoted by d_{i3},

$$\sum_{p=1}^{KW} \delta^{i,p} \le d_{i3} \qquad 1 \le i \le N(N-1) \qquad (5.21)$$

$$\sum_{r=1}^{KW} v^{i,r} \le d_{i3} \qquad 1 \le i \le N(N-1) \qquad (5.22)$$

Best-effort class constraints: these constraints are used only when the best-effort class demands are being solved. For best-effort variation 2 class demands (Eqs. 5.21 and 5.22), no backups are admitted without a primary (i.e. for every node-pair, the number of primaries accepted is equal to or greater than the backups). This constraint is required to ensure that when best-effort variation 2 class demands are admitted, the ILP does not admit more backups than primaries. The topological diversity constraint has to be modified while solving for best-effort class demands. This is because all primaries need not be accepted with backups. Both of these constraints can be stated together as follows:

$$\sum_{p=1}^{W} \delta^{i,p} \ge \sum_{r=W+1}^{KW} v^{i,r} \qquad 1 \le i \le N(N-1) \qquad (5.23)$$

$$\sum_{p=W+1}^{KW} \delta^{i,p} \ge \sum_{r=1}^{W} v^{i,r} \qquad 1 \le i \le N(N-1) \qquad (5.24)$$

In Eq. (5.12), the last term indicates that if a currently working connection ($\chi^{i,p} = 1$) is reassigned in the final solution ($\delta^{i,p} = 0$), then the cost C_w is subtracted from the objective function. Since the objective function is to maximize revenue, it ensures that service is not disrupted unless revenue can be increased. The choice of C_w offers flexibility to the network provider. Although the network would like to avoid service disruption to all connections, there may be customers who are willing to pay more and do not wish to be disturbed. This can be accommodated by modifying C_w to be path specific ($c_w^{i,p}$) and setting a higher cost for disrupting such connections.

Complexity issues. The number of variables $\delta^{i,p}$ and $v^{i,p}$ grow rapidly with network size. This effect is more pronounced with an increase in the number of wavelengths. For a network of size $N = 14$, $W = 32$ and $K = 2$, there are

$K \times W = 2 \times 32$ instances of each variable for every node-pair. Since there are $N \times (N - 1) = 182$ node-pairs, this results in 11 648 $\delta^{i,p}$ variables and 11 648 $v^{i,p}$ variables. The number of equations will be roughly 125 million (11 648²). Thus, the problem is complex even for small networks.

A smarter solution would be to consider only variables that are relevant to the problem at hand. This implies that variables that have a zero value are removed. If a node-pair does not have any demands to be routed between them, then all the variables relating to that node-pair are removed.

Thus, for every node-pair that does not have demands to be routed between them, a reduction of $KW = 2 \times 32$ instances of each variable can be obtained. Thus, if only 10 node-pairs have demands to be routed between them, one needs to deal with only 1320² instead of 11 648² equations.

Additional reductions are possible by considering only links that affect the specific instance of demands to be provisioned. For each link not considered, a further reduction of 248² equations can be obtained. The above discussions suggest that it is necessary to carefully enumerate the constraints.

5.4.1 Demand normalization technique

Another procedure that results in significant problem size reductions is the demand normalization technique. Since all demands between every node-pair source and sink at the same nodes, there is no need to distinguish between each of those requests.

To reduce the solution space, each set of requests between every demand pair is treated as one entity. Since the whole network should have a consistent view of each entity, the demand sets are normalized by finding the greatest common divisor for all the demand requests, and then dividing each demand set by that factor. The capacity on all links is also normalized. This results in a scaled-down version of the original problem which is less difficult to solve.

Since the capacity on each link is normalized, the number of wavelengths W reduces by a factor of m, where m is the greatest common divisor of the demand sets. Considering the network with $N = 14$, $W = 32$ and $K = 2$, if m is say 2, the number of variables reduces by a factor of 2, resulting in a reduction in the number of equations to 660² equations, which is an $O(1/m^2)$ reduction. This technique can yield considerable reductions if m is comparable to W. An appropriate procedure that can be adopted here is to adjust demand requests to obtain a value of m comparable to W, with the resulting solution adjusted accordingly.

5.5 Solution methodology

This section describes the solution methodology for solving the revenue maximization problem for all classes of demands.

Multi-stage approach. As expected, the number of variables grows rapidly with the network size. A multi-stage solution methodology needs to be evolved to solve the combined problem for all classes of demands. At each stage, the problem is solved for one of the classes, and the result is used in successive stages.

Stage 1: in the first stage, the problem is solved for the primary paths of full protection and no protection classes. The following modified maximization problem is solved at this stage.

Maximize

$$\sum_{i=1}^{N(N-1)} \sum_{p=1}^{KW} \delta^{i,p} C_{ND} \tag{5.25}$$

Demand constraint:

$$\sum_{p=1}^{KW} \delta^{i,p} = d_{i1} + d_{i2} \qquad 1 \le i \le N(N-1) \tag{5.26}$$

Link capacity constraint:

$$\sum_{i=1}^{N(N-1)} \sum_{p=1}^{KW} \delta^{i,p} \epsilon_l^{i,p} \le W \qquad 1 \le l \le L \tag{5.27}$$

The solution to the above ILP is a set of primary paths (chosen paths will have $\delta^{i,p} = 1$). For the next stage, for every $\delta^{i,p} = 1$ in the Stage 1 solution, the corresponding $\chi^{i,p}$ variables are set to 1 in Stage 2. Thus, the solution from Stage 1 is fed to Stage 2 as working primary paths.

Stage 2: in this stage, the original problem presented in Section 5.4 is solved. The demand constraints for the full protection class (Eqs. 5.15 and 5.16), the no protection class (Eqs. 5.17 and 5.18) and the best-effort variation 2 class (Eqs. 5.21 and 5.22) are modified as follows:

$$\sum_{p=1}^{KW} \delta^{i,p} \ge d_{i1} + d_{i2} \qquad 1 \le i \le N(N-1) \tag{5.28}$$

$$\sum_{p=1}^{KW} \delta^{i,p} \le d_{i1} + d_{i2} + d_{i3} \qquad 1 \le i \le N(N-1) \tag{5.29}$$

$$\sum_{r=1}^{KW} v^{i,r} \ge d_{i1} \qquad 1 \le i \le N(N-1) \tag{5.30}$$

$$\sum_{r=1}^{KW} v^{i,r} \le d_{i1} + d_{i3} \qquad 1 \le i \le N(N-1) \tag{5.31}$$

There is no distinction between demands from different service classes for a given node-pair i. The results of the ILP are interpreted as follows. The first $d_{i1} + d_{i2}, \delta^{i,p}$ variables which are set to 1, are considered to be the primary paths for the full and no protection classes. Any feasible solution to the ILP has to satisfy this constraint. Similarly, the first d_{i1}, $v^{i,r}$ variables which are set to 1, are considered to be the backup paths for the full protection class. Equation (5.30) ensures that the backup paths for the full protection class demands are chosen in this stage. Any excess primary and backup variables are considered to belong to the best-effort class.

Effect of C_w: the effect of the solution depends on the value of C_w, the higher the value is, the greater the guarantee that the path will remain unaffected. A high value of C_w does not guarantee that the primary path will not be rerouted. Typically, this value is set to be some $\beta = 3, 4$ times the cost of primary paths. This implies that the increase in the objective value for choosing β primary paths is lost by disrupting one existing path.

Complexity: this section presents some insights into a possible reduction in complexity at each stage of the multi-stage solution methodology. To understand the reduction in complexity at each stage, first examine Stage 1 of the solution. Since the interest is only in the primary paths for the full protection and no protection class in Stage 1 (backups are chosen in Stage 2 of the solution), this approach results in a direct reduction in the complexity because $v^{i,p}$ variables do not need to be considered in the formulation. Stage 2 complexity depends on the value of C_w. The higher the value of C_w is, the better the guarantee that the path will remain unaffected in the final solution. Since Stage 2 starts with an initial solution, there may be a decrease in the number of combinations that need to be explored. Hence a faster solution can be obtained. However, a higher value of C_w does not guarantee that the solution is faster because the ILP can choose to reroute any or all of the existing connections in an attempt to maximize the objective function. Although the worst case complexity of Stage 2 is the same as that of solving the combined problem for all classes of demands, typically the solution is obtained much faster.

5.6 Performance evaluation

The combined routing and wavelength assignment problem is known to be NP-complete, and the problems addressed here are expected to be NP-complete. As a result, these formulations are not easily adaptable for real-time reconfiguration in larger and more practical networks, even when the techniques discussed in Section 5.4 for problem size reduction are also employed. Several heuristics and

Table 5.1. *Static optimization stage*

Node-pair	SRC–DST	Alternate routes	Primary paths	Backup paths
1	1–2	1 2	$\lambda_1, \lambda_2, \lambda_3, \lambda_4, \lambda_5$	–
		1 3 2	–	$\lambda_1, \lambda_2, \lambda_3, \lambda_4, \lambda_5$
27	3–1	3 1	$\lambda_1, \lambda_2, \lambda_3, \lambda_4, \lambda_5$	–
		3 2 1	–	$\lambda_1, \lambda_2, \lambda_3, \lambda_4, \lambda_5$
110	9–6	9 4 5 6	$\lambda_1, \lambda_2, \lambda_4, \lambda_8, \lambda_9$	–
		9 12 13 6	–	$\lambda_1, \lambda_2, \lambda_4, \lambda_8, \lambda_9$
142	11–13	11 6 13	$\lambda_1, \lambda_2, \lambda_4, \lambda_8, \lambda_9$	–
		11 10 12 13	–	$\lambda_1, \lambda_2, \lambda_4, \lambda_8, \lambda_9$
167	13–11	13 6 11	–	$\lambda_1, \lambda_2, \lambda_4, \lambda_8, \lambda_9$
		13 12 10 11	$\lambda_1, \lambda_2, \lambda_4, \lambda_8, \lambda_9$	–

decomposition techniques have been explored to significantly reduce the computational complexity of the original problem [36, 42, 52, 188, 284].

A performance evaluation study to analyze the behavior of ILP formulation is studied on a 14-node 21-link NSFnet topology with one fiber per link and 10 wavelengths per fiber. To compare the increase in revenue using the two variations of the best-effort class, results are obtained for various demand sets on the NSFnet topology and the 20-node 32-link ARPANET topology. In an N-node system, nodes are numbered from 1 to N, and the number of a node-pair (i, j) is obtained using the following formula:

$$
\text{node-pair number} =
\begin{cases}
(N-1) \times (i-1) + j & \text{if } j < i \\
(N-1) \times (i-1) + (j-1) & \text{if } j > i
\end{cases}
\tag{5.32}
$$

5.6.1 Capacity minimization

Initial call setup: for a small experiment, a set of 25 demands distributed uniformly across only five node-pairs, as shown in Table 5.1, are considered (the source and destination nodes are represented by SRC and DST, respectively). In the static optimization stage, there are no current working connections and hence the demand matrix is provisioned by providing a primary and backup path for each demand. The resulting routing and wavelength assignments are shown in Table 5.1. The objective value for the ILP is 95.

Long-term reconfiguration: to understand the working of the ILP for long-term reconfiguration, the traffic in Table 5.1 was changed. An example of existing traffic is depicted in Table 5.2. The node-pairs, their alternate routes, and their currently existing primary and backup working paths are shown in Table 5.2. The ILP has to avoid service disruption to the primary paths of the working connections. These

Table 5.2. *Long-term reconfiguration stage*

Node-pair	Alternate routes	Primary paths of working connections (wavelengths)
1	1 2	λ_1, λ_2
	1 3 2	
27	3 1	$\lambda_1, \lambda_2, \lambda_3$
	3 2 1	
110	9 4 5 6	λ_7, λ_8
	9 12 13 6	λ_5
167	13 6 11	
	13 12 10 11	λ_3
32	3 6 5 7	λ_1, λ_2
	3 2 8 7	

Table 5.3. *Long-term reconfiguration stage*

Node-pair	Alternate routes	Primary paths of working connections (wavelengths)
1	1 2	λ_1, λ_2
	1 3 2	λ_5*
27	3 1	$\lambda_1, \lambda_2, \lambda_3$
	3 2 1	$*\lambda_1, \lambda_2*$
110	9 4 5 6	λ_7, λ_8
	9 12 13 6	λ_5
167	13 6 11	$\lambda_5*, \lambda_8*, \lambda_{10}*$
	13 12 10 11	λ_3
32	3 6 5 7	λ_1, λ_2
	3 2 8 7	$*\lambda_3, \lambda_4$

paths are input to the formulation through the $\chi^{i,p}$ variable. The currently working connections are deliberately chosen to demonstrate the operation and selection of the ILP.

Node-pairs 1, 32, 110, and 167 require five connections each and node-pair 27 requires six connections. The total number of connections requested between each node-pair includes those which are currently working.

The ILP is solved for node-pairs shown in Table 5.2 with $C_l = 1$ and $C_w = 4$. The effect of the solution depends on the value of C_w, the higher the value is, the better the guarantee that the working paths remain unaffected. The value of C_w is set to be β times the cost of primary paths, C_{ND}. The value of β is typically set to 3 or 4. For every connection that is disturbed, the objective value is penalized by a factor of C_w.

The connections that are disturbed are denoted in Table 5.3 and are marked by an asterisk (∗). The resulting route and wavelength assignments for the demands are shown in Table 5.4. The objective value for the ILP is 53.

Table 5.4. *Route and wavelength assignment*

Node-pair	Alternate routes	Primaries	Backups
1	1 2	$\lambda_1, \lambda_2, \lambda_3, \lambda_6, \lambda_{10}$	–
	1 3 2	–	$\lambda_1, \lambda_2, \lambda_3, \lambda_6, \lambda_{10}$
27	3 1	$\lambda_1, \lambda_2, \lambda_3, \lambda_6, \lambda_{10}$	λ_9
	3 2 1	λ_9	$\lambda_1, \lambda_2, \lambda_3, \lambda_6, \lambda_{10}$
110	9 4 5 6	$\lambda_3, \lambda_6, \lambda_7, \lambda_8$	λ_5
	9 12 13 6	λ_5	$\lambda_3, \lambda_6, \lambda_7, \lambda_8$
167	13 6 11	–	$\lambda_3, \lambda_6, \lambda_7, \lambda_8, \lambda_{10}$
	13 12 10 11	$\lambda_1\lambda_3, \lambda_6, \lambda_7, \lambda_8$	–
32	3 6 5 7	$\lambda_1, \lambda_2, \lambda_3, \lambda_{10}$	λ_4
	3 2 8 7	λ_4	$\lambda_1, \lambda_2, \lambda_3, \lambda_{10}$

The connections that are disturbed are the ones which use links where backups can be multiplexed. To understand this better, take the case of node-pairs 1 and 27. They share a link (3–2) on one of their routes. Since both the node-pairs have at least one disjoint route, the routes corresponding to link 3–2 can be used for multiplexing the backup paths. Thus the primary paths of connections using wavelength λ_5 on route 1–3–2, and λ_1, λ_2 on route 3–2–1, are reassigned to routes 1–2 and 3–1, respectively.

Recall that in the short-/medium-term reconfiguration stage, the goal is to optimize resource consumption for backup paths. The higher the value of C_w is, the better the guarantee that primary paths of the working connections remain unaffected. In the short-/medium-term reconfiguration phase, the cost of C_w is typically set very high for the primary paths of the working connections. A high value of C_w, however, does not guarantee that the primary path will not be rerouted in the final solution. Hence to avoid disruption to the primary paths of working connections, the capacity consumed by them may be removed and the backup capacity consumption can be optimized.

5.6.2 Revenue maximization

To show the performance for the revenue maximization formulation, it is necessary to have a revenue relationship between a primary path and a backup path to be able to trade between them. Consider the following cost relationship between the primary and backup paths: $C_D = \alpha C_{ND}, 0 \le \alpha \le 1$. The total revenue is calculated as total number of primaries $\times C_{ND}$ + total number of backups $\times \alpha \times C_D$ cost units (cu). The network relies on the best-effort class to increase revenue. The increase

Table 5.5. *Increase in revenue for the two variations of the best-effort class (NSFnet) for α = 1*

| Demand pairs | α = 1 | | | | |
| | Best-effort 1 | | Best-effort 2 | | |
	Primary	Backups	Primary	Backups	Rejected
12	12	8	12	8	0
20	20	16	20	16	0
24	24	12	21	18	3
32	32	20	28	27	4
36	36	22	33	28	3
44	44	30	41	36	3
48	48	32	44	39	4

Table 5.6. *Increase in revenue for the two variations of the best-effort class (NSFnet) for α = 0.5*

| Demand pairs | α = 0.5 | | | | |
| | Best-effort 1 | | Best-effort 2 | | |
	Primary	Backups	Primary	Backups	Rejected
12	12	8	12	8	0
20	20	16	20	16	0
24	24	12	21	18	3
32	32	20	29	26	3
36	36	22	33	28	3
44	44	30	41	36	3
48	48	32	46	36	2

in revenue obtained by the two variations of the best-effort class with a base case are compared with accepting all connections without any protection.

For $C_{ND} = 500$ cu and for two values of α (1 and 0.5), the results for various demand sets on NSFnet and ARPANET topologies are shown in Tables 5.5 and 5.6, and 5.7 and 5.8, respectively. For particular instances of demands, best-effort variation 1 results in a 67% gain in revenue and variation 2 achieves an additional 6% gain, for α = 1. The cases are compared to the revenue generated by accepting all demands without protection number of primaries × C_{ND}. For example, consider the case of 48 demands for α = 1 in Table 5.5. The base case where all demands

Table 5.7. *Increase in revenue for the two variations of the best-effort class (ARPANET) for α = 1*

Demand pairs	α = 1				
	Best-effort 1		Best-effort 2		
	Primary	Backups	Primary	Backups	Rejected
12	12	8	12	8	0
20	20	16	18	18	2
24	24	12	20	20	4
32	32	20	28	28	4
36	36	20	32	28	4
44	44	28	40	37	4
48	48	24	40	40	8

Table 5.8. *Increase in revenue for the two variations of the best-effort class (ARPANET) for α = 0.5*

Demand pairs	α = 0.5				
	Best-effort 1		Best-effort 2		
	Primary	Backups	Primary	Backups	Rejected
12	12	8	12	8	0
20	20	16	20	16	0
24	24	12	20	20	4
32	32	20	28	28	4
36	36	20	32	28	4
44	44	28	41	34	3
48	48	24	41	38	7

are accepted without any protection results in $48 \times C_{\mathrm{ND}} = 24\,000$ cu. The total revenue for variation 1 is $48 \times C_{\mathrm{ND}} + 32 \times C_{\mathrm{D}} = 40\,000$ cu, which is a 66.7% gain. The revenue for variation 2 is $44 \times C_{\mathrm{ND}} + 39 \times C_{\mathrm{D}} = 41\,500$ cu, which is a 72.9% gain. Although both schemes employ backup multiplexing, the first variation has no choice but to choose all the primary paths and then try to accommodate backups, and so it is restricted. The second variation exploits the backup resource consumption better by effectively multiplexing more connections on the same wavelength, thus accepting more connections and resulting in slightly better revenues.

The multi-stage solution methodology is demonstrated on the NSFnet topology. A demand set comprising 48 demands with 12 demands in full protection class,

Table 5.9. *Solution at the end of the third stage*

Node-pair	Class 1	Class 2	Class 3	Primary paths	Backup paths
1	3	3	6	10	10
2	3	3	6	11	10
3	3	3	6	12	8
4	3	3	6	12	8
				45	36

12 demands in no protection class, and 24 demands in best-effort class, distributed uniformly across four node-pairs, needs to be set up. The cost values used are $C_{ND} = 500$, $C_D = 500(\alpha = 1)$, and $C_w = 500(\beta = 1)$.

In the first stage, the problem is solved for full protection demands. It is assumed that there are no currently working connections. Thus, the value of $\chi^{i,p}$ for all the node-pairs is zero. The ILP determines a feasible solution, which is a set of paths with a route and a wavelength associated with each of them for all the 12 demands in the full protection class. This set of paths is fed into the second stage by setting the associated $\chi^{i,p}$ variables to 1. The problem is then solved for the full protection and no protection classes. The 12 paths chosen for the full protection class are assumed to be working paths in the second stage. The ILP assigns primary paths for all full protection and no protection demands. The objective value for the example turns out to be 11 500.

Although the objective value is of no relevance as long as the number of primary and backups selected is known, it is interesting to note how the ILP handles service disruption. Since the ILP determines a feasible solution for all the full protection and no protection demands, the objective value is expected to be 12 000, but the value obtained is 11 500 ($24 \times C_{ND} - 1 \times C_w$). This is due to the fact that one of the primary paths for the full protection demand is reassigned. The objective value incurred a penalty for disturbing the connection. Thus, by appropriately choosing C_w, as explained in Section 5.5, the formulation can be used to try and avoid service disruptions to existing connections in the network.

The set of primary paths is then fed to the third stage. The third stage solves the problem for all classes. The value of C_w is set to 1500 ($\beta = 3$). As explained in Section 5.5, Eq. (5.24) ensures that backups for all demands of the full protection class are chosen. The final solution at the end of the third stage is shown in Table 5.9. The demands rejected are those belonging to the best-effort class. The total revenue generated for provisioning the complete demand set for all classes is $45 \times C_{ND} + 36 \times C_D = 58\,500$ cu.

6

Managing large networks

Several methods discussed for joint working and spare capacity planning in survivable WDM networks in the last chapter considered a static traffic demand and optimized the network cost assuming various cost models and survivability paradigms. The focus here lies in network operation under dynamic traffic. The common framework that captures the various operational phases in a survivable WDM network in a single ILP optimization problem avoids service disruption to the existing connections. However, the complexity of the optimization problem makes the formulation applicable only for network provisioning and offline reconfigurations. The direct use of this method for online reconfiguration remains limited to small networks with a few tens of wavelengths.

6.1 Online algorithm

The goal here is to develop an algorithm for fast online reconfiguration using a heuristic algorithm based on an LP relaxation technique. Since the ILP variables are relaxed, a way is needed to derive a feasible solution from the solution of the relaxed problem. The algorithm consists of two steps. In the first step, the network topology is processed based on the demand set to be provisioned. This preprocessing step ensures that the LP yields a feasible solution. The preprocessing step is based on (i) the assumption that in a network, two routes between any given node-pair are generally sufficient to provide effective fault tolerance, and (ii) an observation on the working of the ILP for such networks. In the second step, using the processed topology as input, the LP is solved. It is interesting to obtain some insights into why the LP formulation may yield a feasible solution to the ILP. This allows the use of the algorithm on realistic sized backbone networks with hundreds of wavelengths per link. The results in this chapter indicate that the run time of the heuristic algorithm is short enough (of the order of seconds) to use for online reconfiguration.

6.1.1 Relaxation methods

LP relaxation of the ILP formulation is one of the most widely used relaxation techniques. In this technique, the integrality constraints of the ILP variables are relaxed. After a solution has been obtained the values of the variables are converted to integers using a heuristic approach. The expectation is that the rounded LP solution will give a value close to that of the optimal solution obtained from ILP.

In [36], the lagrangian relaxation method is used to simplify the integer problem into subproblems for each demand. Since a solution to a relaxed problem may not necessarily be a feasible solution to the original problem, heuristics are employed to extract a feasible solution. In [237], LP relaxation technique is used to derive an upper bound on the carried traffic of connections for any routing and wavelength assignment algorithm. In [52], a randomized rounding technique has been used to convert fractional flows provided by the LP solution to integer flows, and graph coloring algorithms are used to assign wavelengths to the lightpaths. The problem of minimizing the total wavelength mileage, in a network with arbitrary topology, to provide shared line protection has been used in [17]. In [16], an efficient approach for solving the wavelength mileage problem is developed. The algorithm provides a feasible solution, with minimal violation of the design constraints, and a pruning technique of the search space to reduce the problem complexity.

To develop a fast online reconfiguration algorithm, a heuristic algorithm based on the LP relaxation technique is developed in this chapter. The algorithm consists of two steps. In the first step, the network topology is processed based on the demand set to be provisioned. This preprocessing step is done to ensure that the LP yields a feasible solution. The preprocessing step in the algorithm is based on the following two assumptions.

- In a network, two routes between any given node-pair are sufficient to provide effective fault tolerance.
- An observation on the working of the ILP for such networks.

In the second step, using the processed topology as input, the problem is formulated as an LP and solved. Interestingly, the LP relaxation heuristic yields feasible solutions to the ILP in most cases.

6.1.2 Preprocessing step

The purpose of preprocessing is to ensure that the LP yields a feasible solution. The ILP in the formulation decides one of the routes out of the two available to be used for the primary for each node-pair and the other route is used for the backup. Since paths between each node-pair are predetermined, the demands can be classified into one of two categories: (i) if two node-pairs have common links on both of their routes, their backups cannot be multiplexed on the same wavelengths (as they violate the criteria for backup multiplexing) and (ii) if there are two node-pairs and at least one route for each of them is node- and link-disjoint with the other, then backup paths of demands belonging to these node-pairs may or may not be multiplexed depending on the specific instance of traffic that is contending for resources. The following definitions are used.

- A node-pair i is of type 1 on a particular route if it has been assigned exactly the same number of wavelengths as the number of demands on its route.
- A node-pair is of type 2 if it has been assigned more wavelengths than its demand requires.

The preprocessing consists of the following steps.

(i) Identify the bottleneck link for each node-pair as follows.
    (a) The bottleneck link for a node-pair i ($Bl[i]$) is defined as that link on either of its two routes, which is part of the routes of most other node-pairs.
    (b) If multiple links have the same value, the tie can be arbitrarily broken.

(ii) Pre-wavelength set assignment.
    (a) Arbitrarily choose a node-pair i.
    (b) Assign d_i wavelengths on both of its routes. To satisfy d_i demands, a node-pair needs d_i wavelengths on each of its routes for its primary and backup paths. In this case, node-pair i is of type 1 on both of its routes.
    (c) For every node-pair j using $Bl[i]$.
        1. One route of node-pairs i and j already share a common link $Bl[i]$. Without loss of generality, let this be route 1 for both node-pairs. Now, if route 2 of node-pair j is link-disjoint with route 2 of node-pair i, then assign $d_i + d_j$ wavelengths for j on route 1, out of which d_i wavelengths are shared with i. j is of type 2 on its route 1. On the other route of node-pair j, it is assigned exactly d_j wavelengths (j is of type 1 on its route 2).
        2. If node-pairs j and i share link(s) on their other route (route 2), then node-pair j is assigned d_j wavelengths, disjoint to those assigned to i, on both of its routes. In this case, node-pair j is type 1 on both of its routes.
        3. Repeat the procedure for all node-pairs j using $Bl[i]$, comparing with every type 1 node-pair available on the link. These rules are enforced to handle problems arising as a result of the relaxation. This is explained in detail in Section 6.3.
           A. A type 2 node-pair can share wavelengths with only one type 1 node-pair on a link.
           B. Every type 1 node-pair can have exactly one type 2 pair sharing wavelengths with it. If more than one such type 2 pair exists on the link, for every type 1, then the demands belonging to those node-pairs are removed. The problem is solved for only one set of interacting demands at a time.
        4. Once step ii.c is completed, node-pairs which have been assigned wavelengths are marked.

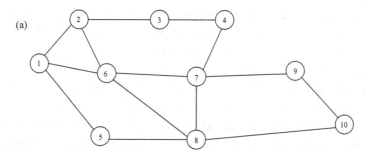

(a)

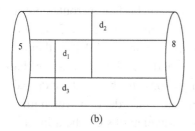

(b)

Pair	Route 1	Route 2
(1, 4)	1–2–3–4	1–5–8–7–4
(2, 10)	2–6–8–10	2–1–5–8–7–9–10
(1, 3)	1–2–3	1–5–8–7–4–3

(c)

Fig. 6.1. An illustrative example to demonstrate the preprocessing step © IEEE. Source: M. Sridharan, M. V. Salapaka, and A. K. Somani, A practical approach to operating survivable WDM networks, in *IEEE Journal of Selected Areas in Communications*: Special Issue on WDM-based Network Architectures, 2002 [188].

5. Arbitrarily choose one of the node-pairs that has been marked, and repeat step ii.c on its bottleneck link.
6. Repeat step ii.c.5 for all marked pairs on link $Bl[i]$.
7. Repeat step (ii) and terminate when all node-pairs that have non-zero demands are marked.

The preprocessing step identifies possible routes for backup multiplexing demands belonging to different node-pairs. Each node-pair is assigned a set of wavelengths based on a set of rules (ii.c.3.A and ii.c.3.B above). It is to be noted that the preprocessing step merely assigns a set of wavelengths to each node-pair and the actual routing and wavelength assignment is performed by the LP formulation as developed in Section 6.3.

6.2 Example

Consider an example network shown in Fig. 6.1(a). The node-pairs of interest and the alternate routes between them are shown in Fig. 6.1(c). Let d_1, d_2 and d_3 be the demand request for each pair 1, 2 and 3, respectively. The links that are of interest are those where more demands belonging to different node-pairs interact. In this example, links $5 \rightarrow 8$ and $8 \rightarrow 7$ fall into that category. An arbitrary link $5 \rightarrow 8$ is chosen for demonstration. Each node-pair is assigned a set of wavelengths. This allocation does not affect the actual routing and wavelength assignment to be performed by the LP formulation. For a node-pair, as long as enough wavelengths (capacity) are allocated to meet the demands, it does not matter what range of wavelengths was assigned to it.

Examining link $5 \rightarrow 8$, node-pair 3 is chosen arbitrarily and wavelength d_3 is assigned on both of its routes. Node-pair 3 is of type 1 on both of its routes. Node-pairs 1 and 3 are arbitrarily chosen and have common links on both of their routes, and hence they cannot be backup multiplexed with each other. This implies that disjoint wavelength sets are needed for these two link-sharing pairs. Node-pair 1 is assigned d_1 wavelengths ($d_3 + 1$ to min($d_3 + d_1$, W)) on both of its routes. Starting from node-pair 1 also gives type 1 on both of its routes. Since node-pair 2 has at least one route (its route 1) that is disjoint from the possible routes of both the other node-pairs, the potential for backup multiplexing exists, and wavelengths may be shared with either of the type 1 pairs available. Arbitrarily choose to share with node-pair 1, since a contiguous set of wavelengths can be assigned. The order of wavelengths can be easily rearranged such that contiguous sets of wavelengths can be allocated to type 2 and 1 node-pairs, which need to share the same set of wavelengths. Node-pair 2 is assigned wavelengths numbered from $d_3 + 1$ to min($d_3 + d_1 + d_2$, W) on its route 2 and d_2 wavelengths on its route 1. Hence, node-pair 2 is of type 2 on its route 2, as it shares wavelengths with a type 1 node-pair 1, and is of type 1 on its route 1. It is assumed here that the LP solves a demand matrix that is feasible for the ILP.

6.3 LP formulation

The LP relaxation of the ILP formulation presented earlier in Section 5.3 is given below. In the formulation $lmin[i, r]$ and $lmax[i, r]$ denote the range of wavelengths assigned for node-pair i on routes $r = 1, 2$.

Minimize

$$\sum_{i=1}^{N(N-1)} \sum_{p=lmin[i,1]}^{lmax[i,1]} \delta^{i,p} \sum_{l=1}^{L} \epsilon_l^{i,p} C_l$$

$$+ \sum_{i=1}^{N(N-1)} \sum_{p=lmin[i,2]}^{lmax[i,2]} \delta^{i,p} \sum_{l=1}^{L} \epsilon_l^{i,p} C_l + \sum_{l=1}^{L} \sum_{\lambda=1}^{W} (-1 \times g_{l,\lambda}) C_l$$

$$+ \sum_{i=1}^{N(N-1)} \sum_{p=lmin[i,1]}^{lmax[i,1]} \chi^{i,p} (1 - \delta^{i,p}) C_w$$

$$+ \sum_{i=1}^{N(N-1)} \sum_{p=lmin[i,2]}^{lmax[i,2]} \chi^{i,p} (1 - \delta^{i,p}) C_w \qquad (6.1)$$

Demand constraints for each node-pair:

$$\sum_{p=lmin[i,1]}^{lmax[i,1]} \delta^{i,p} + \sum_{q=lmin[i,2]}^{lmax[i,2]} \delta^{i,q} = d_i \qquad (6.2)$$

$$1 \le i \le N(N-1)$$

$$\sum_{p=lmin[i,1]}^{lmax[i,1]} v^{i,p} + \sum_{q=lmin[i,2]}^{lmax[i,2]} v^{i,q} = d_i \qquad (6.3)$$

$$1 \le i \le N(N-1)$$

Restoration path wavelength usage indicator constraint:

$$X_{l,\lambda} = \sum_{i=1}^{N(N-1)} \sum_{r=1}^{KW} v^{i,r} \epsilon_l^{i,r} \psi_\lambda^{i,r} \qquad (6.4)$$

$$2g_{l,\lambda} = X_{l,\lambda} \qquad (6.5)$$

$$1 \le l \le L, \quad 1 \le \lambda \le W$$

$$0 \le g_{l,\lambda} \le 1, \quad X_{l,\lambda} \ge 0, \quad 0 \le \delta^{i,p}, \quad v^{i,p} \le 1$$

Primary path wavelength usage constraints:

$$\sum_{i=1}^{N(N-1)} \sum_{p=lmin[i,1]}^{lmax[i,1]} \delta^{i,p} \epsilon_l^{i,p} \psi_\lambda^{i,p}$$

$$+ \sum_{i=1}^{N(N-1)} \sum_{q=lmin[i,2]}^{lmax[i,2]} \delta^{i,q} \epsilon_l^{i,q} \psi_\lambda^{i,q} + g_{l,\lambda} \le 1 \qquad (6.6)$$

$$1 \le l \le L, \quad 1 \le \lambda \le W$$

Constraint to ensure that a type 2 primary never clashes with type 1 backups: for *type 2* demands on a link *l*, the following constraint applies. Node-pair *j* belongs to type 1. Node-pair *i* belongs to type 2, which shares wavelengths with node-pair *j*. *p, r* are those paths on the node-pair routes that use *Bl[i]*,

$$v^{j,r} + \delta^{i,p} \le 1. \qquad (6.7)$$

6.3.1 Feasibility of solution

The LP yields a feasible solution based on the following observation that also provides a basis for further argument. For a given node-pair, the LP formulation has a different cost associated with the primary and backup variables. Also the cost incurred depends only on the route on which the path for the variable is present. For a given node-pair, if all the LP constraints are being met, the LP prefers to route the primary variable of a demand on the route that incurs a lower cost. Also as long as the constraints are being met, the LP allocates all primaries on the same route. The same reasoning holds for the backup variables. Thus it can be expected that the primary variables $\delta^{i,p}$ for a particular node-pair i take non-zero values only on one route. The same is expected of the backup variables. This observation is stated more formally as follows.

Observation 1. The LP has a tendency to group the weights of the variables $\delta^{i,p}$ and $v^{i,p}$ for any given i. As a result, for any i, r, and $lmin[i, r] \le p \le lmax[i, r]$, either all $\delta^{i,p}$ variables have non-zero assignments or all $v^{i,p}$ variables have non-zero assignments.

Based on Observation 1, the following insights into why the LP formulation yields a feasible solution for the ILP is obtained. These claims elucidate the operation of a heuristic based on LP relaxation.

Claim 1. The LP solution guarantees integer (binary) assignments for all type 1 variables.

Indeed, consider Eqs. (6.2) and (6.3). They are of the form $A + B = d_i$ and $C + D = d_i$. Terms A, C represent variables on one route and B, D represent variables on the other route. Based on Observation 1, either A or C is zero. Without loss of generality, let the term $C = 0$. This would force $D = d_i$ and hence $B = 0$. Thus $A = d_i$ and $D = d_i$. Recall that for type 1 variables, $lmax[i, r] - lmin[i, r] = d_i$. Since $0 \le \delta^{i,p}, v^{i,p} \le 1$, all the variables in terms A (primary variables) and D (backup variables) are forced to be assigned 1 and all the variables in the other terms are zero.

Claim 2. The LP solution guarantees integer (binary) assignments for all type 2 δ variables.

The above claim follows from the following argument. Let node-pair i be of type 1 and node-pair j be of type 2. All variables, primary and backup of node-pair i, are guaranteed to be binary (Claim 1). Equations (5.2) and (5.3) are of the form $A + B = d_j$ and $C + D = d_j$. Terms A, C represent variables on one route and B, D represent variables on the other route.

Without loss of generality, let the term $A = 0$. This would force $B = d_j$, $D = 0$ and $C = d_j$. Recall that for type 2 variables, $lmax[i, r] - lmin[i, r] = d_j$ on one of its routes and $lmax[i, r] - lmin[i, r] = d_i + d_j$ on its other route. Let B represent variables on a route where $d_i + d_j$ has been assigned. d_i out of $d_i + d_j$ belong to type 1 variables and are guaranteed to be 1. Equations (6.6) and (6.7) ensure that d_i out of the $d_i + d_j$ variables cannot be used and hence force the remaining δ variables in term B to be 1.

A similar argument can be applied by letting $B = 0$. In this case $A = d_i$ and $lmax[i, 1] - lmin[i, 1] = d_j$ for A and hence the δ variables are forced to be 1.

A similar argument can also be applied for ν variables of type 2. When $C = d_j$ and $lmax[i, 1] - lmin[i, 1] = d_j$ for that route and ν variables in C are forced to be 1. Suppose that if $D = d_j$ and $lmax[i, 2] - lmin[i, 2] = d_i + d_j$, then there are two cases. If d_i are primaries, then Eq. (6.7) forces the variables in D to be 1. However, if d_i are backups, then $d_i + d_j$ variables and d_j capacity is needed to be filled. In this particular case the assignments may be fractional. This case is still acceptable because these violations occur only when type 1 and 2 backups share the link on the route. Since backup multiplexing is allowed, it is possible to reclaim resources by adjusting the fractional flows of type 2 to be 1 and make it coincide with the backups of type 1.

In the ILP formulation, $g_{l,\lambda}$ takes a value of one or zero. To proceed with the LP formulation, a method is needed to identify that a wavelength λ is being used as a backup, otherwise Eq. (6.6) is violated and the primary and the backup path may end up using the same wavelength on a link.

$g_{l,\lambda}$ needs to be appropriately modified for the LP and made to take a higher value whenever a wavelength on a link is used for a backup. Recall rules 2.iii.A and 2.iii.B in the heuristic algorithm that state that every type 1 node-pair can have exactly one type 2 pair sharing wavelengths with it. If more than one such type 2 pair exists on the link, for every type 1, then the demands belonging to those node-pairs are removed. The problem is solved for only one set of interacting demands at a time and the multi-step procedure for such a solution and its implications are discussed in Section 6.4. Since only one type 2 demand is allowed to share wavelengths with a type 1 demand, the value of $X_{l,\lambda}$, which counts the number of backup paths that share a wavelength λ on link l, can be either zero (if the path is not used for backup), 1 (one backup path), or 2 (if two paths share this link l and wavelength λ, as backup). Equation (4.3) of the ILP is modified as shown in Eq. (6.5).

Since $X_{l,\lambda}$ can take the values 0, 1, or 2 (enforced by rules 2.iii.A and 2.iii.B), $g_{l,\lambda}$ in Eq. (5.5) can take the values 0, 0.5, or 1, respectively. In the ILP formulation, $g_{l,\lambda}$ is guaranteed to be 1 or 0. In the LP formulation, this cannot be captured exactly. Since $g_{l,\lambda} = 0.5$ implies that only one backup path uses link l and wavelength λ, $g_{l,\lambda} = 1$ implies that two backup paths share link l and wavelength λ, the objective function can be modified to make it favor cases when $g_{l,\lambda} = 1$. This formulation is not exact, since the cost of two backup paths sharing link l and wavelength λ ($g_{l,\lambda} = 1$) is the same as using two different wavelengths for backup ($2g_{l,\lambda} = 1$). The modified objective function is shown in Eq. (6.1). Equations (4.5) and (4.9), representing the link capacity constraint and the backup multiplexing constraint of the ILP, are no longer constraints in the LP formulation, as they are already ensured in the preprocessing step.

6.4 Solving for excess demands

As explained in the previous subsection, every type 1 node-pair can have exactly one type 2 pair sharing wavelengths with it. If more than one such type 2 pair exists on the link, for every type 1, then the demands belonging to those node-pairs are removed. In such cases, the problem is solved for one set of interacting demands at a time. A multi-stage approach is used for solving this problem. An approach similar to that in Chapter 5 can be used. At each stage, one instance of the problem is solved for one set of interacting demands, and the result is used in successive stages. If the problems are solved independently, the resulting solution may be infeasible, as the same path might be used by multiple primaries or backups. In order to avoid infeasibility, the information concerning one stage is fed to the next through the $\chi^{i,p}$ variable. Typically, this variable is used to feed information concerning existing paths to avoid service disruption. This aspect of the formulation is exploited by feeding the solution of one stage to the next stage. The objective function is modified to include backups chosen during one stage to be fed to the next. This feature is exploited only to make sure that assignments are binary. However, there may be a penalty for this type of solution, first because the problem is solved sequentially and is not shown the full solution space, the result may be suboptimal. Secondly, depending on the solution from one stage, some demands may be blocked.

The above methodology is applied to a 14-node 21-link NSFnet topology and the 20-node 32-link ARPANET topology.

The complexity of the optimization problem makes the ILP solution intractable for large problem instances. This effect is sometimes seen for small problem instances. The solution obtained and the resulting relaxation are depicted in the following sections.

6.5 Quality of the LP heuristic algorithm

Consider the node-pairs and their two alternate routes shown in Table 6.1 to see the effect of the LP relaxation heuristic. Let the number of wavelengths per link be 10. Let the node-pairs, in this example, require five primaries and five backups. Since there is a restriction that only one type 2 node-pair can share wavelengths with a type 1 node-pair on a link, demand requests for node-pair 32 are removed in the first stage. The results of the ILP and the LP are shown Tables 6.2 and 6.3, respectively.

The ILP solution assigns backups for demands belonging to node-pairs 1 and 27 in the route that has the common link $3 \rightarrow 2$ and similarly assigns backups for demands belonging to node-pairs 110 and 167 on the route that has the common link $13 \rightarrow 6$. The primary paths are assigned as shown in the Table 6.2.

Table 6.1. *Illustrative example*

Node-pair	Alternate routes
1	1 2
	1 3 2
27	3 1
	3 2 1
110	9 4 5 6
	9 12 13 6
167	13 6 11
	13 12 10 11
32	3 6 5 7
	3 2 8 7

Table 6.2. *ILP solution (five demand requests/node-pair, 10-wavelengths/link)*

Node-pair	Alternate routes	Primary	Backup
1	1 2	$\lambda_1-\lambda_5$	–
	1 3 2	–	$\lambda_1-\lambda_5$
27	3 1	$\lambda_1-\lambda_5$	–
	3 2 1	–	$\lambda_1-\lambda_5$
110	9 4 5 6	$\lambda_1-\lambda_5$	–
	9 12 13 6	–	$\lambda_1-\lambda_5$
167	13 6 11	–	$\lambda_1-\lambda_5$
	13 12 10 11	$\lambda_1-\lambda_5$	–
32	3 6 5 7	$\lambda_1-\lambda_5$	–
	3 2 8 7	–	$\lambda_1-\lambda_5$

In Table 6.3, since the cost of two backup paths sharing a link and wavelength is the same as using two different wavelengths for backup (refer to discussion on $g_{l,\lambda}$ in Section 6.3), the backup wavelength assignment is different from the ILP assignment. As in the case of the ILP solution, the backups for demands belonging to node-pairs 1 and 27 are assigned on the route that has the common link $3 \rightarrow 2$. But the wavelength assignment for backups is different. The backup paths for node-pair 1 are assigned on route $1 \rightarrow 3 \rightarrow 2$ on wavelengths $\lambda_1-\lambda_5$, and backups for node-pair 27 are assigned on route $3 \rightarrow 2 \rightarrow 1$ on wavelengths $\lambda_5-\lambda_9$. Only one wavelength (λ_5) is used for backup multiplexing, in comparison with all five ($\lambda_1-\lambda_5$) in the ILP solution. However, once the LP provides this feasible solution, the backup routes may be merged to coincide with the backup paths of node-pair 1 and reclaim the wavelengths. Refer to the discussion in Section 6.3 on adjusting

Table 6.3. *LP solution (five demand requests/node-pair,*
10-wavelengths/link)

Node-pair	Alternate routes	Primary	Backup
1	1 2	$\lambda_1-\lambda_5$	–
	1 3 2	–	$\lambda_1-\lambda_5$
27	3 1	$\lambda_1-\lambda_5$	–
	3 2 1	–	$\lambda_5-\lambda_9$
110	9 4 5 6	$\lambda_1-\lambda_5$	–
	9 12 13 6	–	$\lambda_1-\lambda_5$
167	13 6 11	$\lambda_1-\lambda_5$	–
	13 12 10 11	–	$\lambda_1-\lambda_5$
32	3 6 5 7	$\lambda_1-\lambda_5$	–
	3 2 8 7	–	$\lambda_1-\lambda_5$

type 2 backups to coincide with the backup paths of its corresponding type 1. In this case demands of node-pair 1 belong to type 1 and those of node-pair 27 belong to type 2.

For the next set of node-pairs, 110 and 167, primary paths for demands belonging to both pairs are chosen on their first route and backup paths on their second route, as shown in Table 6.3. Hence, no backup multiplexing is done. This is in contrast with the ILP solution that used the route containing the common link $13 \rightarrow 6$ for routing backups and as a result could backup multiplex the demand requests of node-pairs 110 and 167.

In the above example, node-pair 32 has to be solved in the next stage. In such cases, the solution from the first stage is fed to the second stage as the currently working primary and backup paths. In this example, since the LP chooses the backup routes for node-pairs 1 and 27 on the route that uses link $3 \rightarrow 2$, all the requests for node-pair 32 are accommodated with the primary and backup route and wavelength assignments as shown in Table 6.3. Although, the demands for node-pair 32 are accommodated in this example, there is no guarantee that all the demands are accepted for node-pairs that are solved in successive stages. Thus, there may be a penalty for solving the problem sequentially as discussed in Section 6.4.

Now suppose the node-pairs required 10 demands each instead of five demands as in the previous case. The solution for the LP and ILP for this case is the same and is the shown in Table 6.4. The LP in this situation, to accommodate all demand requests, is forced to backup multiplex all possible demands, and thus yields an optimal solution. It is well known that if the LP relaxation to the ILP provides a solution that is an integer vector, then the solution is feasible and hence optimal

Table 6.4. *LP/ILP solution (10 demand
requests/node-pair, 10-wavelengths/link)*

Node-pair	Alternate routes	Primary	Backup
1	1 2	$\lambda_1-\lambda_{10}$	–
	1 3 2	–	$\lambda_1-\lambda_{10}$
27	3 1	$\lambda_1-\lambda_{10}$	–
	3 2 1	–	$\lambda_1-\lambda_{10}$
110	9 4 5 6	$\lambda_1-\lambda_{10}$	–
	9 12 13 6	–	$\lambda_1-\lambda_{10}$
167	13 6 11	–	$\lambda_1-\lambda_{10}$
	13 12 10 11	$\lambda_1-\lambda_{10}$	–

Table 6.5. *Sample results demonstrating
the quality of the LP solution*

Demands	ILP objective	LP objective
10	38	43
20	76	90
30	114	120
40	152	156
50	190	190

to the ILP [42]. This is the reason for the LP providing an optimal and a feasible solution to the ILP in this case, as the LP solution vector is forced to be an integer in such cases. This behavior is demonstrated in Table 6.5. The results are for the NSFnet topology with 10 wavelengths per link, for the example in Table 6.1, with demand requests distributed uniformly across five node-pairs. The LP yields an optimal feasible solution to the ILP as the LP solution vector is forced to be an integer in such cases.

6.6 ILP and LP solution run times

The ILP and LP solution run time comparison is shown in Table 6.6 for one experiment to demonstrate the gap between the time taken by the two approaches. All LP and ILP problems are solved using the software package CPLEX. In the table PT and FT denote partial and full terminations, respectively. CPLEX terminates mixed integer optimizations under a variety of circumstances [2]. CPLEX finds an integer optimal solution and terminates when all nodes have been processed. Optimality

Table 6.6. *Comparing ILP and LP solution run times*

Demands	ILP time (s) (PT)	ILP time (s) (FT)	LP time (s)
22	601 (3.35% mipgap)	>9000	0.12
32	3973 (4.40% mipgap)	>9000	0.11
42	852.31 (4.03% mipgap)	>9000	0.13
52	104.87	104.87	0.14
72	84.00	84.00	0.17
92	20.84 (0.29% mipgap)	8289.76	0.23

Table 6.7. *Results for a 14-node NSFnet topology with 100 wavelengths per link*

Demands	LP constraints	LP variables	LP time (s)
100	14 029	4280	0.52
150	22 029	4520	1.10
200	33 229	4760	2.18
250	47 629	5000	3.89
300	56 429	5160	5.25
400	78 829	5480	21.87
500	107 629	5800	15.75

Table 6.8. *Results for a 20-node ARPANET topology with 100 wavelengths per link*

Demands	LP constraints	LP variables	LP time (s)
100	22 117	9 880	0.70
200	31 767	10 360	1.16
300	47 817	10 840	3.06
400	70 267	11 320	4.94
500	99 117	11 800	8.34
600	116 767	12 120	29.04
700	137 617	12 440	27.26
800	161 667	12 760	35.64
900	188 917	13 080	44.24
1000	219 367	13 400	30.91

in this case is relative to the tolerances and other optimality criteria set by the user. The default relative optimality tolerance is 0.0001, in which case the final integer solution is guaranteed to be within 0.01%. Requiring CPLEX to seek integer solutions that meet a 0.01% tolerance in such cases is a waste of computation time. To make the comparison fair, the problem is terminated when the solution is close to the desired value. As the results show, the LP solution time is considerably lower for this example. In the ILP solution result, the fast run times for the 52 and 72 demands compared with the slower run times for smaller demand requests is not surprising. The solver performs a lot of preprocessing and depending on how close the initial solution is to the final integer optimal solution, the problem can run that much faster.

6.7 Run times for the LP heuristic algorithm

To demonstrate the use of the algorithm on backbone networks of practical size with hundreds of wavelengths per link, consider the results for the NSFnet and ARPANET topologies, with 100 wavelengths per link as shown in Tables 6.7 and 6.8, respectively. All the techniques discussed for problem size reduction are applied before the LP is solved. The complexity of the problem is determined by the number of variables and constraints in the formulation. It is observed from the results that for large demand sets, the run time of the heuristic algorithm is fast (of the order of seconds). This improves the applicability of the solution for online decision making at various phases in survivable WDM network operation.

7

Subgraph-based protection strategy

A subgraph-based routing strategy attempts to provide a passive form of redundancy to optical networks in the event of a given set of failure $\mathcal{F}$ scenarios such as a single link failure or a single node failure. It is passive in that, before a connection is established, it is subjected to the constraints that it can be routed in the network as a fault free of any given set of failures, and is thus guaranteed in the event of a failure from the set. The end user experiences nominal interruption in service due to network state restoration and it is characterized in [194]. The key characteristics of the subgraph-based routing strategy are as follows.

- The redundancy in routing is provided using the resources available in the network.
- Fault recovery network states are maintained throughout the operation of the network.
- Subgraph fault tolerance provides a 100% guarantee for recovery from all possible pre-defined failure states.
- Subgraph routing is a pro-active path-based fault tolerance strategy.
- The network state must be altered to accommodate the new topology of the network caused by a failure state.

The first characteristic highlights one of the most important aspects of the strategy; it does not require the allocation of system transmission resources to ensure recoverability after the detection and location of a link failure. Simply put, there is no explicit link capacity lost due to the routing of backup connections because no active backup connections exist in this strategy. The second characteristic is important because, upon the occurrence of a fault, the network restores itself to a state that eliminates the defective component from consideration, and the network operates as if the failed components had never existed. Thirdly, 100% restoration is guaranteed in the event of a designated failure scenario occurring. Fourthly, subgraph routing is pro-active because when a failure occurs, the network knows how to recover from it. Finally, it must be stressed that there are no active backup paths present in subgraph fault tolerance. This results in reduced capacity needs for connections, but requires connections to possibly be rerouted in the event of a component failure, even if they do not traverse the failed component. This property results in the possible interruption of an altruistic connection as discussed in [62]. However, constrained routing [209], inter-arrival planning [62], and dynamic subgraph routing protection [60] all address this issue of connection reconfiguration, and reduce or eliminate the need for it to occur.

A disadvantage of subgraph-based routing and fault tolerance is that it can potentially require a complete reconfiguration of the network to a predetermined new state. Not all connections may be affected by a network reconfiguration, but no connection is guaranteed to be unaffected by a fault recovery. It also assumes that the probability of suffering any fault outside the given set during network operation is very low. Thus the strategy only guarantees the recovery of the network from any single failure at any given time. It is also assumed that each node knows the entire network state at any given time. This is key because all nodes need to know when and where a fault has occurred so that they can initiate appropriately to adopt the backup network state. Each node is required to maintain all subgraph network state information. Once a failure occurs and is detected and located, all nodes are informed of the location to start the recovery process. Of course, if there is a centralized recovery station, then all such information is part of the centralized recovery server.

7.1 Subgraph-based routing and fault tolerance model

Networks consist of a set of nodes and links that correspond to the various servers, routers, switches, and cables that make up its physical implementation. These nodes and links can be viewed as a set of vertices and edges in a graph. Each graph, G, is defined as a set of V vertices and E edges, or $G = (V, E)$. For the following discussion a set of faults consists of all single-link failures. Therefore the routing strategy presented below is for a single-link failure. It is easy to see how this discussion extends to an arbitrary failure set.

For subgraph-based routing, there exists a set of subgraphs of G, denoted as G_i, where e_i is removed from the graph G, or mathematically, $\forall i : 1 \leq i \leq L, G_i = (V, E - e)$, where L is the cardinality of the set of all edges in graph G. In general L is the cardinality of the fault set, e represents a fault in the fault set, and G_i represents the network graph when components corresponding to the fault i are eliminated from the graph G. Therefore, there exist L subgraphs of graph G, each one missing one of the L edges and the set of L subgraphs of G represents all possible single-link failures in the network. The original full link graph is called the *base network*. The constituent subgraphs of the base network are treated as virtual networks because only their state of utilization is maintained when connections are admitted. When a fault occurs on an edge e, the current set of connections are routed on the routes chosen in subgraph $G - e$.

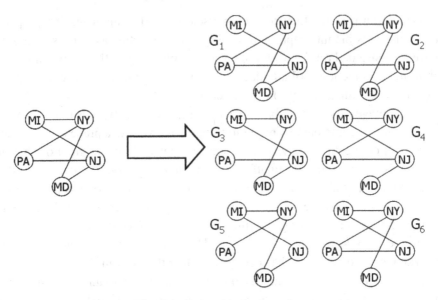

Fig. 7.1. Example of subgraphs.

A graph with five nodes and six edges, as shown in Fig. 7.1, will have six subgraphs. For the purposes of this example, each edge in the base network (and its constituent subgraphs) has a capacity of one and the distance between any pair of adjacent vertices is one.

Subgraph example and routing example. Consider the northeast part of the NSF network subgraph as shown in Fig. 7.2. It has six nodes and six bidirectional links. Let there be a request issued by vertex MI to connect with vertex NY. This connection attempts to find a path from MI to NY on all of the L subgraphs of the base network as shown in Fig. 7.2. The connection request from MI to NY is accepted as a path is available in all subgraphs. Another request from vertex NJ to PA is routed in the same way as the path from NJ to PA is available in all subgraphs (although it may be a different path in different subgraphs). When a third request from NY to MD arrives, it can only find paths in subgraphs G_2, G_3, and G_6. It blocks in subgraphs G_1, G_4, and G_5. Thus the connection request from node NY to node MD now fails due to non-availability of resources, and is consequently not routed in the base network.

7.1.1 Fault tolerance with subgraph-based routing

In the event of a fault, the subgraph-based routed network can fully recover by accepting the subgraph network state corresponding to the located edge failure. For

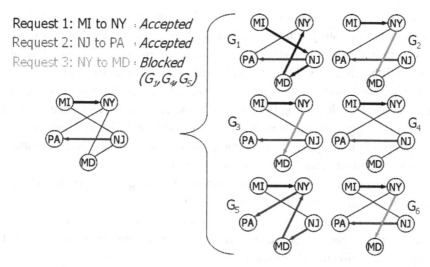

Request 1: MI to NY : *Accepted*
Request 2: NJ to PA : *Accepted*
Request 3: NY to MD : *Blocked*
(G_1, G_4, G_5)

Fig. 7.2. SGR model of the northeast part of the NSF network.

example, assume that there is an arbitrary failure of edge NJ to PA. Suppose the failure is in both directions and for some reason the edge is left non-operational. To recover, the network reroutes all current connections to reflect the network state depicted by subgraph G_5. The fault occurrence and recovery cycle is such that connections from the base network are rerouted to the paths in the selected subgraph (corresponding to the failed link). For the example, the request from MI to NY need not be rerouted as the path taken in the base network as well as in subgraph G_5 are the same. However, the path for the request node NJ to PA now has to be rerouted through the links (NJ, MD), (MD, NY), and (NY, PA) as shown in G_5 of Fig. 7.2.

7.2 Performance of subgraph-based routing

In this section, the performance of the backup multiplexing and subgraph-based routing strategy is compared for link failures. For the purpose of comparison, the shortest path length (in terms of the number of hops) routing strategy is utilized to evaluate the effectiveness of these schemes. Shortest path length routing attempts to dynamically route connections along the path with the least number of links between source and destination nodes. Each link is known as a hop. Wavelength assignment within a link is performed at random. Connections are routed dynamically in that each request is routed based on the network state at the time it enters the network. Dynamic routing and wavelength assignment typically perform better than fixed-path routing, although it requires a higher control overhead because each node must maintain network state information. The no-backup and backup multiplexing routing strategies are compared to provide a reference to measure the effectiveness of subgraph-based fault tolerance. No-backup routing uses the shortest path in terms of hop strategy, and makes no provision for fault tolerance. Backup multiplexing uses shortest-cycle routing. The results laid forth for backup multiplexing are based on the selection of the shortest path for the primary connection. Shortest-cycle routing [36] is used to increase the performance of backup multiplexing by guaranteeing both a potential primary and a backup path are found while attempting to establish a connection and that the primary–backup path pair is the shortest pair of paths from source to destination.

For comparison purposes, the connection requests are generated using a Poisson distribution with arrival rate of λ, where the arrival rates are varied to compare the different schemes. The request hold time follows a negative exponential probability distribution with a parameter value of $\mu = 1.0$. Source and destination nodes for any request are uniformly distributed.

7.2.1 Performance metrics

Several metrics are used to evaluate the effectiveness of the various schemes. These metrics are designed to measure both the efficiency and the feasibility of such a scheme and are only measured in the base network. They include the blocking probability, the average path length, the average shortest path length, the effective network capacity used, and the probability of path reassignment.

Blocking probability. The blocking probability is the most common indicator used to assess network routing and fault tolerance strategies. It is the probability that a request entering the network will be rejected. The blocking probability is the ratio of B and R, where B is defined as the total number of blocked requests and R is the total number of requests, i.e.

$$BP = B/R$$

Average path length. The average path length and the average shortest path length are used to compare how metrics perform within a network. As the ultimate goal in routing a connection is usually to use the shortest path between two points, the average shortest path length provides a way of comparing how effectively a routing strategy performs in a given network configuration. Both the average path length and the average shortest path length are calculated in terms of the number of hops or the number of links along the path.

These averages are calculated for the accepted requests only. The average shortest path is computed by using the shortest possible path in the network,

even if it is not available. The average path length is calculated using the
path length of the path taken (or available for the request) when the request
arrives.

Network utilization. Network utilization metrics are characterized by their inclusion of the ideas of connection and link capacity. They are an indication of how
much of the network is being used over the course of operation and whether there
are enough resources available to handle the request load demands.

Effective utilization refers to the minimum amount of system resources needed
to service all accepted connections if they were to have been routed along the
shortest path. In order for the effective utilization metric to be useful, it first has to
be normalized. The first step is to normalize it to the time duration of the simulation
so that data obtained at different arrival rates can be compared. Dividing by the time
duration of the simulation normalizes utilization. The simulation time is known only
to the network, and can either be obtained by knowing the time that the last request
enters the network, or by calculating it as a function of R/λ, where R is the number
of arriving requests and λ is the arrival rate. Normalizing the utilization with respect
to time yields a value that is bounded on the low side by 0 and on the high side by
the total available capacity in the network. The second step to normalization is to
normalize it by dividing it by the total available capacity of the network, given by
$L \times C$, where L is the total number of links in the network and C is the total available
capacity per link. Thus for utilization computation, $0 \le \lambda/R \times U \le L \times C$, where
U is the utilization. In other words, $0 \le \lambda \times U/(R \times L \times C) \le 1$.

Probability of path reassignment. The effectiveness of a fault tolerance scheme
depends on how the networks recover from a failure. This potentially may require all
connections in the network to be reconfigured to different paths. In order to quantify
the amount of path reassignment taking place, the path reassignment probability
must be measured, which is the probability that the path for a connection on the
base network needs to be changed upon the occurrence of a link failure. Let $P_j(R_i)$
be the probability that the path for request R_i remains the same if network link j
fails. Then the probability of reassignment for a subgraph-based routing strategy is
given by $P(\text{SGR reassignment}) = 1 - \sum_{i=0}^{i=R-B} \sum_{j=0}^{j=L} P_j(R_i)/[L \times (R - B)]$. For
backup multiplexing a path is reassigned only if the failed link is used by a request.
So if L_i is the path length of request i then the probability of reassignment is given
by $P(\text{SGR reassignment}) = \sum_{i=0}^{i=R-B} L_i/[L \times (R - B)]$.

Link load. Link load is a measure of the load placed on each node in the network
at any given time. It is useful in providing a baseline for the comparison of the
effectiveness of routing strategies across different network topologies. Link load,

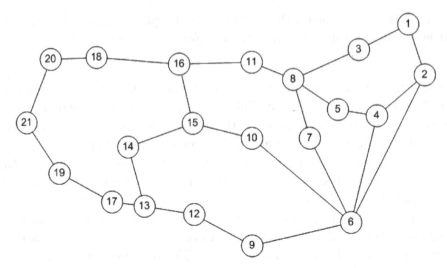

Fig. 7.3. ARPA-2 network.

or γ, is calculated using the following equation, where each duplex link is treated as two links, N is the total number of nodes in the network, and λ_n is the arrival rate per node,

$$\gamma = N \times \lambda_n \times \bar{H}/L$$

where $\bar{H}$ is the expected length of a primary connection in the topology in terms of hops. The link load is expressed in units of Erlangs and is used to compare results among different networks.

7.2.2 Network structures

Three standard network structures are used to assess and compare the performance of different restoration techniques. The networks considered are a 14-node, 21-link NSFnet topology, a 4×4 mesh-torus network, and a 21-node, 26-link ARPA-2 network topology. The ARPA-2 topology is shown in Fig. 7.3. The NSFnet and APRA-2 networks are in use. The 4 × 4 mesh-torus network possesses a high level of connectivity. Each node in a 4 × 4 mesh-torus has a degree of four, resulting in many potential paths between a node-pair.

    All links in the network have one fiber and the number of wavelengths per fiber is 16. In general, the capacity of a link is a direct function of the number of wavelengths on each fiber in the link. The wavelength continuity constraint has to be followed for each request. Each link is a duplex link (as all of the tested networks have this property) and each link is considered as two simplex links operating in opposite

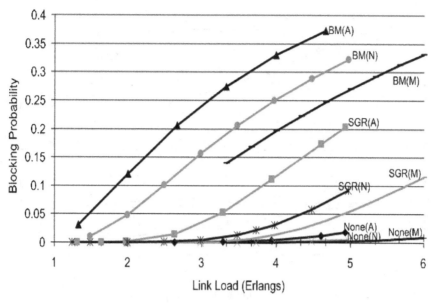

Fig. 7.4. Blocking probability of NSFnet.

directions. In the event of a link failure such as a fiber optic cable being severed, both simplex links are severed.

7.3 Performance results

7.3.1 Blocking probability

The results for subgraph-based fault tolerance and backup multiplexing and no backup strategy for the NSFnet, ARPA-2 and 4×4 mesh-torus topologies are shown in Fig. 7.4. Note that in this figure BM stands for backup multiplexing, SGR stands for subgraph-based routing method and None denotes no backup. Also, A represents ARPA-2 network, N denotes NSFnet and M denotes 4×4 Mesh network. It can be seen that SGR fault tolerance performs much better than backup multiplexing with the same parameters. In the best case (NSFnet), the blocking probability of backup multiplexing is roughly 3.5 times that of the SGR strategy, and is approximately three times higher than SGR in the ARPA-2 and mesh-torus topologies.

One factor in the increasing blocking probability of SGR fault tolerance is the presence of nodes with degree two present in two of the topologies. For example, there are two nodes of degree two in the NSFnet topology, and when subgraphs are formed, these two nodes become nodes of degree one in four of the 21 subgraphs. These isolated nodes are much more difficult to route connections for because of the

Subgraph-based protection strategy

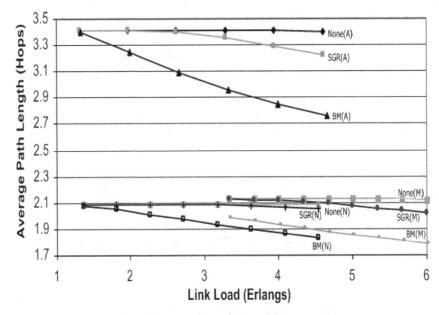

Fig. 7.5. Average path length of three topologies.

severely limited capacity in and out of these nodes. A node with degree two in the base network is referred to as a dead-end node. Dead-end nodes account for 14 of the total 21 nodes in the ARPA-2 topology, and consequently the performance of SGR in that topology is slightly worse when compared to the NSFnet and mesh-torus topologies.

7.3.2 Average path length

The results for SGR fault tolerance and backup multiplexing and no-backup strategy for the NSFnet, ARPA-2 and 4 × 4 mesh-torus topologies are shown in Fig. 7.5. The average path length indicates the average number of hops a connection must contain. The figure shows that the average path length decreases as the arrival rate increases. The large decrease in path length can be attributed to a higher blocking probability and the consequent lower number of connections in the network for a request to have to route around. In general, the average path length decreases as the arrival rate increases. As more requests enter the network at a time, the network becomes more congested and more requests are blocked. The result is that the requests that are accepted as connections are only those that are able to find shorter paths to route on.

The backup multiplexing path lengths are higher across all topologies because the paths are based on shortest-cycle routing, and the primary paths are not necessarily the shortest paths between two nodes. Shortest-cycle routing actually improves

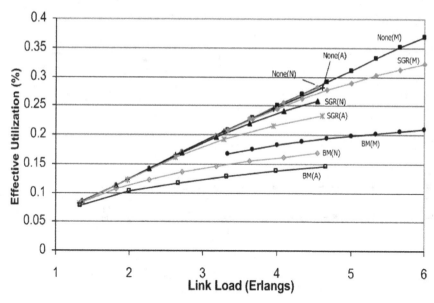

Fig. 7.6. Average effective utilization of three topologies.

performance because, although primary path lengths are longer, a primary–backup pair is almost always found and the total length of the primary–backup pair is the shortest possible.

7.3.3 Effective utilization

Effective utilization measures the minimum amount of network resources needed to accept the connections in the network at any given time for any given link load. In other words, in order for the network to accept the requests it had to provide a minimum amount of resources for the establishment of the connections. The effective utilization figures are shown in Fig. 7.6. For the most part, the effective utilizations of each strategy mirror each other, the exception being under high arrival rates. The difference is again attributed to fewer requests being accepted as the arrival rate increases. Utilization is directly proportional to the sum of the products of the capacity and the path length of each accepted connection; it consequently decreases as fewer requests are accepted. Again the effective utilization is used primarily to compare the performance of both fault tolerance strategies to a no-tolerance strategy.

7.3.4 Comparison between network topologies

An indication of the connectivity of a topology is the number of dead-end nodes that each of the subgraphs of the topology contains. Following this reasoning, the ARPA-2 topology has the worst connectivity with 14 dead-end nodes out of 21,

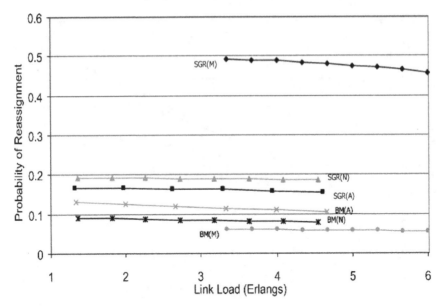

Fig. 7.7. Average probability of path reassignment for the three topologies.

NSFnet is next with two of 14, and the mesh-torus is best with no dead-end nodes. The blocking probabilities confirm this, as the blocking probability is higher per Erlang of link load for the ARPA-2 topology, followed by the NSFnet, and lastly by the mesh-torus. The comparatively lower blocking probability per link load of the mesh-torus indicates that SGR fault tolerance performs much better in topologies with higher connectivity.

7.3.5 Probability of reassignment

In a network recovery situation, SGR fault tolerance requires the network to re-configure to take the state given by the subgraph corresponding to the link failure. This could potentially require all connections in the network to change how they are routed. The probability of reassignment indicates how likely it is that a connection path will change during recovery. In all three topologies, as the arrival rate increases, the probability of reassignment decreases. This observed decrease is due to the higher probability of a request being blocked as the arrival rate increases. As requests enter the network at a higher rate, fewer connections are established. This makes routing the connections that do get accepted on each subgraph easier. Thus there is a much higher probability that a subgraph routes the connection exactly the same as the base network does.

Figure 7.7 shows the probability of reassignment for the NSFnet, ARPA-2, and 4×4 mesh topologies, respectively. As mass connection reassignment is unique to SGR fault tolerance, it is important to show how much more reassignment it requires than backup multiplexing. As the link load increases, the probability of reassignment decreases, indicating that there is less chance of a connection having to be rerouted during a network recovery.

The probability of reassignment for the ARPA-2 and NSFnet topologies remains fairly constant and the probability of reassignment for the mesh-torus topology decreases slightly more as the link load increases. The mesh-torus also has a much higher probability of reassignment overall. This is probably due to the higher connectivity of the mesh-torus. More links means that there are more options to route a path on, and the shortest hop-length routing metric uses this to the fullest. Backup multiplexing requires far less reassignment in the event of a fault occurring.

7.4 Multi-link and other failures

The SGR routing strategy can be easily used to handle multiple link failures, node failures, and specific failure scenarios.

As noted above, link-based sub-graph routing is a strategy for routing dependable connections under arbitrary failures. A connection is accepted only if it is accepted in all subgraphs. In the event of a failure, the network state is restored to the corresponding subgraph state where all connections are guaranteed to be restored for that single-fault scenario. It is also shown that the subgraph routing strategy performs significantly better than the traditional backup multiplexing schemes for routing dependable connections in terms of the blocking probability, network utilization, and redundancy. The following sections demonstrate the more general applications of subgraph routing.

7.4.1 Network and fault model

A network is represented by a graph $G = (V, E)$, where V is the set of nodes, $V = \{v_1, v_2, \ldots, v_N\}$ and $|V| = N$, and E is the set of edges or links, $E = \{e_1, e_2, \ldots, e_L\}$ and $|E| = L$. A failure may be of a single link $e_l \in E$ or a single node $v_n \in V$ or a group of multiple links or multiple nodes, or a combination of links and nodes. Such groups are said to be SRLG as explained earlier. A group failure is represented by $s_m = E \cup V$. A set of all such groups is represented by the set S, $S = \{s_1, s_2, \ldots, s_m\}$. Thus, a set of all possible faults consists of $F = E \cup V \cup S$ and

$$F = \{f_1, f_2, \ldots, f_K\} \tag{7.1}$$

A node failure can be represented equivalently by a set of links incident on it. Thus, failure of a node v_n can be represented by an SRLG s_j, where

$$s_j = \{e_l \mid e_l \text{ is incident on node } v_n\} \tag{7.2}$$

Input: graph $G = (V, E)$, failure set $F = \{f_1, f_2, ..., f_K\}$ and a traffic matrix $T_{N \times N}$.

Output: to route a connection T_{ij}.

Algorithm:

Step 1: create subgraphs $G_k = (V_k, E_k)$ where $V_k = V$ and $E_k = E - f_k, \forall f_k \in F$.

Step 2: attempt routing the connection on each subgraph G_k.

Step 3: If the connection is accepted in all G'_k, then the connection is accepted in the base network.

Else The connection is dropped from the network.

Fig. 7.8. Subgraph routing for all possible failure sets.

Therefore, we do not have to explicitly consider the failure of nodes. Similarly a single-link failure can also be treated as an SRLG failure, but to keep terminologies simple, we will keep single-link and multi-link failure sets represented separately by E and S.

7.4.2 Algorithm description

Subgraph routing is a pro-active fault-tolerant technique that ensures 100% restoration for all failure scenarios, included in set F, for which the network is designed.

A subgraph $G_k = (V_k, E_k)$, derived from a network $G = (V, E)$, is created for each of the failure scenarios $f_k \in F$ by removing all the edges contained within the failure set. Mathematically, $G_k = (V_k, E_k)$, where $V_k = V$ and $E_k = E - f_k$. For a connection entering a subgraph fault-tolerant network, it must be successfully routed in all the subgraphs G'_k. If it cannot be routed for any G_k, then the request is blocked, as it would not be protected against all failure scenarios.

The original graph, $G = (V, E)$, is referred to as the base network. The constituent subgraphs of the base network, $G_k = (V_k, E_k)$, are conceptual graphs as they only maintain a state of the base network.

7.4.3 Complexity analysis for subgraph routing

In order to route a connection R_i in the *base network*, the connection needs to be routed in all the K subgraphs, where K is the cardinality of the failure set. The time complexity of routing a request R_i in a network is governed by the complexity of the routing algorithm. In our case, Dijkstra's shortest-path algorithm can be used for routing the connections in each of the subgraphs. The complexity of this

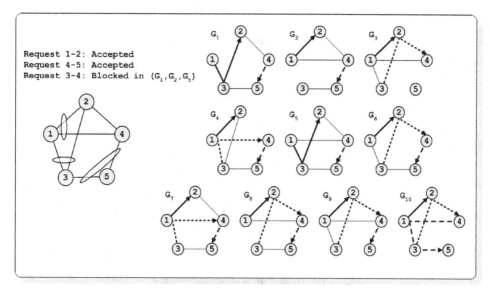

Fig. 7.9. Subgraph routing for tolerating SRLG failures.

algorithm is given by $O(N^2)$ [223], where N is the total number of nodes in the network. Thus, the overall complexity of routing these requests using Dijkstra's shortest-path algorithm is $O(K \times N^2)$. It is important to note that subgraph routing is not limited to using only shortest-path routing, but rather can accommodate any other desired routing metric.

7.4.4 Sub-graph routing for tolerating SRLG failures

Consider the network as shown in Fig. 7.9. There are three shared-risk link groups, each consisting of two links. Each link in the network is assumed to be a unidirectional link of total capacity one unit. The corresponding subgraphs are generated through the removal of individual links, as well as links belonging to each SRLG, and are shown in the same figure. Let there be three requests in the network R_1: $1 \rightarrow 2$, R_2: $4 \rightarrow 5$, and R_3: $3 \rightarrow 4$. Request R_1: $1 \rightarrow 2$ can be routed in all subgraphs and hence it is accepted for routing in the base network. Similarly request R_2: $4 \rightarrow 5$ finds a route in all subgraphs except G_3 and hence is accepted in the base network. A request is accepted on a subgraph which has any node with a degree of zero, i.e. a free node, and that node is either the source or destination of the request. A request is rejected on a subgraph if such a free node appears as an intermediate node in the path of the request. Similarly, request R_3 attempts to find a route on all the subgraphs, but cannot be accepted on subgraphs G_1, G_2, and G_5 because of insufficient capacity. Hence R_3 cannot be routed on the base network.

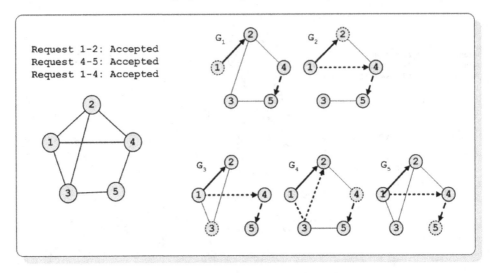

Fig. 7.10. Node-based subgraph routing.

7.4.5 Node-based subgraph routing

Node-disjoint subgraph routing is similar to subgraph routing except that in this case the subgraphs are generated by the removal of each node, $n_i \in N$, one at a time, from the base network. Let us consider the network as shown in Fig. 7.10. Each link in the network is assumed to be a unidirectional link of total capacity one unit. The corresponding subgraphs generated by the removal of each node are shown in the same figure. Let there be three requests in the network R_1: $1 \rightarrow 2$, R_2: $4 \rightarrow 5$, and R_3: $1 \rightarrow 4$. Request R_1: $1 \rightarrow 2$ can be routed in all subgraphs and hence it is accepted for routing in the base network. Similarly, request R_2: $4 \rightarrow 5$ finds a route in all subgraphs except G_4 but is still accepted in the base network. A request is accepted on a subgraph which has any node with a degree of zero, i.e. a free node, if that node is either the source or the destination of the request. Request R_3 is accepted for routing in each subgraph except in G_1 where it cannot be routed because node 1 has a nodal degree of zero; however, since node 1 is the source node of the request, it is accepted in the base network.

Node-disjoint subgraph routing gives a lower bound on the routing performance that can be achieved because it is much more constrained than subgraph routing for tolerating SRLG failures. The subgraphs for tolerating node failures are a special case of the $S + 1$ routing, since it deals with SRLG groups consisting of links that share a common node.

7.5 Constrained subgraph routing

One of the potential drawbacks of incorporating the subsgraph routing scheme as a means of tolerating SRLG failures is the issue of connection re-establishment.

The above-proposed scheme depends on the ability of the network state to change to the state of a subgraph during fault recovery. This potentially requires many connections in the network to be reassigned to different path/trunk combinations as defined in a subgraph. A path is the set of l links that connect the source and destination nodes. A trunk is the specific fiber, wavelength, and time slot that the connection is established on within the path. Similar issues have been discussed in the context of $L + 1$ subgraph routing in [171].

To overcome this limitation we introduce the concept of *constrained subgraph routing*. The constrained subgraph routing minimizes the probability of reassignment during transition from the base network to the final subgraph. There are two levels of constrained subgraph routing. They are:

- *Constraint 1:* a connection is constrained to be routed on the same path as in the base network in all the subgraphs which contain all the l links of the path.
- *Constraint 2:* if constraint 1 is fulfilled, then the connection can be further constrained to be routed along the same trunk in the subgraph as in the base network.

In the case of link-based subgraphs there are L subgraphs. Constraint 1 requires that the connection be routed on $L - l$ subgraphs with the identical path as in the base network. Constraint 2 requires that a subgraph connection not only take the same link path, but also the same trunk along that path as in the base network. In this manner, any connection not directly affected by the failed link will not be interrupted in $L - l$ subgraphs. This is an attempt to avoid as much node reconfiguration as possible by minimizing the probability of reassignment. In our results, we will show that path-constrained routing can actually improve the blocking performance over the unconstrained case. However, trunk-constrained routing significantly degrades network performance in terms of increasing the blocking probability, but realizes a very low probability of reassignment for subgraph routing architectures.

7.6 Example

Three different network topologies, shown in Figs. 7.11–7.13, were simulated to assess the performance of subgraph routing for tolerating SRLG failures. Each of the three topologies consists of links with one fiber per link, 16 wavelengths per fiber, and one timeslot per wavelength. Each link also consists of two unidirectional links that are assumed to be part of the same shared risk link group, meaning that if the link fails in one direction, the link in the opposite direction also fails because they would presumably be physically routed together. No nodes offer any wavelength switching capabilities, thus the *wavelength continuity constraint* is obeyed. The arrival of the requests at a node follow a Poisson process with rate λ, and are equally likely to be destined to go to any other node. The holding time of the requests follows an

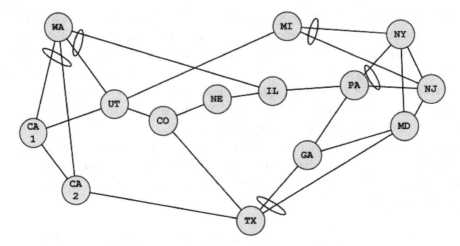

Fig. 7.11. 14-node, 23-link NSFnet network with SRLGs.

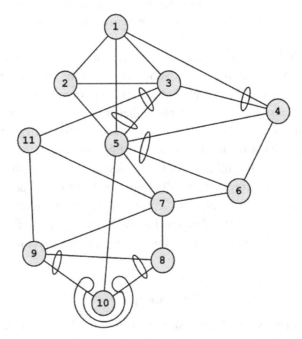

Fig. 7.12. 11-node, 22-link NJLATA network with SRLGs.

exponential distribution with unit mean. The capacity requirement of each request is a unit wavelength.

Three subgraph formation techniques have been assessed: subgraphs based on all physical link failures (link-based subgraph routing), subgraphs based on arbitrarily chosen shared risk link groups, as shown in Figs. 7.11–7.13, and subgraphs based on all single-node failures.

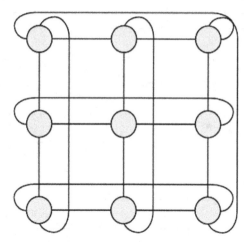

Fig. 7.13. Nine-node, 18-link mesh-torus network.

For the 3×3 mesh-torus, SRLGs were formed as the north and east links leaving a node and the south and west links leaving a node. The number of subgraphs created for each base network is equivalent to the sum of the number of physical link faults and the number of SRLGs. If the source or destination node for a request is a stranded node (one with a degree of zero) in a subgraph, a conditional acceptance on that subgraph is granted as discussed in Section 8.3 (source or destination nodes can be isolated nodes, intermediate nodes cannot be).

Conditional acceptance is allowed, because any connection formed between two nodes automatically incurs the risk of either the source or destination node failing. The intent of providing coverage for all single-node faults is to protect against faults occurring at the intermediate nodes in the path of the connection request.

Using the 14-node, 23-link NSFnet as an example, the previously described subgraph formation techniques will be clarified. In the case of link subgraph generation, an NSFnet base network creates 23 subgraphs, each missing a pair of unidirectional links physically routed together. For SRLG subgraph generation, an NSFnet base network creates six SRLG subgraphs in addition to the 23 subgraphs based on physical links, for a total of 29 subgraphs. Finally, for node subgraph generation, an NSFnet base network creates 14 node subgraphs and 23 link subgraphs for a total of 37 subgraphs. Physical link failures are always considered in each subgraph generation because a physical link can fail in a location where it does not affect the other members of its SRLG.

In all subgraph cases, all single-link faults are 100% guaranteed, and in the case of subgraphs based on shared risk link groups, there is a 100% guarantee for all connections in the event of a shared risk link group fault. In the node-based subgraph case, 100% restoration is guaranteed for all intermediate node and

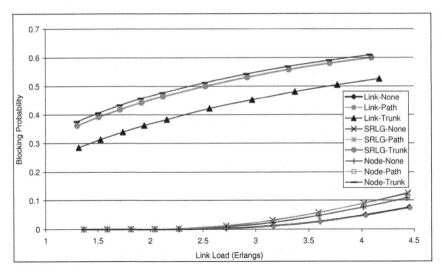

Fig. 7.14. NSFnet blocking probability vs. link load, © IEEE. Source: P. Datta, M. T. Frederick, and A. K. Somani, Sub-graph routing: a novel fault-tolerant architecture for shared-risk link group failures in WDM optical networks, in *Proc. of Design of Reliable Communication Networks DRCN 2003* [209].

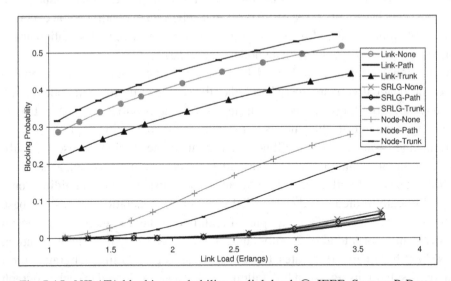

Fig. 7.15. NJLATA blocking probability vs. link load, © IEEE. Source: P. Datta, M. T. Frederick, and A. K. Somani, Sub-graph routing: a novel fault-tolerant architecture for shared-risk link group failures in WDM optical networks, in *Proc. of Design of Reliable Communication Networks DRCN 2003* [209].

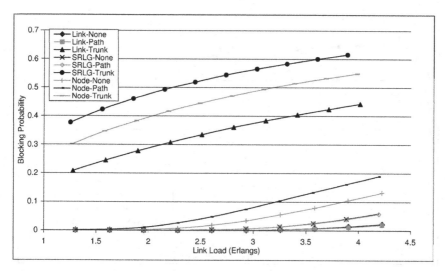

Fig. 7.16. MESH 3×3 blocking probability vs. link load, © IEEE. Source: P. Datta, M. T. Frederick, and A. K. Somani, Sub-graph routing: a novel fault-tolerant architecture for shared-risk link group failures in WDM optical networks, in *Proc. of Design of Reliable Communication Networks DRCN 2003* [209].

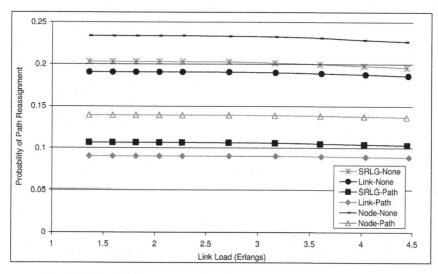

Fig. 7.17. Probability of path reassignment vs. link load for NSFnet.

single-link failures. The blocking probability results for all three topologies are shown in Figs. 7.14–7.16.

To assess the performance of constrained routing, the probabilities of reassignment to a different path and to a different path/trunk combination are shown in Figs. 7.17–7.22. In Figs. 7.17–7.19 the probability of reassignment for

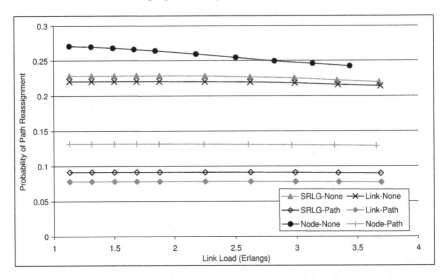

Fig. 7.18. Probability of path reassignment vs. link load for NJLATA.

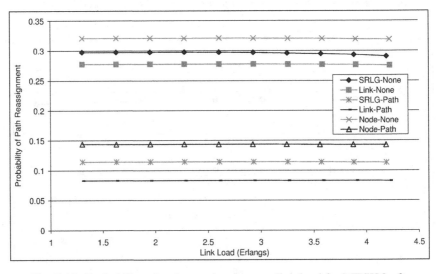

Fig. 7.19. Probability of path reassignment vs. link load for MESH 3×3.

unconstrained routing runs between 18 and 33%. This is in sharp contrast to the probability of reassignment for path-constrained routing which ranges from 8 to 15%. In all cases, path-constrained subgraph routing offers a lower probability of reassignment. According to [171], the calculated probability of path reassignment for backup multiplexing ranges between 8 and 15%, thus making constrained subgraph routing roughly equivalent to backup multiplexing.

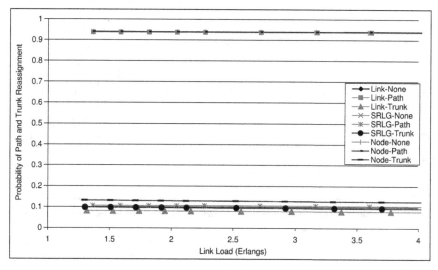

Fig. 7.20. Probability of path and trunk reassignment vs. link load for NSFnet.

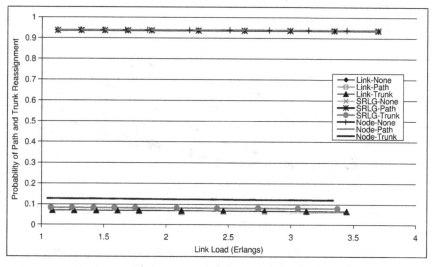

Fig. 7.21. Probability of path and trunk reassignment vs. link load for NJLATA.

In Figs. 7.20–7.22 the probability of path/trunk combination reassignment is about 93% for path and unconstrained subgraph routing. This is significantly higher than the probability of path and trunk reassignment for path/trunk combination constrained subgraph routing, which ranges from 8 to 15%. This is an expected result because connections are more likely to choose same paths, in similar subgraphs on any capacity available. In summary, in a recovery situation, a connection will

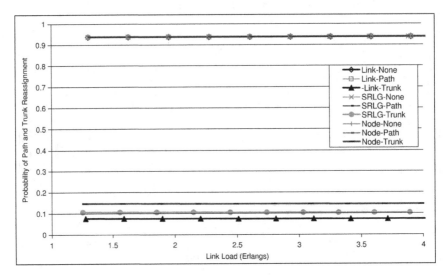

Fig. 7.22. Probability of path and trunk reassignment vs. link load for MESH 3×3.

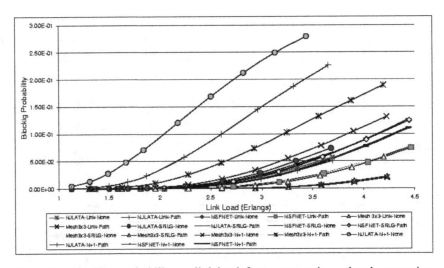

Fig. 7.23. Blocking probability vs. link load, for no-constraint and path constraint,
ⓒ IEEE. Source: P. Datta, M. T. Frederick, and A. K. Somani, Subgraph routing: a
novel fault-tolerant architecture for shared-risk link group failures in WDM optical
networks, in *DRCN 2003* [209].

more likely than not have to change its trunk if path/trunk combination constrained
routing is not imposed.

In Fig. 7.23 we observe that the blocking probability for path-constrained subgraph routing is less than the blocking probability for unconstrained subgraph
routing in all but two cases: NSFnet and mesh-torus 3×3 node-based subgraphs.
In each topology, the blocking probability for node-based subgraphs is higher than

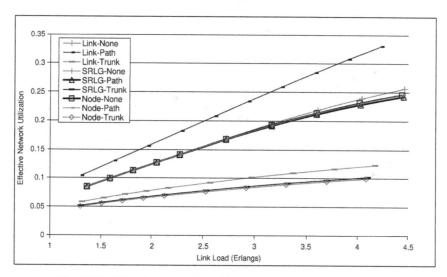

Fig. 7.24. Effective network utilization vs. link load for NSFnet.

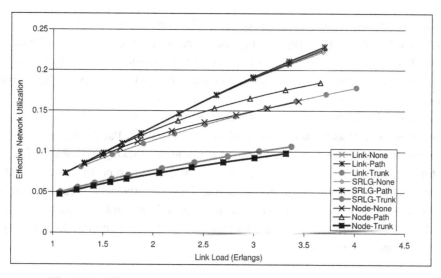

Fig. 7.25. Effective network utilization vs. link load for NJLATA.

that for the SRLG, which in turn is higher than that for link-based subgraphs. The only exception is observed in the results for NSFnet where node-based subgraphs slightly outperformed the SRLG-based subgraphs.

Intuitively, one would expect that, in constrained subgraph routing, forcing the connection to route on the same path on each of the L subgraphs would increase the connection blocking probability. However, in simulation the reverse phenomenon was observed, as shown in Figs. 7.14–7.16. In most cases the blocking

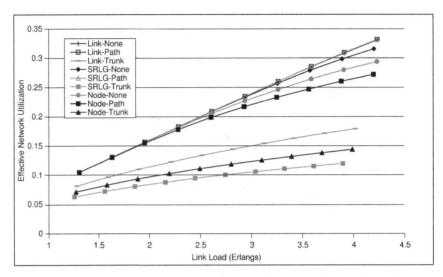

Fig. 7.26. Effective network utilization vs. link load for MESH 3×3.

probability actually decreased slightly when following the path-constrained routing. Path/trunk-constrained routing sharply increased the overall connection blocking probability and is not seen as a viable solution unless minimization of probability of reassignment is more crucial than the minimization of the blocking probability in a particular network.

Figures 7.24–7.26 depict the effective utilization of the network for path-constrained as well as the unconstrained subgraph routing. Path/trunk-constrained routing is not considered due to its drastically higher blocking probability. For all the topologies, the network utilization for the SRLG subgraph routing is slightly lower than that of link-based subgraph routing. However, the network utilization for the SRLG subgraph routing is higher than node-based subgraph routing since node-based subgraph routing offers a higher blocking probability than SRLG subgraph routing for most topologies.

7.7 Observations

The results obtained using path-constrained routing are very interesting. They indicate that constraining subgraph routing to a path actually improves the blocking probability. One of the possible explanations for this phenomenon is the increased resemblance each subgraph takes to the base network. If the path on a subgraph is distinctly different from the base network, the request might have to traverse through links that a different request regularly utilizes. If a request cannot find the

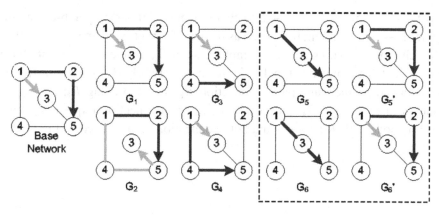

Fig. 7.27. Subgraph shadowing.

necessary resources available on such critical links, it is blocked with a higher frequency. If, however, each subgraph is required to route each connection in the same way that the base network does if the same path exists, the subgraph utilization of critical links more closely resembles that of the base network. This increases the likelihood that an arbitrary request is accepted on all subgraphs, and consequently accepted in the base network. This is referred to as *subgraph shadowing*.

Subgraph shadowing increases the performance of subgraph routing because situations exist where there are several different equidistant paths from a source to a destination node. Each of these paths can be chosen to route the connection in a fewest-hops routing strategy. Constraining the path actually creates subgraph states that more closely resemble, or shadow, the actual state of the base network. It helps to reduce the occurrence of situations where a connection gets blocked because the only possible path where it could have been routed is already occupied by some other connection that should have been routed elsewhere.

A subgraph shadowing situation is depicted in Fig. 7.27. In this figure, the base network consists of six links and subgraphs are created based on single link failures, resulting in subgraphs 1 through 6. Assume that each link in the subgraphs has a total capacity of one unit. Also assume that connection request R_1: $1 \rightarrow 5$ is routed on the base network along R_1: $1 \rightarrow 2 \rightarrow 5$. Similarly, connection request R_2: $1 \rightarrow 3$ is routed on the base network along R_2: $1 \rightarrow 3$. Subgraphs G_1 and G_2 have the option of routing connection R_1, from node 1 to 5, along the paths $1 \rightarrow 2 \rightarrow 5$ or $1 \rightarrow 4 \rightarrow 5$. The path R_1: $1 \rightarrow 2 \rightarrow 5$ is chosen. Connection R_2 gets routed in subgraphs G_1 and G_2. In subgraphs G_3 and G_4, R_1 has the option to select from either of the two paths, R_1: $1 \rightarrow 3 \rightarrow 5$ or R_1: $1 \rightarrow 4 \rightarrow 5$. The path R_1: $1 \rightarrow 4 \rightarrow 5$ is chosen for routing connection R_1 and connection R_2 can be routed without any problem. In subgraphs G_5 and G_6, connection R_1 has paths $1 \to 2 \to 5$ and $1 \to 3 \to 5$ to choose from. Without subgraph-constrained routing, $1 \to 3 \to 5$ might be chosen as the path for routing connection R_1. If this path is chosen for routing R_1 in subgraphs G_5 and G_6, connection request R_2 cannot be accepted and must be blocked by the base network. However, if the routing is constrained on the same path, connection R_1 is routed along R_1: $1 \to 2 \to 5$ as shown in subgraphs G'_5 and G'_6. Furthermore, connection request R_2 can be routed on node path R_2: $1 \to 3$ and both connections can be accepted in the base network.

8

Managing multiple link failures

Although the failure of a single component such as a link or a node is the most common failure scenario, it is possible to have multiple links fail simultaneously. In particular, double-link failures can happen in the following scenarios.

(i) The first link fails. The recovery from the failure of the first link is completed within a few milliseconds to a few seconds. However, it may take a few hours to a few days for the failed physical link to be repaired. It is certainly conceivable that a second link might fail during this period, thus causing two links to be down at the same time. Suppose the link failure is a Poisson process with parameter λ and the repair times are exponentially distributed with parameter μ. Thus the average time to failure is $1/\lambda$ and the average repair time is $1/\mu$. Suppose a link fails at time $t = 0$, then the probability that a failure will occur on a link while the first repair is being carried out is given by

$$FP = \frac{\lambda}{(\lambda + \mu)}[1 - \exp(-\lambda/\mu)] \tag{8.1}$$

For $\mu = 9\lambda$, $FP \cong 0.1$, which is large.

(ii) Two links may be physically routed together for some distance in real situations. A single backhoe accident may lead to the failure of both links.

In order to protect connections from link failures in the network, two paths are often assigned: a primary path on which a connection is established and a backup path on which a connection will be re-established in the case where the primary path fails. Most research to date in survivable optical network design and operation has focused on single-link failures [237]; however, the occurrence of double-link failures is not uncommon in a network topology [63, 89].

Link restoration schemes provide a detour around a failed link that does not necessarily affect the entire source–destination path. Path restoration schemes, in general, attempt to provide a backup path from the source to the destination that is independent of the working path. Path restoration schemes are classified into two categories based on knowledge of the link failure. A backup path that can be used for any link failure on the working path and is link-disjoint with the working path is referred to as *failure-independent path restoration*. Alternatively, a connection may be assigned more than one backup path depending on the failure scenario. Such an approach requires complete knowledge of the failure in the network, hence it is referred to as *failure-dependent path restoration*. Path-based restoration has been established to be a more capacity-efficient approach for mesh-based networks compared to link-based restoration approaches [225, 261].

There are link-based [266, 305] and path-based protection [304] models for surviving double-link failures. For a graph to remain connected after any two edges fail, the graph must be 3-link connected. Thus recovery methods assume the graph to be 3-link connected, i.e. the graph does not become disconnected with fewer than three link failures. For multiple link failures, the graph needs to have higher connectivity.

8.1 Link-based protection for two link failures

Link-based protection for two link failures can be classified into the following two categories.

(i) *Backup paths with link identification*. In this technique, two edge-disjoint paths, a first backup path $b_1(e)$ and a second backup path $b_2(e)$, are precomputed to recover from the failure of each edge e. When edge e fails, the first backup path $b_1(e)$ is used for rerouting. At the same time, all nodes in the network are informed of the failure using appropriate signaling. Suppose the second link f fails at this point. This failure is notified to all nodes as before, and the recovery takes place using an alternate path for the second failure.

Four possible cases may exist for the failure of edges e and f and the two backup paths $b_1(e)$ and $b_2(e)$ (Fig. 8.1).

(a) $b_1(f)$ does not use e, f does not lie on $b_1(e)$: in this case, $b_1(e)$ continues to be used to reroute the traffic on e and $b_1(f)$ is used to reroute the traffic on f.

(b) $b_1(f)$ uses e, f does not lie on $b_1(e)$: in this case, $b_1(e)$ continues to be used to reroute the traffic on e. $b_1(f)$ cannot be used to reroute because link e is still down, therefore $b_2(f)$ is used to reroute the traffic on f.

(c) $b_1(f)$ uses e, f lies on $b_1(e)$: in this case, neither $b_1(e)$ nor $b_1(f)$ can be used as restoration routes. There are two recovery methods to reroute the working traffic on primary links e and f. In Method 1, when f fails, $b_2(f)$ is used to reroute the working traffic on f. When the information concerning the failure of f reaches the end nodes of e, these nodes switch the working traffic originally on e from $b_1(e)$ to $b_2(e)$. Knowledge of which links lie on a backup path is necessary to carry out this process. In Method 2, $b_2(f)$ is used to reroute both the working traffic on f as well as the backup traffic rerouted on $b_1(e)$. Thus, the traffic originally routed on e is now on $(b_1(e) - f) \cup b_2(f)$.

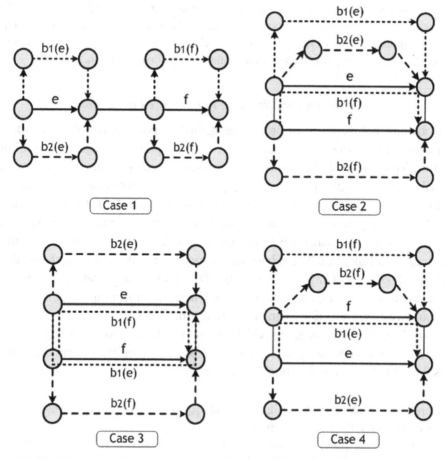

Fig. 8.1. Four cases demonstrating rerouting of traffic on links e and f when they both fail.

(d) $b_1(f)$ does not use e, f lies on $b_1(e)$: as in the case 3, two methods are possible. In Method 1, $b_2(e)$ and $b_2(f)$ are used to reroute the working traffic on e and f, respectively. In Method 2, $(b_1(e) - f) \cup b_1(f)$ is used to reroute the working traffic on e, while $b_1(f)$ is also be used to reroute the traffic originally on f.

(ii) *Backup paths without link identification.* In this method, a single backup path $b(e)$ is precomputed for each link. For every link $f \in b(e)$, a backup path $b(f)$, which does not contain e can be found since the graph is 3-link connected. Suppose e fails first, and then f fails. The working traffic on f and rerouted traffic on f (in this case, the rerouted traffic from e) are both rerouted to $b(f)$ from f. Since $b(f)$ does not use e, this rerouting would be successful. One advantage of this method is that no signaling is necessary to inform the network nodes of the failure of a link. The failure of a link need only be detected at the end nodes of that link. The cost for this is that when two links fail the rerouted traffic may have to traverse many links.

Notice that computing such kinds of backup paths is not trivial. A heuristic algorithm was developed in [266] to compute the backup paths. It works by contracting the graph G according to a set of rules, computing backup paths for the links in the contracted graph, and then mapping these backup paths onto the original graph.

8.1.1 Backup multiplexing in link-based protection

It is obvious that methods for protecting against all possible double-link failures require more backup capacity than the methods for protecting against single-link failures. Thus the efficient utilization of backup capacity is more important. As described above in the case of a backup path with link identification, two backup paths are precomputed and the resources are reserved on these paths at the time of establishing the primary path. An important observation here is that backup paths may not be used simultaneously to reroute the traffic on primary paths at any time when any two links fail. These backup paths can share the wavelengths on their common links without any violation of the 100% protection guarantee. For each of the cases illustrated in Fig. 8.1, the scenarios where backup multiplexing is possible without violating protection guarantees are identified below.

Consider Case 1 in Fig. 8.1 in which for any two links e and f, $b_1(f)$ does not use e and f does not lie on $b_1(e)$. Without loss of generality, assume that link e fails first. The following situations can occur.

(i) Link e fails and then link f fails. $b_1(e)$ and $b_1(f)$ are used as backup paths to reroute the traffic on links e and f, respectively.
(ii) Link e fails first and then one of the links $g \in b_1(e)$ fails. When the information regarding the failure of g reaches the end nodes of e, the rerouted traffic on $b_1(e)$ is switched to $b_2(e)$. Link f cannot fail during this period as only two link failures are being considered.
(iii) Link e fails first and then a link that is not link f and not on either of the two backup paths of link e fails. $b_1(e)$ is used to reroute the traffic on link e. The traffic on the second failed link is rerouted on one of its own backup paths.

As can be seen from the above scenarios, only paths $b_1(e)$ and $b_1(f)$ are used simultaneously (as in the above failure scenario (i)). All other path pairs $b_1(e)$ and $b_2(f)$, $b_2(e)$ and $b_1(f)$, $b_2(e)$ and $b_2(f)$ are not used simultaneously at any time. Thus, if one of the above path pairs, which are not used simultaneously, have any common link(s), then they can share the reserved backup wavelengths on the common link(s). Similar rules of sharing backup wavelengths on common links can be obtained for the other cases as well. Using the above observations, the backup paths can share backup capacity when it is possible. This problem is formulated in Section 8.3.

8.2 Path-based protection

In path-based protection, to provide 100% protection against any two link failures, two link-disjoint backup paths must be provided for each source–destination pair. When a primary path fails, an alarm indication signal is generated by the node that detects the link failure and transferred through the control channel. When the source receives the alarm signal, it prepares to set up the first backup path. The first backup path may also fail due to another link failure. Therefore, a run-time search is needed. Run-time search also detects if the backup capacity on the first backup path is not available due to the sharing, in the case of a shared path scheme. If the source detects the above scenarios using a run-time search, it prepares to set up the second backup path; otherwise, it uses the first backup path to reroute the traffic on the primary. The backup paths may or may not use the same wavelength as the primary path.

8.2.1 Backup multiplexing in path-based protection

Reserving dedicated capacity on two backup paths for every primary path, however, could reserve excessive capacity in some situations. Consider the example network shown in Fig. 8.2 (see Section 7.4): suppose paths $1 \rightarrow 3 \rightarrow 2$ and $4 \rightarrow 5 \rightarrow 1$ are two primary paths p and r, respectively. Let $1 \rightarrow 2$ and $1 \rightarrow 5 \rightarrow 4 \rightarrow 2$ be the two backup paths for p, denoted as b_{p1}, b_{p2}, respectively. Let $4 \rightarrow 3 \rightarrow 1$ and $4 \rightarrow 2 \rightarrow 1$ be the two backup paths for path r, denoted as b_{r1} and b_{r2}, respectively. The only failure scenario that could cause two primary paths to go down simultaneously is when one of the links on path p and one of the links on path r fail at the same time. b_{p1} and b_{r1} can be used to reroute the working traffic on paths p and r, respectively. Thus b_{p2} and b_{r2} are not used at the same time for all possible two link failures, therefore they can share backup capacity for primary paths p and r on link $(4, 2)$.

Backup capacity sharing is not always allowed if a 100% restoration guarantee against any two link failures is required. Suppose the other two primary paths p and r are $2 \rightarrow 3 \rightarrow 1$ and $4 \rightarrow 5 \rightarrow 1$, respectively. The two backup paths for path p are $b_{p1}: 2 \rightarrow 1$ and $b_{p2}: 2 \rightarrow 4 \rightarrow 5 \rightarrow 1$. The two backup paths for path r are, $b_{r1}: 4 \rightarrow 2 \rightarrow 1$, $b_{r2}: 4 \rightarrow 3 \rightarrow 1$. Since paths p and b_{r2} have shared links, and so do path r and b_{p2}, the failure of one link on path p could cause b_{r2} to fail, and the failure of one link on path r could cause b_{p2} to fail. If the above scenario occurs, b_{p1} and b_{r1} are used to reroute the primary traffic on paths p and r, respectively. Therefore they must not share backup capacity even if they have a common link $(2, 1)$.

The backup multiplexing rules are summarized below.

(i) Primary p and r are link-disjoint. Table 8.1 cases 1–6 summarize the topology relationships and backup capacity sharing constraint. The following notation and equations are used to express the topology relations and backup capacity sharing constraint.

    (a) $p \cap r = \phi$: primary path p and r are link-disjoint; otherwise, they have shared link(s).

    (b) $BC(b_{pi})$: backup capacity reserved on backup path b_{pi}.

    (c) $BC(b_{pi}) \wedge BC(b_{rl}) = \phi$: backup paths b_{pi} and b_{rl} must not share backup capacity on their common link(s).

(ii) Primary p and r are not link-disjoint. In addition to the above constraint, the failure of the shared link of p and r causes both p and r to go down simultaneously. The worst-case scenario is that one of the backup paths of p and one of the backup paths of r also have a common link and that link also fails, causing these two backup paths to fail at the same time. If the above failure scenario occurs, the other backup paths for p and r are used to reroute the primary traffic on p and r, respectively. Therefore they must not share backup capacity. This scenario is summarized in case 7 of Table 8.1.

8.3 Formulating two link failures

Integer linear programming (ILP) is used to optimize the capacity utilization. The ILP formulation is used to model shared link and shared path protection schemes and to demonstrate the possible sharing in capacity. Dedicated link and dedicated path protection schemes are also modeled in the same way. In shared link and shared path protection schemes, the backup paths can share wavelengths on their common links, while in dedicated link and dedicated path protection schemes the backup paths do not share the capacity.

The formulation assumes that the following information is given.

- The network topology.
- A demand matrix consisting of the connections to be established.
- Three alternate routes, which are link-disjoint, for each node-pair.
- In link-based protection, two alternate routes (which are link-disjoint) are precomputed and given for each link.
- Each route between every node-pair is viewed as W wavelength continuous paths (lightpaths), one for each wavelength.

The objective is to minimize the total number of wavelengths used on all the links in the network for both the primary and backup paths, measured by the number of wavelength links. One wavelength link represents a wavelength used on a link. The ILP solution determines the primary and backup paths for the demand set and hence the routing and wavelength assignment.

Table 8.1. *The topology relationships, failure scenarios and backup capacity sharing constraint*

Primary paths topology relationships	Backup–primary, backup–backup topology relationships	Backup capacity sharing constraints
$p \cap r = \phi$	1. $b_{pi} \cap r \neq \phi, b_{rl} \cap p \neq \phi,$ i and $l \in \{1, 2\}$	$BC(b_{pi}) \wedge BC(b_{rl}) = \phi$
	2. $b_{pi} \cap r \neq \phi, b_{rl} \cap p = \phi,$ $b_{rm} \cap p \neq \phi,$ $i, l,$ and $m \in \{1, 2\}, l \neq m$	$BC(b_{pi}) \wedge BC(b_{rl}) = \phi$
	3. $b_{pi} \cap r = \phi, b_{pj} \cap r \neq \phi,$ $b_{rl} \cap p = \phi, b_{rm} \cap p \neq \phi,$ $i, j, l,$ and $m \in \{1, 2\}, i \neq j,$ $l \neq m$	$BC(b_{pi}) \wedge BC(b_{rl}) = \phi$
	4. $b_{pi} \cap r = \phi, b_{rl} \cap p = \phi,$ $b_{rm} \cap p \neq \phi,$ $i, l,$ and $m \in \{1, 2\}, l \neq m$	$(BC(b_{p1}) \wedge BC(b_{rl}) = \phi) \|$ $(BC(b_{p2}) \wedge BC(b_{rl}) = \phi)$
	5. $b_{pi} \cap r = \phi, b_{rl} \cap p \neq \phi,$ i and $l \in \{1, 2\}$	$((BC(b_{p1}) \wedge BC(b_{r1}) = \phi) \|$ $(BC(b_{p2}) \wedge BC(b_{r1}) = \phi)) \&\&$ $((BC(b_{p1}) \wedge BC(b_{r2}) = \phi) \|$ $(BC(b_{p2}) \wedge BC(b_{r2}) = \phi))$
	6. $b_{pi} \cap r = \phi, b_{rl} \cap p = \phi,$ i and $l \in \{1, 2\}$	$(BC(b_{p1}) \wedge BC(b_{r1}) = \phi) \|$ $(BC(b_{p2}) \wedge BC(b_{r1}) = \phi) \|$ $(BC(b_{p1}) \wedge BC(b_{r2}) = \phi) \|$ $(BC(b_{p2}) \wedge BC(b_{r2}) = \phi)$
$p \cap r \neq \phi$	In addition to the above primary and backup cases, there is one backup–backup relationship we need to consider 7. $b_{pi} \cap b_{rl} = \phi, i, l \in \{1, 2\}$	$BC(b_{pj}) \wedge BC(b_{rm}) = \phi,$ j and $m \in \{1, 2\}, j \neq i, m \neq l$

8.3.1 Notation used

The following notation is used.

The network topology is represented as a directed graph $G(N, L)$ with N nodes and L links with W wavelengths on each link.

- $k, l, m = 1, 2 \ldots, L$: number assigned to each link in the network.
- $\lambda = 1, 2 \ldots, W$: number assigned to each wavelength.
- $i, j = 1, 2 \ldots, N(N-1)$: number assigned to each s-d pair.
- d_i: demand for node-pair i, in terms of the number of lightpath requests.
- $K = 3$: number of alternate routes between every s-d pair.

- $M = 2$: number of alternate routes for the link l. It is used only in the formulation of link-based protection schemes.
- $p = 1, 2, \ldots, KW$: number assigned to a path for each s-d pair. A path has an associated wavelength (lightpath). Each route between every s-d pair has W wavelength continuous paths. The first $1 \le p \le W$ paths belong to route 1, $W + 1 \le p \le 2W$ paths belong to route 2, and $2W + 1 \le p \le 3W$ paths belong to route 3.
- r: in link-base schemes, $r = 1, 2, \ldots, MW$. It is a number assigned to an *alternate path* for each link. In path-based schemes, $r = 1, 2, \ldots, KW$. It is a number assigned to a path for each node-pair. A path has an associated wavelength (lightpath). Each path has W wavelength continuous paths, as in the definition of notation p.
- (i, p): refers to the pth path for s-d pair i.
- $g_{l,\lambda}$ takes a value of 1 if wavelength λ is used by some restoration routes that traverse link l (binary variable).
- s_l: number of wavelengths used by backup lightpaths, which pass through link l (variable).
- w_l: number of wavelengths used by primary lightpaths, which pass through link l (variable).
- $\delta^{i,p}$: path indicator. It takes a value of 1 if (i, p) is chosen as a primary path and 0 otherwise (binary variable).
- $\psi_\lambda^{i,p}$: wavelength indicator. It takes a value of 1 if wavelength λ is used by the path (i, p) and 0 otherwise (data).
- $\epsilon_l^{i,p}$: link indicator. It takes a value of 1 if link l is used in path (i, p) and 0 otherwise (data).

  The following notation is used only in the formulation of link-based schemes.

- $(k, r), (l, r), (m, r)$: refers to the rth alternate path for the links k, l, m, respectively.
- $v^{l,r}$: path indicator. It takes a value of 1 if (l, r) is chosen as a restoration path and 0 otherwise (binary variable).
- $\epsilon_l^{k,r}$: link indicator. It takes a value of 1 if link l is used in restoration path (k, r) and 0 otherwise (data).
- $\psi_\lambda^{l,r}$: wavelength indicator. It takes a value of 1 if wavelength λ is used by the restoration path (l, r) and 0 otherwise (data).

  The following notation is used only in path-based schemes.

- $v^{j,r}$: path indicator. It takes a value of 1 if (j, r) is chosen as a restoration path and 0 otherwise (binary variable).
- $I_{(i,p)(j,r)}$: link-disjoint indicator. It takes a value of 1 if paths (i, p) and (j, r) share link(s) and 0 otherwise. If $i = j$, then $p \ne q$ (data).
- $\alpha_{(i,p)(j,r)}$: the number of shared links between paths (i, p) and (j, r) (data).

8.3.2 ILP formulation

The problem is formulated in the four schemes in such as way that some constraints are common and some are different. In the following, the constraints are listed and it is noted where they are applied. The four schemes are:

(i) dedicated link protection
(ii) shared link protection
(iii) dedicated path protection
(iv) shared path protection.

The objective function and the following constraints apply to all four schemes.

Objective: the objective is to minimize the total number of wavelengths used on all the links in the network (for both the primary and backup paths).

Minimize

$$\sum_{l=1}^{L}(w_l + s_l) \tag{8.2}$$

Link capacity constraint:

$$w_l + s_l \leq W \qquad 1 \leq l \leq L \tag{8.3}$$

Demand constraint for each node-pair:

$$\sum_{p=1}^{KW} \delta^{i,p} = d_i \qquad 1 \leq i \leq N(N-1) \tag{8.4}$$

Primary link capacity constraint: define the number of primary lightpaths traversing each link.

$$w_l = \sum_{i=1}^{N(N-1)} \sum_{p=1}^{KW} \delta^{i,p} \epsilon_l^{i,p} \qquad 1 \leq l \leq L \tag{8.5}$$

The following constraints differ in different schemes.

(i) *Spare capacity constraint:* a definition of spare capacity is required on link l. A different form of constraint applies in the different cases.

(a) *Dedicated link protection:*

$$s_l = \sum_{r=1}^{MW} \sum_{k=1}^{L} v^{k,r} \epsilon_l^{k,r} \qquad 1 \leq l \leq L \tag{8.6}$$

(b) *This constraint is the same for shared link protection and shared path protection:*

$$s_l = \sum_{\lambda=1}^{W} g_{l,\lambda} \qquad 1 \leq l \leq L \tag{8.7}$$

(c) *Dedicated path protection:*

$$s_l = \sum_{i=1}^{N(N-1)} \sum_{r=1}^{KW} v^{i,r} \epsilon_l^{i,r} \qquad 1 \leq l \leq L \tag{8.8}$$

(ii) *Primary path wavelength usage constraint:* only one primary path can use a wavelength λ on link l, no restoration path can use the same λ on link l. A different form of constraint applies in different cases.

(a) *Dedicated link protection:*

$$\sum_{i=1}^{N(N-1)} \sum_{p=1}^{KW} \delta^{i,p} \epsilon_l^{i,p} \psi_\lambda^{i,p} + \sum_{r=1}^{2W} \sum_{k=1}^{L} v^{k,r} \epsilon_l^{k,r} \psi_\lambda^{k,r} \leq 1 \tag{8.9}$$
$$1 \leq l \leq L, \quad 1 \leq \lambda \leq W$$

(b) *This constraint is the same for shared link protection and shared path protection:*

$$\sum_{i=1}^{N(N-1)} \sum_{p=1}^{KW} \delta^{i,p} \epsilon_l^{i,p} \psi_\lambda^{i,p} + \sum_{\lambda=1}^{W} g_{l,\lambda} \leq 1 \tag{8.10}$$
$$1 \leq l \leq L, \quad 1 \leq \lambda \leq W$$

(c) *Dedicated path protection:*

$$\sum_{i=1}^{N(N-1)} \sum_{p=1}^{KW} \delta^{i,p} \epsilon_l^{i,p} \psi_\lambda^{i,p} + \sum_{j=1}^{N(N-1)} \sum_{r=1}^{KW} v^{j,r} \epsilon_l^{j,r} \psi_\lambda^{j,r} \leq 1 \tag{8.11}$$
$$1 \leq l \leq L, \quad 1 \leq \lambda \leq W$$

(iii) *Demand constraints for link l.* A different form of constraint applies in different cases.

(a) *This constraint is the same for dedicated link protection and shared link protection:* there are two restoration routes for each link l, so that the demand on link l can be met after any double-link failures:

$$\sum_{r=1}^{W} v^{l,r} \psi_\lambda^{l,r} = \sum_{i=1}^{N(N-1)} \sum_{p=1}^{KW} \delta^{i,p} \epsilon_l^{i,p} \psi_\lambda^{i,p} \tag{8.12}$$
$$1 \leq l \leq L, \quad 1 \leq \lambda \leq W$$

$$\sum_{r=W+1}^{2W} v^{l,r} \psi_\lambda^{l,r} = \sum_{i=1}^{N(N-1)} \sum_{p=1}^{KW} \delta^{i,p} \epsilon_l^{i,p} \psi_\lambda^{i,p} \tag{8.13}$$
$$1 \leq l \leq L, \quad 1 \leq \lambda \leq W$$

(b) *This constraint is the same for dedicated path protection and shared path protection:* there are two restoration routes for each primary call. Let $x, y, z \in \{0, 1, 2\}$ and $x \neq y \neq z$; $t, u, v \in \{x, y, z\}$, $t \neq u \neq v$:

$$\sum_{p=tW+1}^{(t+1)W} \delta^{i,p} + \sum_{p=uW+1}^{(u+1)W} \delta^{i,p} = \sum_{r=vW+1}^{(v+1)W} v^{i,r} \tag{8.14}$$
$$1 \leq i \leq N(N-1)$$

(iv) *Restoration path wavelength usage constraint:* this constraint only applies to shared link protection and shared path protection.

(a) *Shared link protection:*

$$g_{l,\lambda} \leq \sum_{r=1}^{2W} \sum_{k=1}^{L} v^{k,r} \epsilon_l^{k,r} \psi_\lambda^{k,r} \tag{8.15}$$

$$1 \leq l \leq L, \quad 1 \leq \lambda \leq W$$

$$2N(N-1)W g_{l,\lambda} \geq \sum_{r=1}^{2W} \sum_{k=1}^{L} v^{k,r} \epsilon_l^{k,r} \psi_\lambda^{k,r} \tag{8.16}$$

$$1 \leq l \leq L, \quad 1 \leq \lambda \leq W$$

(b) *Shared path protection:*

$$g_{l,\lambda} \leq \sum_{i=1}^{N(N-1)} \sum_{r=1}^{KW} v^{i,r} \epsilon_l^{i,r} \psi_\lambda^{i,r} \tag{8.17}$$

$$1 \leq l \leq L, \quad 1 \leq \lambda \leq W$$

$$N(N-1)KW g_{l,\lambda} \geq \sum_{i=1}^{N(N-1)} \sum_{r=1}^{KW} v^{i,r} \epsilon_l^{i,r} \psi_\lambda^{i,r} \tag{8.18}$$

$$1 \leq l \leq L, \quad 1 \leq \lambda \leq W$$

(v) *Backup multiplexing constraints.* These constraints only apply to shared link protection and shared path protection.

(a) *Shared link protection*

Backup multiplexing constraint 1: if link j is not on the alternate routes of link k and link k is not on alternate routes of link j, then the first backup route of link j and the first backup route of link k cannot share wavelength channel on their common links (represents the backup multiplexing rule for case 1):

$$\sum_{r=1}^{W} v^{k,r} \psi_\lambda^{k,r} \epsilon_l^{k,r} + \sum_{r=1}^{W} v^{m,r} \psi_\lambda^{m,r} \epsilon_l^{m,r} \leq 1 \tag{8.19}$$

$$1 \leq k \leq L, \quad k+1 \leq m \leq L, \quad 1 \leq l \leq L, \quad 1 \leq \lambda \leq W$$

Backup multiplexing constraint 2: if link j is not on the alternate routes of link k and link k is on one of the alternate routes of link j, then there should be no wavelength sharing between the backup route of j, which does not pass through link k, and the first backup route of link k (represents the backup multiplexing rule for case (2) and case (4)):

$$\sum_{r=1}^{2W} v^{k,r} \psi_\lambda^{k,r} \epsilon_l^{k,r} \left(1 - \epsilon_m^{k,r}\right) + \sum_{r=1}^{W} v^{m,r} \psi_\lambda^{m,r} \epsilon_l^{m,r} \leq 1 \tag{8.20}$$

$$1 \leq k \leq L, \quad 1 \leq m \leq L, \quad 1 \leq l \leq L, \quad 1 \leq \lambda \leq W$$

Backup multiplexing constraint 3: if link j is on one of the alternate routes of link k and link k is on one of the alternate routes of link j, then there should be no wavelength sharing between the backup route of link j, which does not pass through link k, and the backup route of link k, which does not pass through link j (represents the backup multiplexing rule for case (3)):

$$\sum_{r=1}^{2W} v^{k,r} \psi_\lambda^{k,r} \epsilon_l^{k,r} \left(1 - \epsilon_m^{k,r}\right) + \sum_{r=1}^{2W} v^{m,r} \psi_\lambda^{m,r} \epsilon_l^{m,r} \left(1 - \epsilon_k^{m,r}\right) \leq 1 \quad (8.21)$$

$$1 \leq k \leq L, \quad k+1 \leq m \leq L, \quad 1 \leq l \leq L, \quad 1 \leq \lambda \leq W$$

(b) *Shared path protection*
For the following constraints, let $m, n \in \{0, 1, 2\}$; $s, s' \in \{(m+1) \bmod 3, (m+2) \bmod 3\}$, $s \neq s'$; $t, t' \in \{(n+1) \bmod 3, (n+2) \bmod 3\}$, $t \neq t'$.

$$X_\lambda^{i,m} = v^{i,mW+\lambda} \psi_\lambda^{i,mW+\lambda} \quad (8.22)$$

Constraints for backup multiplexing rules 1–3 in Table 8.1.
if $I_{(i,s)(j,n)} = 1$, $I_{(i,m)(j,t)} = 1$, and $I_{(i,s')(j,t')} = 1$:

$$X_\lambda^{i,s'} + X_\lambda^{j,t'} \leq 1 \quad (8.23)$$

$$1 \leq i < j \leq N(N-1), \quad 1 \leq \lambda \leq W$$

Constraints for backup multiplexing rules 4 and 5 in Table 8.1.
if $I_{(i,s)(j,n)} = 0$, $I(i, s')(j, n) = 0$, $I_{(i,m)(j,t)} = 1$, $I_{(i,s)(j,t')} = 1$, and $I_{(i,s')(j,t')} = 1$:

$$\Pi_{(i,s-min)(j,t)} = min\left(\Pi_{(i,s)(j,t')}, \Pi_{(i,s')(j,t')}\right)$$
$$X_\lambda^{i,s-min} + X_\lambda^{j,t'} \leq 1 \quad (8.24)$$
$$1 \leq i, \quad j \leq N(N-1), \quad 1 \leq \lambda \leq W$$

Constraint for backup multiplexing rule 6 in Table 8.1:
if $I_{(i,s)(j,n)} = 0$, $I(i, s')(j, n) = 0$, $I_{(i,m)(j,t)} = 0$, $I_{(i,m)(j,t')} = 0$, $I_{(i,s)(j,t')} = 1$, $I_{(i,s)(j,t)} = 1$, and $I_{(i,s)(j,t')} = 1$, $I_{(i,s')(j,t)} = 1$, $I_{(i,s')(j,t')} = 1$:

$$\Pi_{(i,s-min)(j,t-min)} = min\left(\Pi_{(i,s)(j,t)}, \Pi_{(i,s)(j,t')}, \Pi_{(i,s')(j,t)}, \Pi_{(i,s')(j,t')}\right)$$
$$X_\lambda^{i,s-min} + X_\lambda^{j,t-min} \leq 1 \quad (8.25)$$
$$1 \leq i < j \leq N(N-1), \quad 1 \leq \lambda \leq W$$

For the following constraint, let $m, n \in \{0, 1, 2\}$; $s, s' \in \{m+1, m+2\}$, $s \neq s'$; $t, t' \in \{n+1, n+2\}$, $t \neq t'$.
Constraint for rule 7 in Table 8.1:
if $I_{(i,m)(j,n)} = 1$, $I_{(i,s)(j,t)} = 1$, and $I_{(i,s')(j,t')} = 1$:

$$X_\lambda^{i,s'} + X_\lambda^{j,t'} \leq 1 \quad (8.26)$$

$$1 \leq i, \quad j \leq N(N-1), \quad 1 \leq \lambda \leq W$$

Table 8.2. *The routes and wavelengths of primary and backup paths under dedicated link protection*

Node-pair	Primary lightpath	Backup lightpath 1	Backup lightpath 2
1	$(1, 2), \lambda_3$	$(1, 3, 2), \lambda_3$	$(1, 5, 4, 2), \lambda_3$
5	$(2, 1), \lambda_3$	$(2, 3, 1), \lambda_3$	$(2, 4, 5, 1), \lambda_3$
13	$(4, 3, 1), \lambda_1$	$(4, 2, 3), \lambda_1$	$(4, 5, 3), \lambda_1$
		$(3, 2, 1), \lambda_1$	$(3, 5, 1), \lambda_1$
20	$(5, 4), \lambda_2$	$(5, 3, 4), \lambda_2$	$(5, 1, 2, 4), \lambda_2$

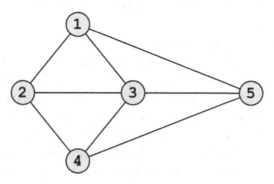

Fig. 8.2. A five-node eight-link network.

8.4 Examples and comparison

To compare and demonstrate the performance of these methods using the above formulation, consider a simple five-node network as shown in Fig. 8.2. Assume the network has one fiber per link and three wavelengths per fiber.

Suppose that four node-pairs are requesting one lightpath between them. In an N-node network, there are $N(N - 1)$ node-pairs. They are numbered sequentially. The source node number i and the destination node number j for node-pair r are determined by the following relationships:

$$i = \lceil r/(N - 1) \rceil$$
$$\text{let } k = r - (i - 1) \times (N - 1)$$
$$\text{then } j = k, \text{ if } k < i, \text{ else } j = k + 1.$$

8.4.1 Link-based protection

The routes and wavelengths of primary and backup lightpaths for the dedicated link protection are illustrated in Table 8.2. The routes and wavelengths of primary

Table 8.3. *The routes and wavelengths of primary and backup paths*
under shared link protection

Node-pair	Primary lightpath	Backup lightpath 1	Backup lightpath 2
1	$(1, 2), \lambda_3$	$(1, 3, 2), \lambda_3$	$(1, 5, 4, 2), \lambda_3$
5	$(2, 1), \lambda_2$	$(2, 3, 1), \lambda_2$	$(2, 4, 5, 1), \lambda_2$
13	$(4, 5, 1), \lambda_3$	$(4, 3, 5), \lambda_3$	$(4, 2, 1, 5), \lambda_3$
		$(5, 3, 1), \lambda_3$	$(5, 4, 2, 1), \lambda_3$
20	$(5, 4), \lambda_2$	$(5, 3, 4), \lambda_2$	$(5, 1, 2, 4), \lambda_2$

and backup lightpaths for the shared link protection for the same demand set are illustrated in Table 8.3.

For each link on the primary path, two backup paths are provided and wavelengths are reserved on these paths. In Table 8.2, each reserved wavelength on a link of backup paths is dedicated to a link on a primary path. For example, λ_3 on link $(2, 4)$ is reserved and dedicated to link $(2, 1)$, which is a link on the primary path $2 \rightarrow 1$. Similarly, λ_2 on link $(2, 4)$ is reserved and dedicated to link $(5, 4)$, which is a link on primary path $5 \rightarrow 4$. In contrast, in Table 8.2, λ_2 on link $(2, 4)$ is shared by backup path $2 \rightarrow 4 \rightarrow 5 \rightarrow 1$ and backup path $5 \rightarrow 1 \rightarrow 2 \rightarrow 4$. The path $2 \rightarrow 4 \rightarrow 5 \rightarrow 1$ is the second backup path for link $(2, 1)$ on primary path $2 \rightarrow 1$, while the path $5 \rightarrow 1 \rightarrow 2 \rightarrow 4$ is the second backup path for link $(5, 4)$ on primary path $5 \rightarrow 4$. Therefore one backup wavelength is saved by sharing the wavelength on the common link in the shared link protection scheme.

An interesting observation is that the primary path for node-pair 13 in shared link protection is different from the primary path for node-pair 13 in the dedicated link protection scheme. The reason is that routing a primary for request 13 on path $4 \rightarrow 5 \rightarrow 1$ rather than on $4 \rightarrow 3 \rightarrow 1$ has better wavelength sharing on the backup paths. This leads to minimum capacity utilization for this demand. The shared link protection scheme utilizes a total of 23 wavelength links, while the dedicated link protection scheme utilizes a total of 28 wavelength links for this demand. Shared link protection saves about 18% capacity.

Even in larger networks with more wavelengths and requests, an improvement of 10% or larger is noticed.

8.4.2 Path-based protection

The routes and wavelengths of primary and backup lightpaths for the dedicated path and shared path protection schemes for the example network are given in Table 8.4.

In the dedicated scheme, each reserved wavelength on a backup path is dedicated to a primary path. In contrast, in the shared scheme, λ_3 on links $(5, 1)$ and $(2, 4)$

Table 8.4. *The routes and wavelengths of primary and backup paths*
under dedicated path and shared path protection

Scheme	Node-pair	Primary lightpath	Backup 1	Backup 2
Dedicated	1	$(1, 2), \lambda_3$	$(1, 3, 2), \lambda_2$	$(1, 5, 4, 2), \lambda_2$
	5	$(2, 1), \lambda_3$	$(2, 3, 1), \lambda_1$	$(2, 4, 5, 1), \lambda_3$
	13	$(4, 2, 1), \lambda_1$	$(4, 3, 1), \lambda_2$	$(4, 5, 1), \lambda_2$
	20	$(5, 4), \lambda_3$	$(5, 3, 4), \lambda_3$	$(5, 1, 2, 4), \lambda_1$
Shared	1	$(1, 3, 2), \lambda_3$	$(1, 2), \lambda_3$	$(1, 5, 4, 2), \lambda_1$
	5	$(2, 3, 1), \lambda_3$	$(2, 1), \lambda_3$	$(2, 4, 5, 1), \lambda_3$
	13	$(4, 5, 1), \lambda_2$	$(4, 2, 1), \lambda_1$	$(4, 3, 1), \lambda_1$
	20	$(5, 3, 4), \lambda_3$	$(5, 4), \lambda_1$	$(5, 1, 2, 4), \lambda_3$

is shared by backup paths $2 \rightarrow 4 \rightarrow 5 \rightarrow 1$ and $5 \rightarrow 1 \rightarrow 2 \rightarrow 4$. λ_1 on links $(5, 4)$ is shared by backup paths $1 \rightarrow 5 \rightarrow 4 \rightarrow 2$ and $5 \rightarrow 4$. λ_1 on links $(4, 2)$ is shared by backup paths $4 \rightarrow 2 \rightarrow 1$ and $1 \rightarrow 5 \rightarrow 4 \rightarrow 2$. λ_3 on link $(1, 2)$ is shared by backup paths $1 \rightarrow 2$ and $5 \rightarrow 1 \rightarrow 2 \rightarrow 4$. Backup paths $2 \rightarrow 1$ and $4 \rightarrow 2 \rightarrow 1$ cannot share backup capacity on their common link $(2, 1)$ because of backup multiplexing constraint 3 in Table 8.1. The routings for the primary paths under dedicated and shared path schemes are different because the routing under the shared path scheme yields maximum saving for total capacity. The shared and dedicated path protection schemes use a total of 19 and 24 wavelength links for this demand matrix, respectively. The shared path protection saves about 21% capacity.

For larger networks and more requests, the shared path scheme requires between 22.3% and 33.4% lower total capacity than the dedicated path scheme.

8.5 Dual-link failure coverage of single-failure protection schemes

As noted in the last section, double-link failures, i.e. any two links in the network failing in an arbitrary order, are becoming critical in survivable optical network design. Multi-link failure scenarios can arise out of two common situations. First, an arbitrary link may fail in the network, and before that link can be repaired, another link fails, thus creating a multi-link failure sequence. Secondly, it might happen in practice that two distinct physical links may be routed via the same common duct or physical channel. A failure at that shared physical location creates a logical multiple-link failure. Such instances where separate fiber optic links share a common failure structure is often referred to as a shared-risk link group (SRLG) [120, 121, 215]. Simultaneous link failures can be treated as an arbitrarily ordered sequential failure with no latency. In this chapter, it is assumed that there are two independent link failures, where recovery for the two failures is undertaken in a sequential order, i.e. the system recovers from one failure and then attempts recovery from the second.

A significant finding is that designs offering complete dual-failure restorability require almost triple the amount of spare capacity. In this chapter an alternative approach is presented. This approach is based on a paradigm in which networks are designed to achieve 100% restorability under single-link failures, while maximizing coverage against any second-link failure in the network. In the event of a link failure, the restoration model attempts to dynamically find a second alternate link-disjoint end-to-end path to provide coverage against a sequential overlapping link failure.

In order to achieve efficient utilization of network resources, multiplexing of resources across primary and/or backup paths may be employed. More than one backup path may share resources as long as any failure in the network causes, at most, one of the corresponding working connections to fail. This is often referred to as shared-mesh protection or backup multiplexing. Shared-mesh protection is employed to optimize the capacity utilization and to provide a 100% protection guarantee for all single-link failure scenarios. It is of interest here to analyze the performance of backup multiplexing to ascertain what percentage of second-link failure scenarios can be dynamically tolerated after a single physical link failure occurs.

Protection paths may also be provided by deconstructing the network into multiple subgraphs to mimic each failure scenario [171]. A connection is established if it can be accepted in all subgraphs. This method does not require the explicit allocation of backup resources in the network, but it does require the network to reconfigure to a state corresponding to the fault that occurred. Reconfiguration can be performed according to the scheme presented in [194]. Subgraph routing [171] can also be extended to provide dual-failure restorability for a network provisioned to tolerate all single-link failures.

Since networks are provisioned to tolerate all single-link failure scenarios using both shared-mesh protection and subgraph routing, additional link faults are assessed. In the case of shared-mesh protection, additional backup paths are dynamically calculated for all affected backup or primary connections. In a shared-mesh protection scheme, this approach is similar to the protection-reconfiguration approach taken in [259]. With respect to subgraph routing, the approach is a recursive one in that the set of subgraphs fails the link that has failed in the original network. Attempted rerouting of the connections affected by the failed link is performed in order to obtain second-link failure protection. The scheme is recursive because it can be extended to an indefinite number of successive link failures as long as the network remains adequately connected.

While node, link and shared-risk link group based subgraph routing [209] have the ability to pro-actively protect against a wide variety of multiple link failures,

Table 8.5. *Primary and backup paths before failure*

Requests	Primary lightpath	Backup lightpath
R_1 (1→2)	(1→2), λ_1	(1→3→2), λ_1
R_2 (2→3)	(2→3), λ_1	(2→1→3), λ_1
R_3 (1→4)	(1→5→4), λ_1	(1→3→4), λ_1

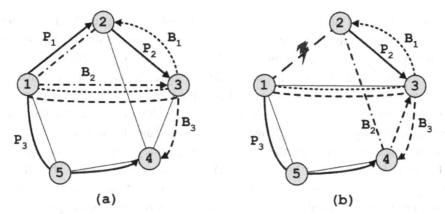

Fig. 8.3. (a) Routing of requests using shared-mesh protection. (b) Rerouting of requests on failure of link 1 → 2.

they cannot protect against all possible multiple-link failure scenarios. Pro-active subgraph routing has the advantage of providing protection for 100% of the failure scenarios for which subgraphs are designed. Unfortunately, in the case of link fault tolerant subgraphs, there are $L(L - 1)$ possible dual-link failure states in a network with L links. The problem grows with the number of failure scenarios.

The flexibility of the subgraph strategy also allows it to be used in the reactive tolerance of link failures. The ability to tolerate a high percentage of multiple overlapping sequential or unrelated simultaneous link failures, using the merits of subgraph fault tolerance in a reactive manner, is the subject of this chapter.

8.6 Dual-link failure coverage using shared-mesh protection

Let the primary path of a request R_i be denoted by P_i and the two link-disjoint end-to-end backup paths for tolerating two independent link failures e and f be $B_i(e)$ and $B_i(f)$. Figure 8.3(a) depicts the primary and backup routes and wavelength assignment using backup multiplexing or shared-mesh protection for three requests (R_1–R_3). The routes and the wavelengths assigned for each request are also shown in Table 8.5.

Table 8.6. *Primary and backup paths after failure of link* 1 → 2

Requests	Primary lightpath	Backup lightpath
R_1 (1→2)	(1→3→2), λ_1	No routes possible
R_2 (2→3)	(2→3), λ_1	(2→5→3), λ_1
R_3 (1→4)	(1→5→4), λ_1	No routes possible

The primary and backup connections for request R_1 are given by 1 → 2, and 1 → 3 → 2, respectively, and are assigned wavelength λ_1. Similarly request R_2 is assigned primary and backup routes 2 → 3 and 2 → 1 → 3, respectively. Since the primary routes of the requests R_1 and R_2 do not share common links their backup paths can share wavelength λ_1. Connection request R_3 is routed using primary and backup routes 1 → 5 → 4 and 1 → 3 → 4 on wavelength λ_1. The backup route for R_3 can be assigned the wavelength λ_1 and it can share this wavelength with the two other backup paths, since the corresponding primary paths are link-disjoint.

Let there be a failure at link e (1 → 2) as shown in Fig. 8.3(b). The affected primary route is $P_1 : 1 → 2$. The affected backup connections which were multi-plexed on the affected link e correspond to requests R_2 and R_3, the primary and backup path combinations of which are shown in Table 8.5.

After the failure of link e, $B_1(e): 1 → 3 → 2$ restores P_1 and P_2, respectively, as shown in Tables 8.3 and 8.6. A new alternate backup connection corresponding to P_1 should be found on the graph so that the connection can tolerate a second failure in the network. This backup connection is referred to as $B_1(f)$, since it guarantees restoration for the second failure f in the network. Assuming each request is of unit capacity and each link is a bidirectional link having a capacity of one unit in each direction, in the above example $B_1(f)$ does not exist. Thus, this request cannot be restored in the event of a second failure overlapping in time and incident on one of the links of $B_1(e)$.

P_2 corresponding to the backup connection $B_2(e)$, which was multiplexed on the failed link 1 → 2, needs to find $B_2(f)$. Since P_2 remains unaffected by the link failure, it can potentially reroute its backup such that $B_2(f)$ can also be mul-tiplexed. However, this is constrained by the available capacity on a link and more importantly, the availability of an alternate backup path in the first place because the failure of a critical link may cause a partition of the graph. It is important to note that primary and backup connections that are unaffected by any link failures remain uninterrupted in their service and are not rerouted.

The second alternate backup path for the connection R_2 is given by $B_2(f): 2 →$ 4 → 3. Moreover, the routing of request R_1 along $B_1(e)$ would force R_3 to search for a new alternate backup path to tolerate a second link failure, due to the capacity constraint on link $1 \rightarrow 3$. However, request R_3 fails to find $B_3(f)$, and hence cannot be recovered in the event of a second overlapping link failure along its primary path.

8.7 Dual-link failure coverage: subgraph routing

In this section, the capability of the subgraph routing scheme to tolerate sequential overlapping link failures in the network is studied. A network composed of nodes and links can be viewed as a graph G, consisting of a set of V vertices and E edges, or mathematically $G = (V, E)$. There exists a set of subgraphs of G, denoted by G_i, where one edge e_i is removed from the graph, $G_i = \{V_i = V, E_i = E - e\}$, where e is an edge in the graph and L is the cardinality of the set of edges in the graph G. Therefore there exist L subgraphs of graph G, one for each missing link e.

The set of L subgraphs represent all possible single-link failures in the network. The original graph is referred to as the *base network*. The constituent subgraphs of the base network are not referred to as networks because they correspond to a base network state reached through failure of any one link.

In the subgraph routing strategy, a connection request is accepted in the base network only if it can be routed in all the subgraphs. Hence, the accepted connections are guaranteed restorability against all single-link failure scenarios. If any link, e_i, fails in the network, the network transitions to the state of the subgraph, G_i, and some of the connections directly unaffected by the failure in the base network are potentially rerouted, corresponding to the routing of the requests in subgraph G_i. Suppose link $1 \rightarrow 3$ fails. The network tries to restore all existing connections by migrating to the subgraph G_4 as shown in Fig. 8.4. In order to ensure that these requests also have complete coverage against a second link fault in the network, consider the situation where the corresponding failed link e_i is deleted from the other $L - 1$ sub-graphs, and route the compromised connections in G_4 on the remaining subgraphs. Connections that are accepted in all remaining subgraphs satisfy complete 100% coverage against all overlapping dual-failure scenarios. However, connections that are unsuccessful in being rerouted in the remaining subgraphs are guaranteed restorability only against the initial single-link failure or those second failures for which they may be rerouted. It is expected that connections that are protected against all single-link failures, hopefully also have high coverage for all possible sequential overlapping dual-link failures.

8.7.1 Dual-link subgraph routing complexity

The following section analyzes the complexity of the sub-graph routing algorithm while ensuring complete dual-failure coverage. In order to route a connection R_i

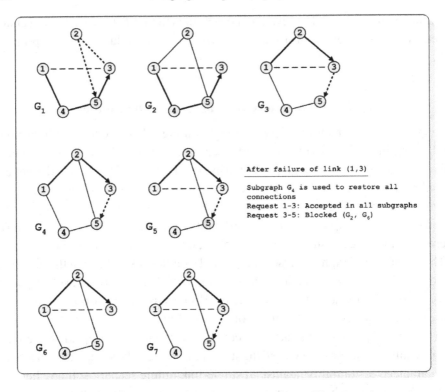

Fig. 8.4. Rerouting of requests upon failure of link 1 → 3.

in the *base network*, the connection needs to be routed in all of the $L = |E|$ sub-graphs. The time complexity of routing a request R_i in a network is governed by the complexity of the routing algorithm. Dijkstra's shortest-path algorithm can be used for routing the connections in each of the L subgraphs. The complexity of Dijkstra's shortest-path algorithm is $O(N^2)$ [223], where N is the total number of nodes in the network. Thus, the overall complexity of routing these requests using this algorithm is $O(L \times N^2)$. Notice that subgraph routing is not limited to using only shortest-path routing, and can accommodate any other desired routing metric.

After the failure of the first link e, the network makes a transition from the base network to the G_e^{th} subgraph. To ensure restorability against failure of another link f, a subset of connection requests need to be rerouted on the remaining $L - 1$ subgraphs, that is all subgraphs except graph G_e. The requests that need to be rerouted on all the $L - 1$ subgraphs in order to tolerate a second fault would require an additional worst-case computation of $O(L \times N^2)$. Thus the worst-case complexity for routing each request for dual-failure survivability is $O(L^2 \times N^2)$.

A variation of subgraph routing is called constrained subgraph routing. Constrained subgraph routing requires that each subgraph containing all of the links of the selected path of a request in the base network use the same path if the path exists. In other words, if the path of a request contains l links, that request, if accepted, will be routed on the same path in the base network as well as $L - l$ subgraphs. Constrained subgraph routing increases the network performance by decreasing the blocking probability and the probability of reassignment when a failure occurs. The probability of reassignment is the probability that an active connection in the network is reassigned in the event of a failure. Constrained subgraph routing also lowers the time complexity of routing in subgraph routed networks. Instead of a complexity of $O(L \times N^2)$, assuming use of Dijkstra's shortest-path algorithm, an overall time complexity of $O(l \times N^2)$ is obtained, where l is the length of the path. A path only needs to be selected for the base network and the l subgraphs that do not contain all of the links of the path selected in the base network. Other $L - l$ subgraphs use the same path as in the base network.

As a result of subgraph routing being a recursive technique, the time complexity of connection recovery from a link failure also changes to $O(l \times N^2)$. This, along with the original routing time complexity discussed in the previous paragraph, yields an overall time complexity for single overlapping sequential link failures of $O(l^2 \times N^2)$ instead of the $O(L^2 \times N^2)$ time needed in unconstrained subgraph routing.

This formulation can be extended in a recursive fashion for any number of sequential overlapping link failures. For r link failures $O(l^r)$ graphs need to be considered, yielding a complexity of $O(l^r \times N^2)$ for constrained subgraph routing.

8.8 Coverage computation

The performance of both the backup multiplexing and the subgraph routing schemes are evaluated through simulation. Three topologies are used: the 14-node, 23-link NSFnet; the 11-node, 22-link NJLATA; and a standard 9-node, 18-link 3×3 mesh-torus.

The blocking probabilities for both schemes are computed in the absence as well as the presence of faults in the system. A subgraph routing strategy offers considerably lower blocking in the absence of faults compared to the backup multiplexing scheme. For the purposes of the simulation, each link is assumed to be composed of two unidirectional links, each with only one fiber. The total number of wavelengths used is $W = 16$ for each fiber in each unidirectional link.

The performance of the network in the presence of faults is assessed in two ways. In the first, an arbitrary link is failed randomly and repaired during the simulation time frame as a fault would occur in a real world situation. In the second scenario,

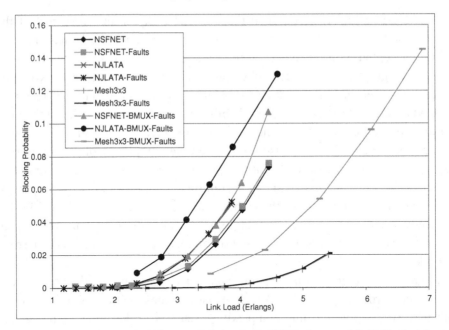

Fig. 8.5. Blocking probability vs. link load, © IEEE. Source: M. T. Frederick, P. Datta, and A. K. Somani, Evaluating dual-failure restorability in mesh-restorable WDM optical networks, in *ICCCN, 2004* [172].

the simulation is paused periodically, and the network state is tested against all possible link failures. The simulation then continues without the network state being altered by the occurrence of any fault.

The arrival of a request at a node follows a Poisson process with rate λ and is equally likely to be destined to any other node. The holding time of the requests follows an exponential distribution with unit mean. The capacity requirement of each request is a full wavelength. The random link faults are assumed to occur following a Poisson distribution.

The link load is a measure of the load placed on each link in the network at any given time. It is useful in providing a baseline for the comparison of the effectiveness of routing strategies across different network topologies. The link load can be calculated using the formula $\gamma = N\lambda H/L$, where N is the number of nodes in the network, λ is the arrival rate of the requests per node, H is the average hop length of each connection, and L is the total number of links in the network.

The blocking probability is illustrated in Fig. 8.5, which shows how the backup multiplexing and subgraph routing perform with and without the presence of random faults. The blocking probabilities of the subgraph routing strategy are extremely low compared to the backup multiplexing scheme and are reasonably close to the blocking probability achieved in the presence of no faults.

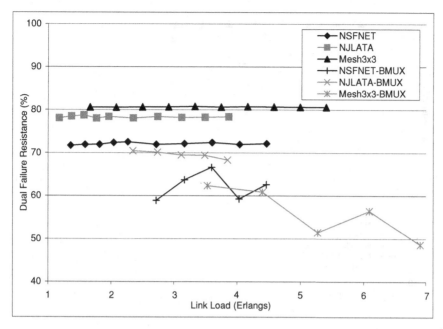

Fig. 8.6. Automatic dual-failure coverage vs. link load for periodic testing, ©
IEEE. Source: M. T. Frederick, P. Datta, and A. K. Somani, Evaluating dual-
failure restorability in mesh-restorable WDM optical networks, in *ICCCN, 2004*
[172].

Figure 8.6 depicts the *automatic sequential overlapping fault coverage*, indicating that around 72–81% of connections are automatically covered for all possible
dual failures (across different topologies) without being rerouted in the subgraph
scheme as compared to 49–70% for the backup multiplexing routing strategy. The
automatic dual-failure coverage in the subgraph routing strategy is calculated as
the number of connections in the final subgraph G_e (reached by a failure in link e),
that do not need to be rerouted in the other subgraphs, and hence are automatically
covered for two link failures, the first failure being on link e.

High automatic coverage is important because it means that fewer connections
are rerouted in the event of a single link failure, in order to provide protection against
a second overlapping failure. Additionally, higher automatic coverage means that
active connections also have a better chance of being protected against a second
link failure because the connection does not have to attempt rerouting. The total
capacity reservation for tolerating a single fault in the backup multiplexing scheme
has been shown to be around 150–160% [305]. Thus the probability of reserving
a second path, in the event of a fault, to tolerate a second failure is extremely low
either due to lack of capacity in the network or due to graph disconnection caused
by the first failure.

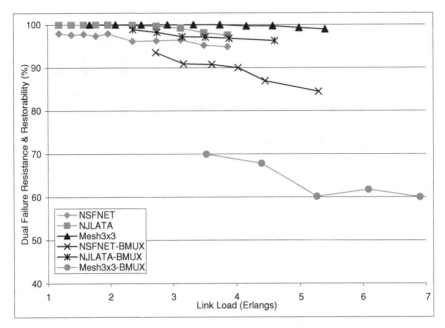

Fig. 8.7. Total restorability vs. link load for random testing, © IEEE. Source: M. T. Frederick, P. Datta, and A. K. Somani, Evaluating dual-failure restorability in mesh-restorable WDM optical networks, in *ICCCN, 2004* [172].

Dual-failure restorability in the presence of random faults is shown in Fig. 8.7 and is an indication of the degree of restorability that can be achieved by both algorithms in the event of random faults in the system. The subgraph routing strategy achieves a significantly higher degree of restorability compared to the backup multiplexing scheme. Although the backup multiplexing scheme is able to achieve total restorability varying between ∼60–97% across different topologies, the subgraph routing scheme achieves a restorability of over 95% for all topologies.

Dual-failure restorability in the presence of periodic faults, depicted in Fig. 8.8, indicates the complete network-wide dual-failure restorability achieved by both the restoration algorithms. In the presence of periodic faults, restorability is computed by successively failing each link in the network, computing the coverage for a second failure, and averaging it over all possible dual-failure scenarios. The network is left in its original state. The total dual-failure restorability achieved by both the algorithms is quite high (∼62–96%) except for MESH3×3 where the subgraph routing strategy outperforms the backup multiplexing scheme.

Subgraph routing provides a passive form of redundancy without any physical allocation of any redundant capacity in the network, by maintaining the state information of L distinct subgraphs of the network. Effectively, there is a trade off between the physically redundant capacity that needs to be stored in the network to

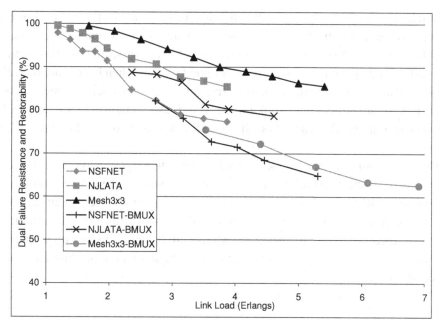

Fig. 8.8. Total restorability vs. link load for periodic testing, © IEEE. Source: M. T. Frederick, P. Datta, and A. K. Somani, Evaluating dual-failure restorability in mesh-restorable WDM optical networks, in *ICCCN, 2004* [172].

achieve fault tolerance in the case of backup multiplexing, and the reconfiguration and redundant network states that need to be maintained in the subgraph routing strategy. However, since network state information is always cheaper to maintain than physical allocation of spare capacity, the subgraph routing strategy is a viable alternative routing methodology in WDM optical networks.

8.9 Observations

Pro-active subgraph fault tolerance has the ability to protect against all possible multi-link failures for which its subgraphs are designed. It also has the advantage of having predetermined subgraph states that the base network can emulate in the event of a failure. However, pro-active fault tolerance has the drawbacks of not being able to handle all possible multiple-link failure situations, as well as sequential overlapping link failures.

Reactive subgraph fault tolerance addresses some of these pitfalls by employing a recursive method for tolerating numerous sequential overlapping failures. It can also tolerate simultaneous multiple link failures simply by serializing the handling of each individual fault. One of the drawbacks of reactive fault tolerance is that it can rarely provide 100% protection for all connections against subsequent link failures, although simulation results have shown that protection against a subsequent link failure is of the order of 75–96%. Another drawback is that it cannot simply begin reconfiguring the network as soon as a multi-link fault occurs, but must first attempt to reroute all compromised connections for one fault, and then handle another. For example, if an SRLG were to fail in a reactive subgraph fault-tolerant network, each of the link failures in the multi-link failure would have to be handled sequentially. Network reconfiguration cannot occur until all faults are processed and recovered from.

The best solution to multi-link fault tolerance is to employ a hybrid of reactive and pro-active subgraph fault tolerance. Initially subgraphs are defined taking into account link, SRLG [61, 70], or node failures, and could, in the event of an unrelated or subsequent multi-link failure, incorporate the reactive form of subgraph fault tolerance. Incorporating a hybrid approach to fault tolerance using subgraph fault tolerance provides a complete solution to the problem of multiple link failures in optical networks.

9

Traffic grooming in WDM networks

Data traffic in ultra-long-haul WDM networks is usually characterized by large, homogeneous data flows, and metropolitan area WDM networks (MAN) have to deal with dynamic, heterogeneous service requirements. In such WAN and MAN networks, equipment costs increase if separate wavelengths are used for each individual service. Moreover while each wavelength offers a transmission capacity at gigabit per second rates (e.g., OC-48 or OC-192 and on to OC-768 in the future), users may request connections at rates that are lower than the full wavelength capacity. In addition, for networks of practical size, the number of available wavelengths is still lower by a few orders of magnitude than the number of source-to-destination connections that may be active at any given time. Hence, to make the network viable and cost-effective, it must be able to offer subwavelength level services and must be able to pack these services efficiently onto the wavelengths. These subwavelength services, henceforth referred to as *low-rate traffic streams*, can vary in range from, say, STS-1 (51.84 Mbit/s) capacity up to the full wavelength capacity. Such an act of multiplexing, demultiplexing, and switching of lower-rate traffic streams onto high-capacity lightpaths is referred to as *traffic grooming*. WDM networks offering such subwavelength low-rate services are referred to as *WDM grooming networks*. Efficient traffic grooming improves the wavelength utilization and reduces equipment costs.

In WDM grooming networks, each lightpath typically carries many multiplexed lower-rate traffic streams. Optical add–drop multiplexers (OADMs) add/drop the wavelength for which grooming is needed and electronic SONET-ADMs multiplex or demultiplex the traffic streams onto the wavelength. The act of switching the traffic streams from one wavelength to another is performed by a cross-connect at a node. That is, a traffic stream occupying a set of time slots on a wavelength on a fiber can be switched to a different set of time slots on a different wavelength on another fiber. However, this *traffic stream switching* capability comes at the cost of increased cross-connect complexity. In addition to the space-switching of wavelengths, the

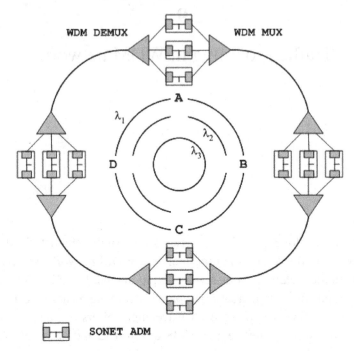

Fig. 9.1. No traffic grooming.

cross-connect may have to be provided with wavelength conversion and time-slot interchange equipment. Currently, all-optical wavelength conversion and all-optical time-slot interchange devices are still not commercially available and it is more economical to use electronic methods of implementation to incorporate these features into the network. In the future, it may be possible to perform all-optical traffic grooming.

To illustrate the importance of efficient traffic grooming using a simple example, consider a WDM unidirectional path-switched ring (UPSR) with four nodes inter-connected by fiber optic links with three wavelengths on each fiber. Each wavelength has capacity of 2.5 Gbit/s (or OC-48 capacity). Suppose the traffic requirement is for two OC-12 circuits between each pair of nodes in the ring. Then the total traffic requests are 12 OC-12 circuits on each link or three OC-48 rings (thus at least three wavelengths). Consider the following two assignments of traffic on the wavelengths of the ring:

Assignment 1

$\lambda_1 = A \leftrightarrow D, C \leftrightarrow B$

$\lambda_2 = A \leftrightarrow B, C \leftrightarrow D$

$\lambda_3 = A \leftrightarrow C, D \leftrightarrow B$

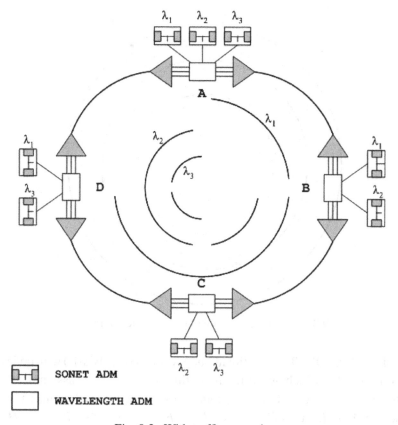

Fig. 9.2. With traffic grooming.

Assignment 2

$$\lambda_1 = A \leftrightarrow B, B \leftrightarrow D$$
$$\lambda_2 = A \leftrightarrow C, B \leftrightarrow C$$
$$\lambda_3 = A \leftrightarrow D, C \leftrightarrow D$$

Assignment 1 of the traffic to wavelengths would result in each node in the ring requiring a SONET ADM for each wavelength as every wavelength is either for receiving or transmitting at each node. This requires a total of 12 SONET ADMs. This is equivalent to that number of SONET ADMs required in a point-to-point WDM ring (Fig. 9.1) to support the same traffic. On the other hand, in Assignment 2 traffic is groomed efficiently so that only two SONET ADMs are needed at three nodes (B, C, and D), while node A needs three SONET ADMs in the ring. Figure 9.2 depicts this scenario, giving a total requirement of only nine SONET ADMs.

Using another example, it is shown that the minimum number of ADMs is not necessarily achieved with the minimum number of wavelengths. Consider a

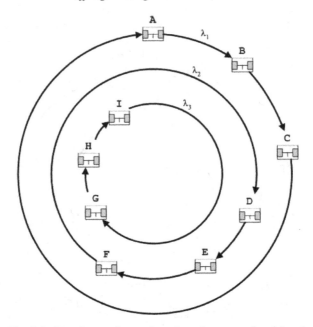

Fig. 9.3. Routing and wavelength assignment algorithm 1.

nine-node WDM/SONET bidirectional line-switched ring (BLSR) with nodes numbered from *A* to *I* in which two different routing and wavelength assignment (RWA)
algorithms have been used to route a common set of demands. Let the set of traffic
demands be $A \leftrightarrow B$, $A \leftrightarrow C$, $B \leftrightarrow C$, $D \leftrightarrow E$, $E \leftrightarrow F$, $F \leftrightarrow D$, $G \leftrightarrow H$, $H \leftrightarrow$
I, and $I \leftrightarrow G$. Further, assume that each traffic demand is for a part wavelength
capacity. The traffic can be groomed using the following two different algorithms.
Figures 9.3 and 9.4 depict the two solutions.

RWA algorithm 1

$$\lambda_1 = A \leftrightarrow B, B \leftrightarrow C, C \leftrightarrow A \text{ (three ADMs)}$$
$$\lambda_2 = D \leftrightarrow E, E \leftrightarrow F, F \leftrightarrow D \text{ (three ADMs)}$$
$$\lambda_3 = G \leftrightarrow H, H \leftrightarrow I, I \leftrightarrow G \text{ (three ADMs)}$$

RWA algorithm 2

$$\lambda_1 = B \leftrightarrow C, C \leftrightarrow A, E \leftrightarrow F, F \leftrightarrow D, G \leftrightarrow H \text{ (eight ADMs)}$$
$$\lambda_2 = A \leftrightarrow B, D \leftrightarrow E, H \leftrightarrow I, I \leftrightarrow G \text{ (seven ADMs)}$$

The first RWA algorithm uses nine ADMs and three wavelengths to route the
traffic demands. The second RWA algorithm uses 15 ADMs and only two wavelengths to route the traffic demands. This shows that the RWA algorithm that uses
additional capacity uses fewer ADMs while that which uses additional ADMs may
route the traffic demands in a smaller number of wavelengths.

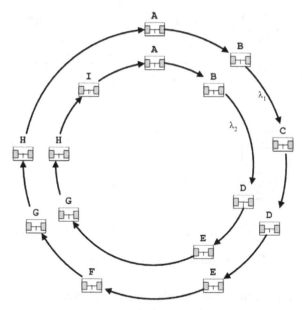

Fig. 9.4. Routing and wavelength assignment algorithm 2.

9.1 Traffic grooming in WDM rings

Most applications of traffic grooming have been concentrated on providing efficient network designs in SONET/WDM rings for improving the overall network cost. In a SONET/WDM ring network, each lightpath carries several low-rate traffic streams. The nodes in the ring network are equipped with OADMs and SONET ADMs. These OADMs terminate only those lightpaths for which the traffic streams need to source or sink at the node. The remaining lightpaths optically pass through the node without any processing. The terminated lightpaths are converted to electronic form and are each processed by a SONET ADM. The SONET ADM adds or drops low-rate traffic streams from the channel stream and sends the channel back to the OADM in optical form. The OADM then multiplexes the wavelength with other wavelengths in the outgoing fiber. The goal is to reduce the need for higher-layer electronic processing equipment (SONET ADMs), the cost of which dominates over the cost of the optical equipment (number of wavelengths) in the WDM ring network. Traffic grooming can be achieved based on whether they address static or dynamic traffic.

9.2 Static traffic grooming in rings

The static traffic grooming problem is defined as follows. Given the traffic demand set of low-rate connections needed between the node-pairs on the ring, assign the

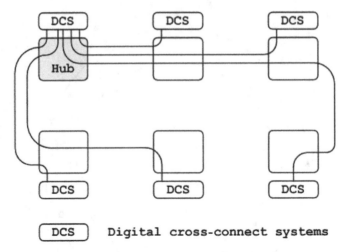

Fig. 9.5. An example of a single-hub ring.

traffic to wavelengths such that the number of SONET ADMs are minimized. Since a set of lightpaths are established on the WDM network, this problem is a special instance of the virtual topology design problem.

In [204] traffic grooming for different ring architectures is discussed. It is assumed that electronic processing equipment at the nodes included SONET ADMs and digital cross-connect systems (DCS) for cross-connecting and switching traffic streams from one wavelength to another. In the best cost scenario, all the nodes in the ring network are equipped with DCSs and each link has a one-hop lightpath to the neighboring nodes. This ring is called a *point-to-point WDM ring* (PPWDM), and uses the minimum number of wavelengths, but is the most expensive in terms of electronic processing cost. On the other extreme, a *fully-optical ring* has no electronic cost since all the traffic streams are directly connected between nodes through connecting lightpaths. However, it requires the maximum number of wavelengths.

A *hub ring* is one that has a node designated as a hub. The hub is the only node to contain a DCS and has lightpaths connecting it directly to all other nodes. This ring is wide-sense non-blocking, which implies that it can also support dynamic traffic. Although this ring has the least non-zero electronic cost, it still requires many wavelengths. Figure 9.5 gives an example of a *hub ring*.

A *double-hub ring* has two hubs with lightpaths connecting them to all of the other nodes. Although this ring is rearrangeable and non-blocking, it is reasonably efficient in the number of wavelengths and the electronic cost. Figure 9.6 depicts an example of a *double-hub ring*.

A *hierarchical* ring is composed of two PPWDM subrings and performs as well as a PPWDM ring at the expense of a few extra wavelengths. An example of a *hierarchical ring* is shown in Fig. 9.7.

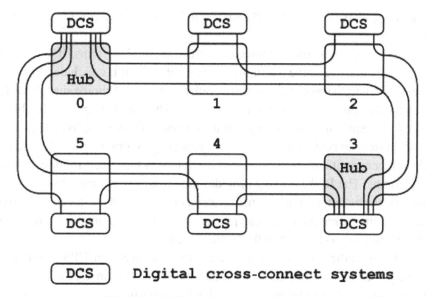

Fig. 9.6. An example of a double-hub-ring.

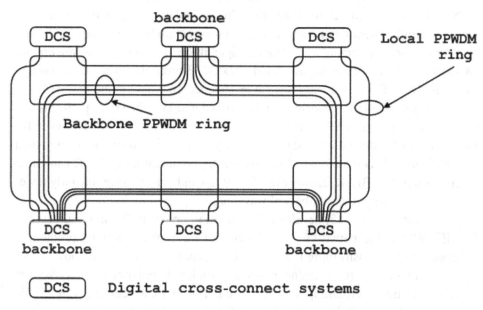

Fig. 9.7. An example of a hierarchical ring.

An *incremental ring* is a ring network that is recursively defined from smaller sections of the ring. It contains two types of nodes, namely *backbone* nodes and *local* nodes. All wavelengths are terminated by transceivers at backbone nodes and only a certain number of wavelengths are terminated at local nodes, which are called *transit* wavelengths. The incremental ring is designed in such a way that it is equivalent to a PPWDM ring with the same number of wavelengths for incremental traffic.

The various different ring architectures are compared in terms of wavelength cost, transceiver cost, and the maximum number of hops. It is concluded that given sufficient wavelength resources, the single-hub ring is a good choice since it has low transceiver costs and can support dynamic traffic. The double ring is preferable for dynamic traffic since it requires only half the number of wavelengths and has about the same transceiver cost. When the wavelengths are precious, the PPWDM, hierarchical, and incremental rings are reasonable choices as they use a minimal number of wavelengths. The PPWDM ring provides the most efficient use of wavelengths for dynamic traffic but has maximum transceiver cost. Therefore, given some spare wavelengths, the hierarchical ring can potentially reduce the transceiver cost, while the incremental ring is best when the traffic is static.

In [200] the traffic grooming issues are further analyzed in UPSR and BLSR WDM rings. The lower bounds on the number of ADMs in UPSR and BLSR WDM rings are established using static and uniform traffic patterns. Ring interconnection is also allowed in which low-speed traffic could be routed through different SONET rings from source to destination. In particular, each pair of nodes has r low-speed traffic streams between them. Each wavelength has a line speed indicated by a parameter g (for granularity). For example, if the low-speed traffic streams are OC-3 then each pair of nodes has r full duplex OC-3 communication. If each wavelength is a SONET OC-48 ring, then $g = 16$ since $16 \times$ OC-3 rate = $1 \times$ OC-48 rate. Typically, BLSR/2 (a two-fiber bidirectional line-switched ring) performs significantly better than UPSR in terms of both wavelengths and ADMs. However, for finer low-speed traffic (static and uniform), that is if r is much smaller than g, then the two bounds become closer. If ADMs are the dominant cost, UPSR cannot be much costlier than BLSR/2. However, BLSR/2 can be used to significantly lower the secondary cost of the numbers of wavelengths.

The research in [14, 15] develops algorithms to groom traffic in unidirectional SONET/WDM ring networks. It is shown that the traffic grooming problem is NP-complete by transforming the bin-packing problem (which is already known to be NP-complete) into the traffic grooming problem in polynomial time. Several scenarios for traffic grooming are considered. The first case considers the traffic grooming problem for a ring with uniform, all-to-one node traffic. The lower bounds on the number of wavelengths W_{min} is developed from the aggregated low-speed traffic L_{max} at an egress node in a unidirectional ring network, $W_{min} = \lceil L_{max}/g \rceil$. Different lower bounds on the number of ADMs are hence developed for the special cases: (i) a uniform traffic pattern, (ii) a uniform traffic pattern with minimizing ADMs subject to the minimum number of wavelengths, and (iii) all-to-all uniform traffic. Heuristics are designed and demonstrated to perform close to the lower bound for the all-to-all uniform traffic pattern.

In another scenario, the distance-dependent traffic model is considered where the amount of traffic between node-pairs is inversely proportional to the distance separating them and a set of loose lower bounds and heuristics for the minimum number of ADMs are developed. For all-to-all uniform traffic, it is proved that any solution not using a hub can be transformed into a solution with a hub using fewer or the same number of ADMs (a single-hub ring with a DCS at the hub).

The research in [310] proposed a suite of six heuristic algorithms for traffic grooming and wavelength assignment under uniform and non-uniform traffic in both unidirectional and bidirectional SONET/WDM rings. The problem is split into two phases: (i) grooming the low-speed traffic into high-speed lightpaths and (ii) grouping the lightpaths into SONET BLSR rings. The study shows that effectively separating the wavelength assignment from traffic grooming helps to simplify both problems and to obtain efficient solutions.

In the first phase, all the traffic demands are packed into as few rings (each ring corresponds to a wavelength) as possible, where each circle has capacity equal to the tributary rate and contains non-overlapping demands. In the second phase, multiple circles are groomed together so as to maximize the overlap of end nodes. An end node is a node that terminates a traffic connection in a circle. The maximization of the overlap of end nodes is reflected as a minimization in the number of SONET ADMs needed to satisfy the set of traffic demands. An optimal circle construction algorithm for uniform traffic and a heuristic for circle construction for non-uniform traffic have been developed. A generic circle grooming algorithm applicable to both unidirectional and bidirectional rings is also developed. The results obtained from the simulations demonstrate that the designed algorithms perform very well in reducing the number of SONET ADMs as well as minimizing the number of wavelengths. An example in [200] shows that the two-step approach in [204] can lead to a 20% more cost-efficient design.

A similar two-phase approach for grooming arbitrary traffic in BLSR rings has been proposed in [217]. In the first stage, primitive rings from the traffic requests (which are denoted as arcs) are generated. The cost of each primitive ring is the number of nodes in it. The objective in this phase is to obtain a set of primitive rings such that the total cost of all primitive rings is minimum. In the next phase, the primitive rings are grouped together such that each group forms a logical SONET ring. The associated ADM cost of each group is the number of nodes contained in this group and the total ADM cost is the sum of all the ADM costs for the groups. Since both of these phases are NP-hard problems in themselves, heuristics are devised for the generation of primitive rings and various approximation algorithms

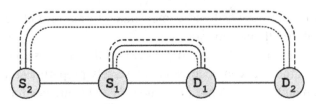

Fig. 9.8. Example of grooming streams for the same node-pair.

are presented for solving each subproblem in the optimal grooming of rings. Either an exact or estimated approximation ratio for each algorithm is derived and proved in [217]. The traffic grooming in WDM rings has also been studied in [12, 13, 117, 119, 130, 134, 142, 201, 202, 216, 220, 221, 222, 227, 228, 238, 307].

9.3 Dynamic traffic grooming in WDM networks

Most of the SONET traffic is groomed and set up in a static scenario. It is provisioned once and remains in place for a long period of time. In the future, as IP becomes the prevailing protocol over WDM and SONET, ADMs may no longer be required to pack traffic onto the wavelengths. In this scenario, it is the responsibility of the IP layer to effectively multiplex traffic onto wavelengths. These IP-over-WDM networks are likely to be arranged in a mesh topology rather than a ring. More importantly, the traffic requirements of IP are bound to change much faster than the legacy traffic based on SONET. In such a scenario, it is important that dynamic traffic grooming is employed so that the networks are able to efficiently accommodate changes in traffic. Minimizing equipment costs in such a dynamic traffic grooming scenario for SONET/WDM rings is an important consideration in [51].

It is possible to restrict traffic grooming in such a way that all traffic streams that are groomed on a path originate and terminate at the same node-pair. For example, in Fig. 9.8, traffic streams between node-pair (S_1, D_1), and traffic streams between node-pair (S_2, D_2) are groomed on their respective paths. In another case, it is possible that traffic streams between different node-pairs share a path. For example, in Fig. 9.9 traffic streams between node-pair (S_1, D_1), and traffic streams between node-pair (S_1, D_2) can share a link. Similarly, traffic streams between node-pair (S_2, D_1) and traffic streams between node-pair (S_2, D_2) can share a link.

A WDM network may include two types of nodes: *wavelength-selective cross-connect* (WSXC) nodes and *wavelength-grooming cross-connect* (WGXC) nodes. Figures 9.10 and 9.11 show the node architecture of a WSXC and a WGXC node, respectively. These network nodes are interconnected by fiber-optic links which can be either bidirectional or unidirectional.

WSXC nodes have OXCs or OADM (if the node degree is two), which switch full wavelengths from an input port to an output port, and add/drop wavelengths

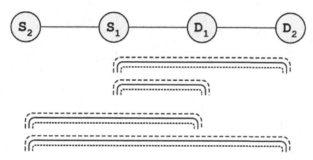

Fig. 9.9. Example of grooming streams for a different node-pair.

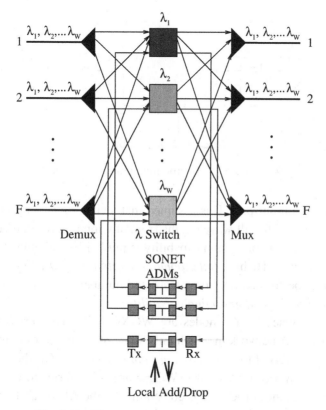

Fig. 9.10. Wavelength select cross-connect node.

for electronic processing. All nodes have SONET-ADMs which groom the traffic streams onto the added/dropped wavelengths. However, WSXC nodes cannot switch traffic streams between wavelengths.

WGXC nodes, in addition to having the functionality of a WSXC, are capable of time-slot interchange and can switch lower-rate traffic streams from a set of time slots on one wavelength on an input port to a different set of time slots on another

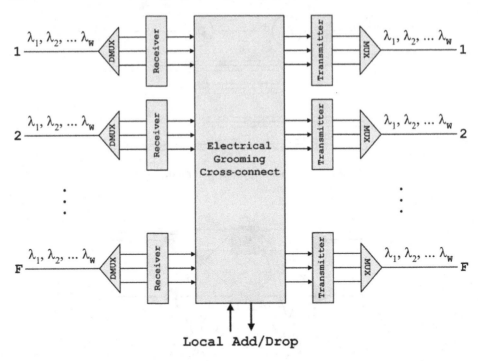

Fig. 9.11. Wavelength grooming cross-connect node.

wavelength on an output port. It is assumed that switching is fully non-blocking and can be performed for all wavelengths from any input port to any output port. Hence, full wavelength conversion capability is implicitly available at the WGXC node. Such a node is said to have *full grooming capability*. If switching of lower-rate traffic streams is performed only on a restricted number of wavelengths, then the node is said to have *limited grooming capability*.

A network in which all the nodes are WGXC nodes is referred to as a *full grooming network*. A network in which only some of the nodes of the network are WGXC nodes is referred to as a *sparse grooming network*. On the other hand, a network with only WSXC nodes and no WGXC nodes is referred to as a *constrained grooming network*, since grooming is constrained to the ADMs at the nodes.

9.3.1 Connection setup and release

The connection setup and release procedure in traffic grooming networks is different from the lightpath establishment process of conventional wavelength routing networks. A low-rate traffic session is routed along a path traversing through intermediate WSXC and WGXC nodes between the source and destination. If the path traverses through one or more intermediate WGXC nodes then the traffic session

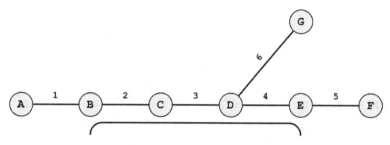

Fig. 9.12. A wavelength grooming network, example A.

involves more than one lightpath. Lightpaths between the source, destination, or intermediate WGXC nodes satisfy the wavelength continuity constraint, that is, the traffic stream occupies the same wavelength on all the links of the path between the source, destination, or intermediate WGXC nodes. However, lightpaths between WGXC nodes can be routed on different wavelengths. In this manner, each lightpath typically carries many multiplexed lower-speed traffic streams. During connection setup, it only needs to confirm whether the lightpaths, that have been established earlier, have the required amount of capacity before they can be used to accommodate the new traffic session.

9.3.2 Grooming example

Consider an example of a subnetwork, shown in Fig. 9.12, which is part of a bigger mesh network. Assume a single wavelength is currently available on the path from A to F. Let the capacity of the wavelength be C. Further assume that all the nodes on the path are WSXC nodes. Suppose a request arrives for a connection from node B to E for a line capacity of $C/4$. This is established immediately on the available wavelength by configuring the OADMs at nodes B, C, and E, and by configuring the OXC at node D (see Fig. 9.12). Note that the wavelength is added/dropped only at nodes B and E, and not at nodes C and D.

Let a second request arrive for a connection from node A to node F for a line capacity of $C/4$. This is also established on the same wavelength by setting up lightpaths from A to B and from E to F, and by using the same lightpath on the wavelength between B to E that was established for the first connection (see Fig. 9.13).

In this process, the first traffic stream is not disturbed. The wavelength is now added/dropped at four nodes, namely, A, B, E, and F.

Let a third request arrive for a connection from A to G for line capacity $C/4$. However, this connection cannot be established on the wavelength and is blocked. The reason is that the path for the third connection request from A to G deviates

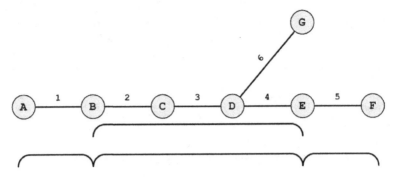

Fig. 9.13. A wavelength grooming network, example B.

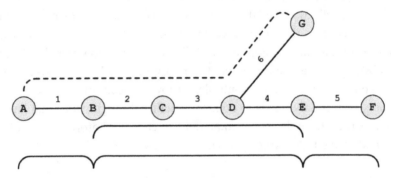

Fig. 9.14. A wavelength grooming network, example C.

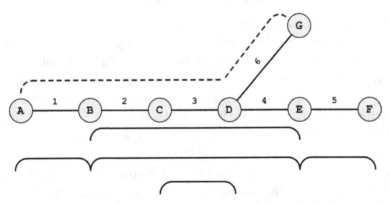

Fig. 9.15. A wavelength grooming network, example D.

away from the path of the lightpath on the wavelength and node D is only a WSXC node (see Fig. 9.14).

Let a fourth request arrive for a connection from node C to D for a line capacity of $C/4$. At this point, there are two options. (i) It can be assumed that any lightpath that has been established should not be disturbed. Therefore, the traffic stream cannot be established on the same wavelength and is blocked. (ii) However, if a temporary disturbance to the lightpath is acceptable, then the third call can be established on the same wavelength by adding/dropping the wavelength at nodes C and D. The lightpath is now split into three parts. This temporary disturbance is made possible by the presence of fast reconfigurable OADMs and OXCs at the nodes. For the network model, case (ii) is assumed and the nodes have fast OADMs and OXCs. On the other hand, if nodes C and D happen to be WGXC nodes, then it is possible to satisfy all the call requests.

When a call leaves the network, the lightpaths that are used to hold the traffic connection release the capacity used by the traffic stream. However, the lightpaths themselves might continue to operate over the wavelengths since they might have other traffic streams multiplexed over them. If the traffic stream was the sole one to have used the lightpath, then the lightpath itself can be released and the wavelengths on the links can be freed.

10

Gains of traffic grooming

The focus of this chapter is to provide an analytical framework and to obtain some insight into how traffic grooming affects performance in terms of the call blocking probability in different network topologies. Specifically, the performance of constrained and sparse grooming networks are compared using simulation-based studies. Constrained grooming corresponds to the case where grooming is performed only at the SONET-ADMs on an end-to-end basis. Sparse grooming corresponds to the case where, in addition to grooming at the SONET-ADMs, the cross-connects at some or all of the nodes are provided with a traffic stream switching capability. The goal is to develop techniques to minimize electronic equipment costs and to provide solutions for efficient WDM network designs.

It has been established that wavelength conversion, that is, the ability of a routing node to convert one wavelength to another, reduces wavelength conflicts and improves the performance by reducing the blocking probability [149]. Lower bounds on the blocking probability for an arbitrary network for any routing and wavelength assignment algorithm are known [237]. It is further shown that the use of wavelength converters results in a 10–40% increase in wavelength reuse. A reduced load approximation scheme to calculate the blocking probabilities for the optical network model for two routing schemes, fixed routing and least-loaded routing, has been used in [9]. This model does not consider the load correlation between the links. Analytical models of networks, using fixed routing and random wavelength assignment, taking wavelength correlation into account, have been developed in [219]. This work also studies the effects of path length, switch size, and hop number on the blocking probability in networks with and without wavelength conversion.

Sparse wavelength conversion and its effects on blocking performance have been studied in [271]. Their analytical model takes into account both wavelength correlation and the dynamic nature of the traffic. A new analytical technique based on the inclusion–exclusion principle for networks with no wavelength conversion and random wavelength assignment has been developed in [185]. Other routing algorithms have been also studied in [6, 46, 50, 58, 59, 77, 86, 113, 115, 128, 148, 151, 163, 167, 179, 226, 230, 231, 232, 246, 257, 262, 276, 306].

There has also been considerable interest in studying the performance of networks which utilize both TDM and WDM. Multi-wavelength TDM networks in which time slots on a wavelength are dedicated to each source–destination pair is developed in [195]. In [146] the performance improvements offered by wavelength converters and time-slot interchanges are described for shared-wavelength TDM networks and dedicated-wavelength TDM networks. However, the analysis is restricted to all calls having uniform bandwidth requirements and occupying one time slot on a wavelength.

Developing an analytical framework using the blocking performance of grooming networks is the subject of this chapter. This model takes into account the capacity distribution on a wavelength and the correlation on neighboring links given that incoming calls are of varying capacity. Prior research has concentrated on the case where the wavelength or a single time slot on a wavelength is considered as the basic unit of bandwidth.

10.1 Network parameters

WDM grooming networks can be described using the following parameters.

 (i) The network consists of V nodes with L links. Each link is bidirectional and consists of a pair of fibers with W wavelengths each in each direction.
 (ii) Each wavelength has capacity C and a parameter g (C is assumed to be divisible by g) referred to as the *granularity*. A lightpath that traverses a wavelength can support a maximum of g traffic streams. In other words, at most g low-rate traffic streams can be multiplexed onto the lightpath. The capacity of a traffic stream can vary from C/g (line-speed 1) to the full wavelength capacity C (line speed g). A *line-speed j* traffic stream is defined as a traffic stream occupying a capacity jC/g on the lightpath.
 (iii) Calls arrive at a node according to a Poisson process with rate λ. Each call is equally likely to be destined to any of the remaining $V - 1$ nodes. The arrival rate of calls λ_{sd} for a node-pair (s, d) is then $\lambda/(V - 1)$. Each call can request a line-speed j, where $1 \leq j \leq g$. The arrival rate of calls at a source–destination pair and requesting a line-speed j is $\lambda_{sd}(j)$. Each set of calls of line-speed j from a source to a destination requests equal total capacity of calls in its line-speed class. In other words, if the combined capacity of calls to a node-pair is say, Kg, then each line-speed class contributes a capacity of K through its call arrivals. For example, line-speed 1 traffic will have K call arrivals, line-speed 2 will have $K/2$ call arrivals, and similarly line-speed j traffic will have K/j call arrivals. Therefore, the probability, r_j that a call is of

line-speed j is

$$r_j = \frac{1/j}{\sum_{i=1}^{g} 1/i} \tag{10.1}$$

The expected value of j, $E\{j\}$ is then given by

$$E\{j\} = \sum_{j=1}^{g} j r_j = \frac{g}{\sum_{i=1}^{g} 1/i} \tag{10.2}$$

The *arrival rate per unit line-speed per s-d pair* is now defined as

$$\hat{\lambda}_{sd} = \lambda_{sd} E\{j\} \tag{10.3}$$

Here the term "unit line-speed" refers to the capacity of the lowest-granularity traffic stream that can be groomed on to the lightpath. $\hat{\lambda}_{sd}$ is essentially the arrival rate of calls at s-d pairs in the network if all call requests are of the lowest granularity, i.e. of line-speed 1.

(iv) Fixed-path routing is assumed, i.e. each call uses a prespecified path. If the path cannot accommodate the call, then the call is assumed to be blocked and is lost.

(v) Call requests cannot be split up among wavelengths on a link. Specifically, it is assumed that a traffic request can occupy only one wavelength on a link in the path. As explained in the previous section, it is assumed that fast reconfigurable OADMs and OXCs are available in the nodes. Hence a lightpath can be disturbed to multiplex a traffic stream into it.

(vi) The duration of each call request is assumed to be exponentially distributed with unit mean.

(vii) It is also assumed that there are enough receivers/transmitters at the nodes to handle the traffic that originates from the nodes. The traffic from a node is then limited by the degree of the node, the number of wavelengths on the fibers, and the capacity of the wavelengths.

10.2 Modeling constrained grooming networks

An exact analysis of grooming networks is a hard problem. Therefore an approximate analytical model with reasonable computational requirements for constrained grooming networks is developed and presented here. The analytical models provide the blocking probabilities of calls for different capacities and can be applied to networks with arbitrary topologies using the parameters described in the previous section. The model assumes random wavelength assignment (RWA) for analysis. In this strategy, the wavelength to be assigned to a call is chosen randomly from the set of available wavelengths on the path. Other algorithms such as first-fit and max-sum provide better performance. However, they are considerably more complex to analyze.

10.2.1 Single-wavelength link

Consider a single-link, single-wavelength system with wavelength capacity C and granularity g. Let $\lambda_l(j)$ be the link arrival rates of traffic streams of line-speed j (capacity jC/g). Let n_j be the number of traffic streams of line-speed j multiplexed into the wavelength. The wavelength can contain $(n_1, n_2, \ldots, n_j, \ldots, n_g)$ traffic streams of line-speeds from 1 to g provided:

$$0 \leq n_j \leq \left\lfloor \frac{g}{j} \right\rfloor \tag{10.4}$$

$$\sum_{j=1}^{g} j n_j \leq g \tag{10.5}$$

Let the call holding time be exponentially distributed with parameter μ. The above system can be modeled as a Markov chain with a state space given by $(n_1, n_2, \ldots, n_j, \ldots, n_g)$ where $n = \sum_{j=1}^{g} n_j$ is the total number of calls in the system and values of n_j satisfy constraints (10.4) and (10.5). The generator matrix Q^* of the Markov process governing the system can be formed according to the following rules. For a state transition:

(i) From state $(n_1, n_2, \ldots, n_j, \ldots, n_g)$ to state $(n_1, n_2, \ldots, n_j + 1, \ldots, n_g)$, provided constraints (10.4) and (10.5) are satisfied. The arrival of a call of line-speed j and the transition rate is given by $\lambda_l(j)$.

(ii) From state $(n_1, n_2, \ldots, n_j, \ldots, n_g)$ to state $(n_1, n_2, \ldots, n_j - 1, \ldots, n_g)$, provided $n_j > 0$. Denote the departure of a call of line-speed j. The transition rate is given by $n_j \mu$.

(iii) The diagonal elements of the matrix are negative such that the sum of all the elements of any row in the matrix equals zero. Specifically their value is given by $-\lambda_s - n\mu$, where λ_s is the sum of only those λ_j rates, with valid single-call arrivals so that constraints (10.4) and (10.5) would not be violated.

(iv) For all other state transitions, the transition rate is zero.

The stationary probability vector X at an arbitrary time for the generator Q^* is the unique solution to the equations:

$$XQ^* = 0, \quad Xe = 1 \tag{10.6}$$

where e is a column vector of ones. Hence the elements of vector X are $x(n_1, n_2, \ldots, n_j, \ldots, n_g)$ which correspondingly provide the probability that the system is in state $(n_1, n_2, \ldots, n_j, \ldots, n_g)$. Once the steady-state probability vector X is calculated, we obtain the blocking probability $Q(j)$ of the class-j traffic stream by

$$Q(j) = 1 - \sum_{i=0}^{g-j} X_c(i) \tag{10.7}$$

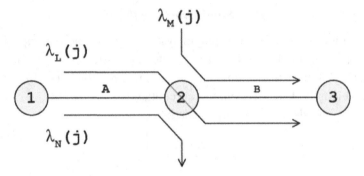

Fig. 10.1. A two-hop single-wavelength system.

where $X_c(i)$ is the sum of only those probability values $x(n_1, n_2, \ldots, n_j, \ldots, n_g)$ of X such that the corresponding state $(n_1, n_2, \ldots, n_j, \ldots, n_g)$ yields a capacity of i, i.e. $\sum_{k=1}^{g} k n_k = i$.

10.2.2 Analysis of a single-wavelength two-hop path

Consider the two-hop, single-wavelength system as shown in Fig. 10.1. Let $\lambda_N(j)$ be the arrival rates of traffic streams of line-speed j using only link A, i.e. the traffic stream enters at node 1 and leaves at node 2. Such traffic streams are denoted as type-N traffic streams. Let $\lambda_M(j)$ be the arrival rates of traffic streams of line-speed j using only link B, i.e. the traffic stream enters at node 2 and leaves at node 3. Such traffic streams are denoted as type-M traffic streams. Let $\lambda_L(j)$ be the arrival rates of traffic streams of line-speed j which use both links A and B, i.e. it denotes the arrival of traffic that enters at node 1, continues through node 2, and leaves at node 3. Such traffic streams are denoted as type-L traffic streams. Type-L traffic streams are not a part of type-N or type-M traffic streams.

Let n_j and m_j be the number of traffic streams of line-speed j multiplexed onto the wavelength at links A and B, respectively. Let l_j correspond to the number of traffic streams of line-speed j which continue on both links A and B. Hence the wavelength at link A contains $N = (n_1, n_2, \ldots, n_j, \ldots, n_g)$ and $L = (l_1, l_2, \ldots, l_j, \ldots, l_g)$ traffic streams and link B contains $M = (m_1, m_2, \ldots, m_j, \ldots, m_g)$ and $L = (l_1, l_2, \ldots, l_j, \ldots, l_g)$ traffic streams provided:

$$0 \leq n_j, m_j, l_j \leq \left\lfloor \frac{g}{j} \right\rfloor$$

$$0 \leq \sum_{i=1}^{g} j \cdot n_j + \sum_{k=1}^{g} k \cdot l_k \leq g$$

$$0 \leq \sum_{i=1}^{g} j \cdot m_j + \sum_{k=1}^{g} k \cdot l_k \leq g \qquad (10.8)$$

Let the call holding time be exponentially distributed with parameter μ. The above system can be modeled as a Markov process with a state space given by $(n_1, n_2, \ldots, n_j, \ldots, n_{g-1}, n_g, m_1, m_2, \ldots, m_j, \ldots, m_{g-1}, m_g, l_1, l_2, \ldots, l_j, \ldots, l_{g-1}, l_g)$. Let $n = \sum_{j=1}^{g} n_j$, $m = \sum_{j=1}^{g} m_j$, and $l = \sum_{j=1}^{g} l_j$ be the total number of traffic streams of types N, M, and L, respectively with the state space variables satisfying constraints in Eq. (10.8). The generator matrix Q^* of the Markov process governing the system can be formed according to the following rules. For a state transition:

(i) From state $(n_1, n_2, \ldots, n_j, \ldots, n_g, M, L)$ to state $(n_1, n_2, \ldots, n_j + 1, \ldots, n_g, M, L)$ provided constraints in Eq. (10.8) are satisfied denotes the arrival of a type-N call of line-speed j. The transition rate is given by $\lambda_N(j)$. Similar transition rates apply for type-M and type-L call arrivals.

(ii) From state $(n_1, n_2, \ldots, n_j, \ldots, n_g, M, L)$ to state $(n_1, n_2, \ldots, n_j - 1, \ldots, n_g, M, L)$ provided $n_j > 0$ denotes the departure of a type-N call of line-speed j. The transition rate is given by $n_j \mu$. Similar transition rates apply for type-M and type-L call departures.

(iii) The diagonal elements of the matrix are negative such that the sum of all the elements of any row in the matrix equals zero. Specifically their value is given by $-\lambda_s - (n + m + l)\mu$, where λ_s is the sum of only those λ_j rates for all N-, M-, and L-type traffic streams, with valid single-call arrivals that will not violate constraints in Eq. (10.8).

(iv) For all other state transitions, the transition rate is zero.

The stationary probability vector for the two-hop path Y at an arbitrary time for the generator Q^* is the unique solution to the equations

$$YQ^* = 0, \quad Ye = 1 \tag{10.9}$$

The elements of vector Y, $y(n_1, n_2, \ldots, n_j, \ldots, n_g, m_1, m_2, \ldots, m_j, \ldots, m_g, l_1, l_2, \ldots, l_j, \ldots, l_g)$ give the probability that the system is in the corresponding state specified in the parentheses. The blocking probabilities $Q_N(j)$, $Q_M(j)$ and $Q_L(j)$ of traffic streams of types N, M, and L, respectively, on the two-hop path can be obtained in a manner similar to Eq. (10.7).

10.2.3 A multi-hop single-wavelength path

A multi-hop single wavelength path (referred to as a wavelength-path) may be modeled in two ways to calculate the blocking probability. The first model, based on link independence, assumes the wavelength capacity distribution on the wavelength on a link is independent of that of neighboring links. In the second model, wavelength capacity correlation is also considered between wavelengths on successive links of the path. The probability of blocking on a multi-hop single wavelength-path uses the results for the two-hop single wavelength-path derived in the previous section.

10.2.3.1 Independence model

In this model, the load on the wavelengths of the link is assumed to be independent of the loads of wavelengths on other links. The independence model has been shown to give reasonable estimates for densely connected network topologies or for networks with low end-to-end traffic loads. Consider an h-hop wavelength-path $v_0, v_1, v_2, \ldots, v_{h-1}, v_h$, where $\lambda_l^{(i)}$, $1 \leq i \leq h$, is the arrival rate at link (v_{i-1}, v_i). Using the results of Section 10.2.1, the values of $Q_i(j)$ for the ith hop in the path are obtained. The end-to-end blocking probability $Q_{v_0 v_h}(j)$ that a class-j call is blocked on the path is simply given by the product form

$$Q_{v_0 v_h}(j) = 1 - [1 - Q_{v_0 v_1}(j)][1 - Q_{v_1 v_2}(j)]$$
$$\cdots [1 - Q_{v_{h-1} v_h}(j)] \tag{10.10}$$

However, the above model is inaccurate for sparse topologies and in the case of high traffic on multi-hop calls. In this case, one needs to consider a wavelength correlation model.

10.2.3.2 Capacity correlation model

A Markovian correlation model based on the wavelength capacity distribution to calculate the blocking probability is more appropriate here. On this single wavelength-path, given the load of hop $1, 2, \ldots, i - 1$, the load on hop i is dependent only on the load on hop $i - 1$. Consider an h-hop path $v_0, v_1, v_2, \ldots, v_{h-1}, v_h$ from source v_0 to destination v_h. Let $\lambda_l^{(1)}, \lambda_l^{(2)}, \ldots, \lambda_l^{(h)}$ be the link arrival rates at hops $1, 2, \ldots, h$, respectively. Specifically, the calls corresponding to $\lambda_l^{(i)}$ at link i enter the path at node $i - 1$ and leave at node i, where $1 \leq i \leq h$. Let $\lambda_s^{(1)}, \lambda_s^{(2)}, \lambda_s^{(3)}, \ldots, \lambda_s^{(h)}$ be the segment arrival rates at wavelength segments of length $1, 2, 3, \ldots, h$, respectively where each segment starts at the source node v_0. Specifically, calls corresponding to $\lambda_s^{(i)}$ for segment i enter the path at the source node v_0 and leave the path at node i. It is to be noted that $\lambda_s^{(1)}$ is the same as $\lambda_l^{(1)}$.

A call of line-speed j on this wavelength-path can be established if each link of the wavelength-path has adequate capacity to accommodate the call. Due to the correlation assumption, the blocking probability of a call of line-speed j on the h-hop wavelength-path can be calculated, using the results from the two-hop path in the previous section. This is done as follows.

(i) Divide the path into the following set of two-hop paths:

$$(v_0, v_1, v_2), (v_0, v_2, v_3), \ldots,$$
$$(v_0, v_i, v_{i+1}), \quad 1 \leq i \leq h - 1$$
$$\ldots, (v_0, v_{h-2}, v_{h-1}), (v_0, v_{h-1}, d) \tag{10.11}$$

(ii) For each two-hop path, (v_0, v_i, v_{i+1}), $1 \leq i \leq h - 1$ in the above set, its associated arrival rates are $\lambda_N = \lambda_s^{(i)}$, $\lambda_M = \lambda_l^{(i+1)}$, and $\lambda_L = \lambda_s^{(i+1)}$.

(iii) Using the arrival rates, obtain the steady-state probability vector $X_i(N, M, L)$, $1 \leq i \leq (h-1)$ for each two-hop path (v_0, v_i, v_{i+1}), using the procedure described in Section 10.2.2.

(iv) For a given wavelength, calculate the end-to-end blocking probability $Q_{v_0 v_h}(j)$ of a call of line-speed j (or capacity jC/g) on the h-hop wavelength-path from v_0 to v_h as

$$
\begin{aligned}
Q_{v_0 v_h}(j) = 1 - &Pr\{\text{capacity } jC/g \text{ is available}\\
&\text{on the two-hop path } (v_0, v_{h-1}, v_h)\}\\
= 1 - &Pr\{\text{cap. } jC/g \text{ is available on link } h|\\
&\quad \text{cap. } jC/g \text{ is available on segment } (v_0, v_{h-1})\}\\
&\times Pr\{\text{cap. } jC/g \text{ is available on segment } (v_0, v_{h-1})\}
\end{aligned}
$$

The above formulation leads to a recursive function by viewing the segment (v_0, v_{h-1}) as a two-hop path with the middle node as v_{h-2}. In general, a segment (v_0, v_{i+1}) can be viewed as a two-hop path with node v_i as the middle node. That is, the first i spans are viewed as the first hop and the $(i+1)$th span as the second hop. Using the chain rule of probability, the following is obtained:

$$
\begin{aligned}
&Pr\{jC/g \text{ is available on segment } (v_0, v_{i+1})\}\\
&= \sum_{\forall \text{valid } N^i} \sum_{\forall \text{valid } L^i} \left\{ \sum_{\forall \text{valid } M^i} X^i(N^i, M^i, L^i) \right.\\
&\qquad \left. \times Pr\{\text{capacity is available on segment } (s, v_i)|L^{i-1} = L^i \oplus N^i\} \right\} \quad (10.12)
\end{aligned}
$$

$$
\begin{aligned}
0 \leq C_{N^i} + C_{L^i} \leq g - c\\
0 \leq C_{M^i} + C_{L^i} \leq g - c\\
1 \leq i \leq (h-1) \quad (10.13)
\end{aligned}
$$

N^i, M^i and L^i are the set of N-, M-, and L-type call vectors for the two-hop path of (v_0, v_i, v_{i+1}), respectively. For example, $X^i(N^i, M^i, L^i)$ gives the steady-state probability values for the two-hop system of (v_0, v_i, v_{i+1}) when the call combinations are N^i, M^i, and L^i. The variables C_{N^i}, C_{M^i}, and C_{L^i} in Eq. (10.13) are the total capacity of N-, M-, and L-type calls, respectively, on the two-hop path of (v_0, v_i, v_{i+1}). The *valid* vector combinations of N^i, M^i, and L^i for Eq. (10.12) are defined so that the inequalities in Eq. (10.12) are satisfied.

The operation $L^{i-1} = L^i \oplus N^i$ is defined such that the number of calls, $l_j^{(i)}$, of capacity j, that use both the first and second links of the two-hop path (v_0, v_i, v_{i+1}), and the number of calls, $n_j^{(i)}$, of capacity j that use only the first link of the two-hop path (v_0, v_i, v_{i+1}), are added together, $l_j^{(i)} + n_j^{(i)}$, to yield $l_j^{(i-1)}$, which is the number of calls of capacity j that use both the first and second links of the two-hop path (v_0, v_{i-1}, v_i) on the channel.

10.2.4 Network analysis

Using the end-to-end blocking probability for a single wavelength-path, the path blocking probability can be calculated using the wavelength independence, assuming that the distribution of capacity on a wavelength on a link is independent of the distribution of capacity on the other wavelengths on the same link. From the blocking probability $Q_p(j)$ of line-speed j calls for a wavelength-path, the path blocking probability for path p with W wavelengths is given by

$$P_p(j) = [Q_p(j)]^W \tag{10.14}$$

In this case, $Q_p(j)$ is calculated using the procedure given in the previous subsections. The analysis in the previous sections assumes that the single wavelength arrival rates at the links and segments of the network are known. Hence, if $\lambda_{(l)}(j)$ is the arrival rate of line-speed j calls at the link, the arrival rate at the wavelength $\lambda_{(l)}^w(j)$ can be estimated by

$$\lambda_{(l)}^w(j) = \frac{\lambda_{(l)}(j)}{W} \tag{10.15}$$

Typically, the traffic in a network is specified in terms of a traffic matrix A which specifies the offered load between pairs of stations. There are j traffic matrices, with each traffic matrix A_j corresponding to traffic loads of line-speed j. The offered loads between node-pairs are used to estimate the load at the links and segments of the network. However, the offered load between node-pairs is not entirely carried by the links as some of the calls are blocked. The extent of call blocking is in turn dependent on the offered load. This interdependence between the blocking probability and the offered load leads to a set of coupled non-linear equations called the Erlang map [19] and its solution is called the Erlang fixed-point solution. The holding time of all the calls between the node-pairs are exponentially distributed with unit mean. Thus the offered loads at the links and nodes are equal to their respective arrival rates.

Hence if the probability of blocking of a line-speed j call on a link l on the path is $P_l(j)$, the probability of blocking of a line-speed j call on path p is $P_p(j)$, and the arrival rate of a line-speed j call on path p is $\lambda_p(j)$, then a good approximation [19] for the arrival rate of a class-j call at link l is given by

$$\lambda_l(j) = \sum_{\forall p|l\in p} \lambda_p(j) \frac{1 - P_p(j)}{1 - P_l(j)} \tag{10.16}$$

A similar formula also applies to determining the arrivals of class-j calls, $\lambda_s(j)$ at a segment s of a path,

$$\lambda_s(j) = \sum_{\forall p|s \in p} \lambda_p(j)\frac{1 - P_p(j)}{1 - P_s(j)} \qquad (10.17)$$

Of course, $P_p(j)$ is in turn determined by the link and segment arrival rates and is calculated by using the procedure given in Subsection 10.2.3.2 and using Eqs. (10.14) and (10.15). The average blocking probability for a line-speed j call in the network is then given by

$$P_B(j) = \frac{\sum_{\forall p} \lambda_p(j)P_p(j)}{\sum_{\forall p} \lambda_p(j)} \qquad (10.18)$$

The probability that a call is of line-speed j is r_j, hence the blocking probability of a call, P_B, irrespective of its line-speed, is

$$\sum_{j=1}^{g} P_B(j)r_j \qquad (10.19)$$

But the expected capacity of a call is $E\{j\}C/g$, where $E\{j\}$ is as defined in Eq. (10.2). Therefore, the average capacity lost, C_L, due to call blocking can be expressed as

$$C_L = P_B E\{j\}/g \qquad (10.20)$$

This gives rise to a system of non-linear equations for the constrained grooming network. The following iterative procedure is used to solve for $P_B(j)$ for all paths. Define $\lambda_l^{(i)}(j)$, $\lambda_s^{(i)}(j)$, $Q_p^{(i)}(j)$, $P_p^{(i)}(j)$, $P_B^{(i)}(j)$, $P_l^{(i)}(j)$, and $P_s^{(i)}(j)$ as the values obtained at the end of the ith iteration for the respective variables without the superscript (i). First set $Q_p^{(0)}(j)$, $P_p^{(0)}(j)$, $P_B^{(0)}(j)$, $P_l^{(0)}(j)$, and $P_s^{(0)}(j)$ to zero and i to 1 as part of the initial conditions. Then follow the iterative method specified in the following.

(i) Determine the link arrival rates, $\lambda_l^{(i)}(j)$, using Eq. (10.16). Using the wavelength capacity correlation, the segment arrival rates, $\lambda_s^{(i)}(j)$, are determined using Eq. (10.17). The respective wavelength arrival rates are obtained using Eq. (10.15).
(ii) In the case of wavelength correlation, the methods given in Subsections 10.2.2 and 10.2.3.2 used to calculate $Q_p^{(i)}(j)$, $P_l^{(0)}(j)$, and $P_s^{(0)}(j)$ can be used. Otherwise, for wavelength independence the methods specified in Subsections 10.2.1 and 10.2.3.1 are used to calculate $Q_p^{(i)}(j)$ and $P_l^{(0)}(j)$.
(iii) Calculate $P_p^{(i)}(j)$ using Eq. (10.14).
(iv) Calculate $P_B^{(i)}(j)$ using Eq. (10.18).

(v) If the absolute percentage difference between $P_B^{(i)}(j)$ and $P_B^{(i-1)}(j)$ is smaller than a preselected threshold value, ϵ (say, $\epsilon = 1 \times 10^{-3}$), then terminate. Otherwise, increment i and go to step (ii).

It should be mentioned that the above iterative procedure does not always converge to a solution. In particular, the procedure for the correlation model case is highly sensitive to input traffic data and is prone to failing to converge. Although there exist methods such as Newton's method that are guaranteed to converge, this method is used since it is computationally efficient and simpler, and generally converges within a few iterations for most cases.

10.3 Sparse grooming network

In a sparse grooming network, only some (say K) of the N nodes of the network are provided with full grooming capability. Instead of calculating the blocking performance of each of the $\binom{N}{K}$ placement combinations, a probabilistic approach similar to [268] can be used to obtain an ensemble average. Let q be the probability that a node is equipped with full grooming capability such that on average there are $K = Nq$ WGXC nodes in the network. The probability q is referred to as the *grooming factor*, and results in a binomial distribution of WGXC nodes in the network with mean Nq. Consider an h-hop path $s, 1, 2, \ldots, i, \ldots, h-1, d$ and the blocking probability of a call of line-speed j on the path is computed. Using a recursion formula that uses the principle that node i on the path is the last WGXC node from s and there are no WGXC nodes after node i until node d is used, a relation for the blocking probability can be obtained. A call of line-speed j is not blocked if (i) it is not blocked on the first i hops of the path and (ii) a capacity jC/g exists on the last $h-i$ hops of the path. Assuming that the wavelength occupancy on a link is independent of the wavelength occupancy of the other links on the paths, the two events (i) and (ii) can be considered to be independent. Using this, the blocking probability of a call of line-speed j on an h-hop path can be calculated as

$$
\begin{aligned}
P_p^{(h)}(j) = \ &P(\text{the call is blocked} \mid \\
&\quad \text{no WGXC nodes are on the path}) \\
&\times P(\text{no WGXC nodes are on the path}) \\
&+ \sum_{i=1}^{h-1} P(\text{Blocking} \mid \text{node } i \text{ is the last WGXC node} \\
&\qquad\qquad\qquad \text{on the path from } s) \\
&\times P(\text{node } i \text{ is the last WGXC node})
\end{aligned}
$$

which is given by

$$P_p^{(h)}(j) = \left[Q_{(sd)}^w(j)\right]^W (1-q)^{h-1}$$

$$+ \sum_{i=1}^{h-1} \left[1 - \left[1 - P_p^{(i)}(j)\right]\{1 - \left[Q_{(id)}^w(j)\right]^W\}\right]q(1-q)^{h-i-1} \quad (10.21)$$

$Q_{(ik)}^w(j)$, where i, k are nodes on the path p, is calculated using the single-hop wavelength link model specified in Subsection 10.2.1. Note that the correlation model cannot be directly applied here as the event of blocking on the last $h - i$ hops is not independent of the event of blocking on the first i hops. To obtain the overall blocking in an arbitrary network, the iterative procedure in the previous section is used, where line 3 of the procedure needs to be replaced as:

3. Calculate $P_p^{(h)}(j)$ using Eq. (10.22).

10.4 Validation of the model

The analytical model in the previous section is used and blocking probabilities for two network topologies, a 16-node mesh-torus and an eight-node ring, under constrained and sparse grooming conditions are calculated. To verify the accuracy of the proposed analytical models, simulation studies are used to compare the results. A simulation study for a six-node network based on a non-blocking centralized switch configured as a WGXC and WSXC is performed. Simulations are run in batches of 10^6 calls each, until the blocking values between successive call batches differ by less than 0.1%. Similarly, the iterative procedure for analysis is stopped when the performance values between successive iterative steps differ by less than 0.1%.

In the case of a 16-node mesh-torus under constrained (with all WSXC nodes) and sparse grooming (with all WGXC nodes) conditions, the independence model provides reasonable estimates for the blocking probability at different line-speeds as shown in Figs. 10.2 and 10.3. For the constrained grooming case in Fig. 10.2, the correlation model provides more accurate estimates than the independence model. For the case of the ring, shown in Figs. 10.4 and 10.5, the correlation model provids more accurate estimates than the independence model for high line-speed connections. In both models, accuracy of the estimates decrease as the line-speed of the call decrease. The large differences in magnitude between the blocking for high and low line-speed connections makes it more difficult for the models to accurately predict the blocking at both high and low magnitudes. In fact, this effect is more pronounced in the case of the ring network due to the high load correlation of links.

Compared to constrained grooming, sparse grooming offers an order of magnitude decrease in blocking probability for high line-speed connections and a multiple

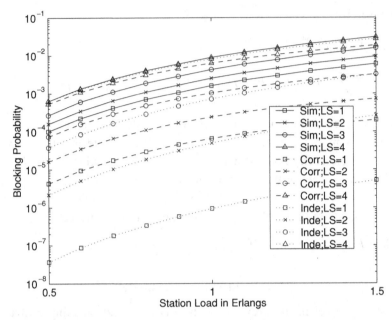

Fig. 10.2. Blocking probability vs. load per station in Erlangs for a constrained grooming 4 × 4 mesh-torus network with $W = 5$ and $g = 4$. (LS: line-speed, Sim: simulation, Corr: correlation model, Inde: independence model.) © IEEE. Source: S. Thiagarajan and A. K. Somani, A capacity correlation model for WDM networks with constrained grooming capabilities, in *ICC 2001* [277].

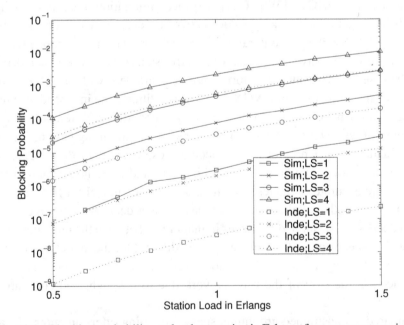

Fig. 10.3. Blocking probability vs. load per station in Erlangs for a sparse grooming 4 × 4 mesh-torus network with $W = 5$ and $g = 4$. (LS: line-speed, Sim: simulation, Inde: independence model.)

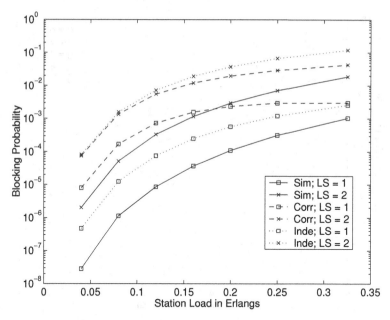

Fig. 10.4. Blocking probability vs. load per station in Erlangs for a constrained grooming eight-node ring network with $w = 5$ and $g = 2$. (LS: line-speed, Sim: simulation, Corr: correlation model, Inde: independence model.)

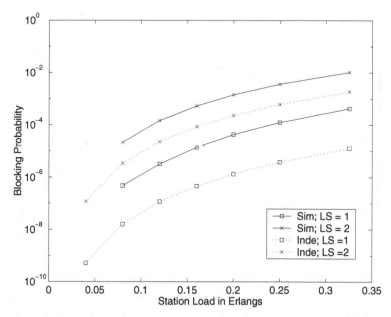

Fig. 10.5. Blocking probability vs. load per station in Erlangs for a sparse grooming eight-node ring network with $W = 5$ and $g = 2$. (LS: line-speed, Sim: simulation, Inde: independence model.)

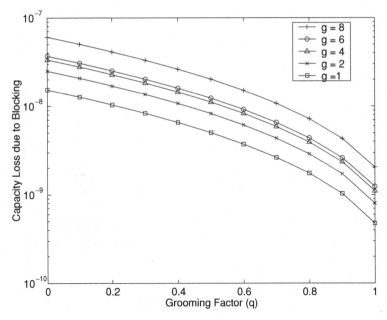

Fig. 10.6. Capacity loss due to blocking vs. grooming factor, q for a 4×4 mesh-torus network with $W = 5$ for different g.

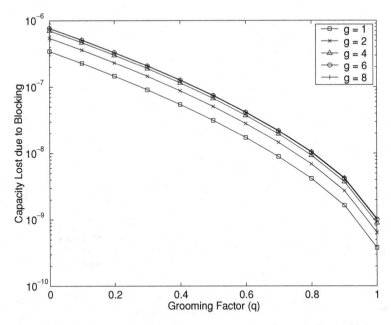

Fig. 10.7. Capacity loss due to blocking vs. grooming factor q for an eight-node ring network with $W = 5$ for different g.

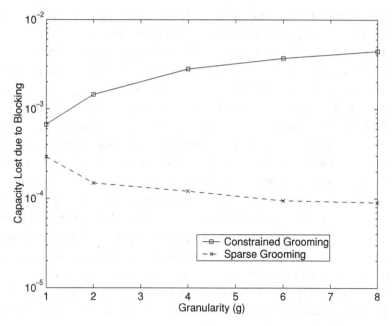

Fig. 10.8. Capacity loss due to blocking vs. granularity g for a non-blocking centralized switch with constrained and sparse grooming and attached to six access station nodes, $W = 5$

orders of magnitude decrease in blocking for low line-speed connections. The reason for this is in part due to the fact that smaller connections groom better and switch easier into wavelengths and also in part due to the assumption of random wavelength assignment for traffic streams. This shows that sparse grooming offers a significant improvement in blocking performance even for random wavelength assignment and better performance for other wavelength assignment schemes can be obtained, as compared to the random wavelength assignment scheme.

In Figs. 10.6 and 10.7, the effects of sparse grooming and granularity on the capacity loss, C_L, are observed, due to blocking of calls in the mesh-torus and the ring under low link loads (0.01 for the ring and 0.05 for the mesh). Here, the offered load at the nodes in terms of wavelength capacity is kept constant over different g. As is obvious, in both networks, C_L decreases as q is increased. Here the case for $q = 0$ corresponds to a constrained grooming network and that for $q = 1$ corresponds to a sparse grooming network with all WGXC nodes. However, it is interesting to note that the rate of decrease in C_L for the ring is more than that of the mesh. C_L increases with an increase in granularity (g). This increase in C_L is more due to the ineffective packing of traffic streams by the random wavelength assignment algorithm and can possibly be improved by using a better wavelength assignment scheme. Finally, the effects of g on a non-blocking centralized switch network connected to six access stations in both constrained and sparse grooming conditions are considered. Studying such a network provides the lower bound on the blocking performance as blocking occurs only at the links leading to the access stations. Here interestingly, as g is increased, the capacity loss due to blocking of a WSXC node increases while that of a WGXC node decreases. This suggests that in some cases, increasing the granularity might be beneficial for a sparse grooming network.

Based on the above analysis, the following observations are made. (i) The correlation model captures the system behavior better than the independence model. (ii) Both of the models are accurate in providing estimates for high line-speeds. However, they do not exhibit the same level of accuracy for low line-speeds in sparse networks. (iii) Compared to constrained grooming, sparse grooming offers at least an order of magnitude decrease in blocking probability for high line-speed connections and a multiple orders of magnitude decrease in blocking for low line-speed connections. (iv) The performance improvement is not equal for traffic of different line-speeds. (v) At low link loads in a network, increasing the granularity of traffic on a lightpath results in an increase in capacity loss due to blocking for constrained grooming networks but results in an improvement in some cases for sparse grooming networks.

11

Capacity fairness in grooming

In the last chapter, the characteristics of traffic grooming WDM networks with arbitrary topologies were studied from the perspective of blocking performance. It has been shown that the blocking performance is not only affected by the link traffic and the routing and wavelength assignment strategy, it is also affected by the arrival rates of different low-rate traffic streams, their respective holding times and more importantly, the capacity distribution of the wavelengths on the links. In such networks, call requests arrive randomly and can request for a low-rate traffic connection to be established between the source and the destination. Under dynamic traffic conditions, call requests that ask for capacity nearer to that of the full wavelength experience a higher probability of blocking than those that ask for a smaller fraction. In fact, the difference in blocking performance between the high- and low-capacity traffic streams becomes more significant as the traffic stream switching capability of the network increases. This difference in blocking performance for different capacities is directly affected by the routing and wavelength assignment policy that is used to route the call request. Hence, it is important that a call request is provided with its service in a *fair* manner commensurate with the capacity it requests. This *capacity fairness* is different from the fairness measure based on hop count that has traditionally been addressed in the literature [11].

In optical networks without wavelength conversion, due to the wavelength continuity constraint there is an increase in probability of a call request being blocked. As the path length from the source to the destination increases, the blocking probability of the corresponding call request further increases. Hence in a network with no wavelength conversion, long paths have a higher blocking probability than short paths. Wavelength conversion can be employed at the network nodes to reduce the blocking of longer-hop connections. Wavelength conversion for WDM networks was studied in [149, 219, 237, 268]. It has been shown that wavelength converters reduce wavelength conflicts and improve the performance by reducing the blocking probability. But an increase in the wavelength conversion capability of the network

201

results in the network admitting longer-hop paths that consume more network resources. This increases the blocking probability of shorter-hop paths, compared to their performance in networks with no wavelength conversion.

11.1 Managing longer paths

Several techniques have been used to handle requests that require longer path lengths in the network [11].

One way to handle such requests while improving the fairness is to use a reservation technique. In this case, some wavelengths are exclusively reserved for longer-hop paths on every link. The longer-hop paths can also compete with the shorter-hop paths for other wavelengths. This partly solves the problem.

In another method, called the protecting threshold technique, the traffic on long paths is protected from the traffic on short paths by admitting the short-path traffic only when the link utilization is below a given threshold. This leads to fairness in blocking.

Alternatively, a limited alternate routing method also improves fairness. In this method, long paths have a larger number of alternate paths than short paths. By limiting the number of alternate connections in short paths, more long-path connections can be accommodated.

All of these methods introduce fairness by regulating the admission of connection requests based on the path length at the expense of an increase in the overall blocking probability. They can be grouped into the general category of algorithms called *connection admission control* algorithms.

Usually, the common metric in evaluating the performance of dynamic routing and wavelength assignment (D-RWA) algorithms is the blocking probability. However, a good D-RWA algorithm for traffic grooming networks should treat all call requests in a "fair" manner while ensuring efficient utilization of the network. Usually, the problem of D-RWA is separated into the subproblems of routing and wavelength assignment and solved independently. Some of the dynamic wavelength assignment (D-WA) algorithms, such as first-fit (FF) [112], random assignment (R) [179], most-used (MU) [24] and max-sum (MS) [273], were designed in a network scenario where the full wavelength was the basic unit of bandwidth. However, in WDM networks capable of traffic grooming, the basic unit of bandwidth is a traffic stream, the capacity of which can be less than that of a wavelength.

In the next section, a definition for fairness is presented. The fairness performance of traditional D-WA algorithms in terms of capacity in a traffic grooming WDM network is also studied. It is observed that these algorithms do not treat call requests of different capacities in a fair manner. This motivates the need for a good mechanism to provide capacity fairness. A new connection admission control scheme is then developed, which can be used along with existing wavelength assignment algorithms to attain fairness in capacity. This algorithm achieves fairness in capacity while not over-penalizing the network blocking performance.

11.2 Capacity fairness

In a network, where every user pays for the bandwidth that is requested and consumed, it is important that every user receives the same type of services as any other. The network system that offers the service must be fair and should not have any inherent bias against a particular subset of users. Although this concept of fairness is simple to understand, the exact definition of fairness is, however, extremely case-dependent and goal-dependent upon the networking issues involved. In addition, it is usually the case that any efforts by a control mechanism to ensure fairness on an issue results in the degradation of performance in other qualities of the network.

The *capacity fairness* is defined as follows. The connection requests arrive randomly at a node-pair. A traffic stream of line-speed j is defined as a j traffic stream. A connection that requests a j traffic stream is referred to as a j call request. Suppose each class of traffic streams generates the same amount of combined capacity. (Otherwise, the model needs to be generalized to handle a variable amount of traffic for different rates.) In such a case, it can be expected that the calls of high capacity are blocked more often than those of small capacity if no specific measures are taken. In fact, as the number of WGXC nodes increases, i.e. the traffic stream switching capability in the network, there is more than an order of magnitude difference in blocking probability between the calls of highest and lowest capacity. A user who has knowledge of this unfairness can request the total required capacity in smaller traffic units rather than as a whole. This is unfair to those users who are either ignorant of the unfairness and/or cannot request their total capacity in splittable flows. To prevent this, i.e. to achieve capacity fairness, the blocking probability of a high-capacity call (say of line-speed m) should equal the combined blocking performance of m calls of line-speed 1. Hence, the following definition of fairness is used.

Definition. *Capacity fairness is achieved when the blocking performance of m calls of line-speed n is equal to the blocking performance of n calls of line-speed m.*

At this point, it is assumed that a user who requests capacity in terms of a smaller number of calls relinquishes all accepted calls immediately, even if one of them is blocked. The scenario, where the user can request a set of calls and accept or reject a subset of the accepted set is not considered. Therefore, if p_m is the blocking probability of a class-m call and p_n is the blocking probability of a class-n call, then to achieve capacity fairness

$$1 - (1 - p_m)^n = 1 - (1 - p_n)^m \quad \forall m, n : 1 \leq m, n \leq g \qquad (11.1)$$

In addition, an algorithm should achieve capacity fairness while keeping the overall blocking probability to an acceptable level. The overall blocking performance of the network can be defined in terms of the blocking probability per unit line-speed of the call requests. When capacity fairness is achieved, according to Eq. (11.1), the blocking probability, p_j, of a class-j call is the same as the blocking performance value of j class-1 calls, whose blocking probability is p_1, i.e.

$$p_j = 1 - (1 - p_1)^j \tag{11.2}$$

or

$$p_1 = 1 - \sqrt[j]{1 - p_j} \tag{11.3}$$

Hence using Eq. (11.3), an estimate of p_1 from p_j can be obtained. This estimate, $\hat{p}_j$, is referred to as the blocking probability per unit line-speed of a class-j call. Now the overall network blocking probability per unit line-speed, $\hat{P}$, is given by

$$\hat{P} = \frac{\sum_{j=1}^{g} \hat{p}_j}{g} \tag{11.4}$$

Recall that $\hat{\lambda}_{sd}$ should be the equivalent arrival rate of calls at s-d pairs in the network if all call requests are of the lowest granularity, i.e. of class 1. For a given physical topology, all incoming call requests are of class 1 and their arrival rate per node-pair is $\hat{\lambda}_{sd}$, then the corresponding network blocking performance Q obtained is the best estimate for $\hat{P}$ when capacity fairness is achieved.

It is usually the case that the unfairness in an algorithm affects calls of the highest or the lowest capacity more than calls of intermediate capacity. Keeping this in mind, a good estimate of fairness can be provided by using just the blocking performance of the highest- and lowest-capacity calls. Hence, the *fairness ratio* F_r for an algorithm running a network can be defined as the ratio of blocking probability per unit line-speed of the call with the highest line-speed (g), $\hat{p}_g$, to the blocking probability per unit line-speed of the call with the lowest line-speed (1), $\hat{p}_1$, or

$$F_r = \frac{\hat{p}_g}{\hat{p}_1} \tag{11.5}$$

If the value of F_r is greater than 1, then the algorithm is said to favor high-capacity call requests over low-capacity call requests and vice versa. Therefore, if F_r for an algorithm is close to 1, then it can be reasonably assumed that the algorithm is also fair to calls of *all* capacities. Therefore a good admission control algorithm should ensure that $\hat{P}$ is close to Q and at the same time ensure capacity fairness using Eq. (11.1) and have a fairness ratio close to 1.

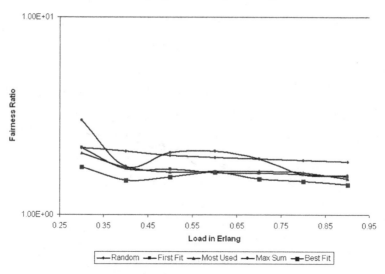

Fig. 11.1. The fairness ratio versus node load in Erlangs for a bidirectional 6×6 mesh-torus with $g = 4$ and $W = 5$.

11.3 Fairness performance of RWA algorithms

A simulation study of the call blocking performance in a 6×6 bidirectional mesh-torus network to study the capacity fairness property of various wavelength assignment algorithms is performed in [278, 280]. Several wavelength assignment schemes, including random, first-fit, most-used, max-sum, and best-fit wavelength assignment schemes are evaluated for the grooming networks. In the best-fit wavelength assignment algorithm, among the available wavelengths for the traffic request, the traffic stream is assigned to that wavelength which has the least free capacity remaining when the incoming traffic stream is accommodated.

The mesh-torus network is selected for comparison over the other topologies such as a ring, hypercube, etc. because of various interesting topological properties. Compared to the ring, the mesh-torus has more connectivity, which can help to generate a good amount of traffic switching at the nodes. Also compared to the hypercube, the average hop length is larger in the mesh-torus network, which also gives rise to a good amount of load correlation between the links.

Only a small number of wavelengths would really be used for wavelength-level grooming as the other wavelengths are likely to be used to carry full wavelength capacity traffic. Therefore, for all the evaluation cases, the number of wavelengths (W) per fiber is assumed to be 5. The granularity (g) of the wavelength is assumed to be 4. This means that a traffic stream can ask for a minimum of a quarter of the capacity of the wavelength.

Figure 11.1 depicts the fairness ratio versus the node load in Erlangs. It is observed that for low node loads, the fairness ratio is high indicating that high-capacity

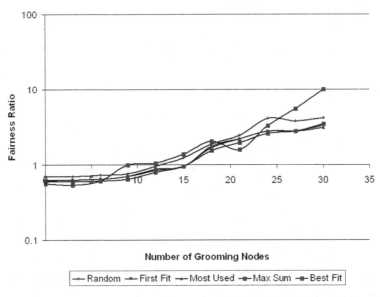

Fig. 11.2. The fairness ratio versus the number of WGXC nodes for a bidirectional 6×6 mesh-torus with $g = 4$ and $W = 5$.

calls are favored more than low-capacity calls. But as the node load is increased, the network traffic as a whole is increased. This increases the blocking probability of both low- and high-capacity connections. Hence all calls irrespective of whether they are high or low capacity start to experience blocking. It is observed that the best-fit performs the best with respect to fairness and provides the fairness ratio that is closest to 1. On the other hand, the max-sum algorithm, which provides the least overall network blocking compared with the other wavelength assignment algorithms considered, has the highest fairness ratio at low loads in the network. This means that it performs poorly when it comes to the fairness.

11.4 Connection admission control for fairness

The fairness ratio increases as the number of grooming nodes, i.e. WGXC nodes, in the network is increased (see Fig. 11.2). Initially when no WGXC nodes are present in the network, the connections of high capacity have a lower blocking probability per unit line-speed than those with low capacity. But as the grooming capability of the network is increased, a reversal is observed and connections of high capacity have a higher blocking probability per unit line-speed than low-capacity connections. The reason for this is that low line-speed connections groom better and fit more easily into wavelengths than high line-speed connections. It is interesting to note that when the number of grooming nodes is around 10–15, the fairness ratio is close to 1. The max-sum algorithm also has the highest fairness ratio when the grooming nodes are high, and the lowest fairness ratio (away from one) when there are no grooming nodes in the network. This motivates the need for effective schemes to achieve capacity fairness, in particular when the grooming nodes are present (and required) in the network.

The following algorithm works along with any routing and wavelength assignment algorithm and introduces capacity fairness by exercising connection admission control using run-time blocking performance information. *Connection admission control* (CAC) is simply defined as the set of actions that are to be taken upon a call arrival in order to establish whether to accept or reject the connection request. Connection admission control relies on two factors. The first is that incoming call requests can specify their network requirements and the second is on the ability of the system to measure, monitor, and update the global state of the network, which in this case is the blocking performance of calls of various capacities at the nodes. Since the algorithm makes its decision to accept or reject a call based only on the current blocking values, the CAC algorithm is effective only after the network has been up and running for quite some time. This is because initially the blocking performance of the network may be inaccurate or may not be known. Due to the dependence of the algorithm on such a "warm-up" period, the algorithm should not be relied upon to provide capacity fairness during this period before attaining an accurate and stable run-time blocking probability. Hence the CAC procedure is assumed to be carried out when the network is in a state where there are previously established traffic streams and wavelengths are already assigned to those traffic streams in the network. The CAC algorithm uses the run-time blocking probabilities p_j of calls of class j in the network. The CAC algorithm is independent of the routing and wavelength assignment scheme and can work along with any routing and wavelength assignment scheme. The procedure for the CAC algorithm is as follows.

Assume a new call arrives for a node-pair (s, d) and requests a capacity j to be established from s to d.

(i) Check if a traffic stream of capacity j can be established on the path, i.e. whether enough capacity exists in terms of wavelengths to carry the traffic stream. If the path cannot be established, reject the call.

(ii) Obtain an estimate of the overall network blocking probability per unit line-speed, $\hat{P}$, from Eq. (11.4) and the blocking probability per unit line-speed of the class j calls, $\hat{p}_j$, from Eq. (11.3).

(iii) If $(\hat{p}_j \geq \hat{P})$ then accept the call and go to step vi.

(iv) Let $q_m = (\hat{P} - \hat{p}_j)/\hat{P}$.

(v) Reject the call with probability q_m.

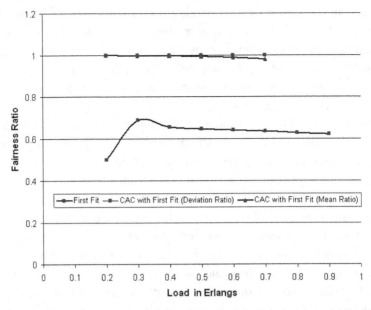

Fig. 11.3. The fairness ratio versus the node load in Erlangs for a bidirectional 6×6 mesh-torus with $g = 4$ and $W = 5$.

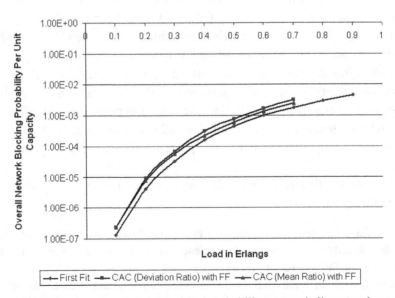

Fig. 11.4. The overall network blocking probability per unit line-speed versus node load in Erlangs for a bidirectional 6×6 mesh-torus with $g = 4$ and $W = 5$.

(vi) If the call is not rejected, start the wavelength assignment algorithm for the set of available wavelengths on the path to establish the connection. Update the blocking performance parameters.

The factor q_m is referred to as the *mean ratio*. Essentially, q_m is the rejection probability for the call. Another estimation for the rejection probability can be obtained using the standard deviation D of the blocking probability per unit line-speed values, $\hat{p}_j$, $1 \leq j \leq g$. In this case q_D is given by

$$q_D = (\hat{P} - \hat{p}_j)/(D) \tag{11.6}$$

The factor q_D is referred to as the *deviation ratio*. q_D can be substituted for q_m in the algorithm. The fairness performance for the two cases of the algorithm is shown later on. As the network services the calls that arrive, due to changes in network topology or traffic, it might happen that the current blocking of the network might differ significantly from the average blocking performance calculated over the lifetime of the network for the algorithm. To ensure accuracy in the estimation of blocking, a rolling window of the most recent set of call arrivals can be used and the blocking performance can be estimated using only those call arrivals rather than considering the complete set of calls since the network started operation.

11.4.1 Performance of the CAC algorithm

The performance of the connection admission control scheme is studied using the same simulation as described earlier. The network topology is a 6×6 mesh-torus. The performance is compared when the CAC algorithm is used along with the first-fit wavelength assignment scheme and when the FF scheme is used without the CAC scheme. The network scenario when there are no grooming nodes in the network is also considered. In the fairness ratio graph shown in Fig. 11.3, it is observed that both of the CAC schemes (*mean ratio* and *deviation ratio*) achieve excellent fairness where the fairness ratio is equal to 1 when compared to the first-fit algorithm, where the fairness ratio is less than 1. This confirms that the CAC algorithm works well under high-load scenarios in the network.

The overall network blocking probability per unit line-speed that is required to achieve the fairness is shown in Fig. 11.4. It is observed that there is a relatively small and consistent increase in blocking performance of the CAC-FF scheme when compare to just the FF scheme. This shows that the CAC can achieve capacity fairness with little increase in overall network blocking probability per unit line-speed.

12

Survivable traffic grooming

As mentioned in earlier chapters, due to the high bandwidths involved, any link failure in the form of a fiber cut has catastrophic results unless protection and restoration schemes for the interrupted services form an integral part of the network design and operation strategies. Although network survivability can be implemented in the higher layers above the optical network layer (e.g., self-healing in SONET rings and the ATM virtual path layer, fast rerouting in MPLS and changing routes using dynamic routing protocols in the IP layer), it is advantageous to use optical WDM survivability mechanisms since they offer a common survivability platform for services to the higher layers. For example, it is possible that several IP routes may eventually be routed through the same fiber. Hence the failure of a single fiber may affect multiple routes, possibly alternative paths for an IP route. Thus, protection at the IP layer requires complete knowledge of the underlying physical fiber topology.

As discussed earlier, a variety of optical path protection schemes can be designed using concepts such as disjoint dedicated backup paths, shared backup multiplexing, and joint primary/backup routing and wavelength assignment. Lightpath restoration schemes, on the other hand, do not rely on prerouted backup channels but instead dynamically recompute new routes to effectively reroute the affected traffic after link failure. Although this saves bandwidth, the timescale for restoration can be difficult to specify and can be of the order of hundreds of milliseconds. Hence in a dynamic scenario, path protection schemes are likely to be more useful and practical than path restoration schemes.

Lightpath protection schemes for WDM networks with grooming capabilities when the traffic demand is dynamic and consists of low-rate traffic streams are of interest as they offer a unique challenge [39, 48, 123, 140, 275]. Specifically, protection against any *single-link failure* at the optical network layer using a one-to-one optical path protection scheme is preferred. The following questions are of interest.

210

- How to optimally multiplex the primary and backup traffic streams onto a wavelength on a link?
- How to extend the multiplexing to traffic stream paths in the network?
- Is it better to multiplex the primary and backup streams together onto the same wavelength or is it better to segregate them to different wavelengths?
- What is the effect of such a grooming policy on the blocking performance?
- How does the topology and the RWA algorithm affect the choice of grooming?
- Evaluation of the effectiveness of the grooming policies on different topologies with different RWA algorithms.

Every physical link is assumed to have a fixed number of wavelengths and the blocking probability of traffic stream connections is considered as the primary performance metric. Calls for a source–destination pair arrive randomly and can request a low-rate dependable traffic connection to be established between the node-pair for the duration of the call. The call request is served by first establishing a primary (working) *traffic stream path* (TSP) and then establishing a link-disjoint backup (protection) TSP. If either the backup or the primary TSP cannot be set up, then the traffic stream request is assumed to be blocked. Indeed, the backup TSP does not carry any information, but takes over the role of the primary in the case of a link failure on the primary TSP. Hence the backup path can also be used to carry preemptable low-priority traffic, which is interrupted when the original primary path fails. Both the primary and the backup TSPs can traverse through the intermediate WSXC and WGXC nodes between the source and destination. A TSP can traverse more than one lightpath on its path from the source to the destination. Thus each lightpath typically carries many multiplexed lower-speed traffic streams, each of which can be either a primary TSP or a backup TSP.

In addition to the above, to reduce the overhead of backup traffic streams, the bandwidth sharing technique, namely *backup multiplexing* [256] is used. In this technique, only a small fraction of the link resources, i.e. a fraction of the wavelength, is reserved for all the backup traffic streams going through the link. The basic idea is the same as in the case of full wavelength routing, i.e. the two backup TSPs can share part of the wavelength capacity if their corresponding primary (working) TSPs do not fail simultaneously. This happens when the primary paths are link-disjoint, for a single-link failure model.

In this chapter, the dynamic establishment of primary and backup traffic streams and their grooming onto the WDM network are considered in detail.

12.1 Traffic stream multiplexing on a single wavelength link

Consider a wavelength w on a fiber link l with capacity C. Typically, many traffic stream paths of varying capacity can be groomed onto the wavelength. These TSPs can be either primary or backup traffic streams. Let C_P denote the total capacity required for all primary TSPs on the link. Let C_B denote the total capacity required for all backup TSPs on the link. The free (unused) capacity, C_F, is then given by $C_F = C - C_B - C_P$. Primary TSPs are distinctly multiplexed on the wavelength so that they occupy their individual capacities on the wavelength. Hence, if $P = \{p_1, p_2, p_3, \ldots, p_n\}$, with respective capacities $\{c_1^P, c_2^P, c_3^P, \ldots, c_n^P\}$, is the set of primary TSPs that traverse the link, then $C_P = c_1^P + c_2^P + c_3^P + \cdots + c_n^P$. On the other hand, backup TSPs share the capacity using the resource sharing technique of *backup multiplexing* [256]. Let $B = \{b_1, b_2, b_3, \ldots, b_n\}$ denote the set of backup TSPs traversing the wavelength w on link l. Let their respective capacities be $\{c_1^b, c_2^b, c_3^b, \ldots, c_n^b\}$ and their respective primary TSPs be denoted by the set $Q = \{q_1, q_2, q_3, \ldots, q_n\}$. Hence a primary TSP q_i with capacity c_i^b has a corresponding backup TSP b_i. Under the single-link failure model, the exact total backup capacity C_B needed to handle all possible cases of single-link failures can be calculated using Algorithm 12.1(b).

Algorithm 12.1(a). Algorithm to calculate C_B for wavelength w for link l with capacity C.

1: For each link i, $i \in L$. $i \neq l$
2: *SpareCapacity* $(i) = 0$
3:      For each primary TSP. $q_j \in Q$
4:          If q_j contains link i then
5:              *SpareCapacity* $(i) = SpareCapacity(i) + c_i^j$
6:          EndIf
7:      EndFor
8: EndFor
9: $C_B = \max\{SpareCapacity(i)\}, \forall i \neq l$

Algorithm 12.1(b). Dynamic algorithm to calculate $C_B^{\max}$ for wavelength w for link l with current backup capacity $C_B^{\max}$ and a new arrival of backup TSP b_{sd}, where $l \in b_{sd}$, with capacity c_{sd} and the primary TSP can be set up on p_{sd}.

1: For each link i, $i \in p_{sd}$, $i \neq l$
2: *SpareCapacity* $(i) = c_{sd}$
3:      For each primary TSP, $q_j \in Q$
4:          If q_j contains link i then
5:              *SpareCapacity* $(i) = SpareCapacity(i) + c_i^j$
6:          EndIf
7:      EndFor
8: EndFor
9: $C_B^{\max} = \max\{C_B^{\max}, SpareCapacity(i)\}, \forall i \neq l$

The algorithm can be modified to serve the dynamic case where there is already precomputed backup capacity, C_{prev}, serving the set of backup paths, B. Let a new backup traffic stream request b_{sd} for a capacity c_{sd} arrive at the wavelength channel. Assume that its primary traffic stream path p_{sd} can be established in the network. In this case, one only needs to calculate the spare capacity for each link j of the primary path p_{sd}, where $j \in p_{sd}$. The new backup capacity C_{new} is max$\{C_{prev}, s_j\}$, where s_j are the spare capacities obtained for each link j of the primary path p_{sd}. If $C_{new} \leq C_{prev}$, then no extra capacity on the channel need be reserved as backup capacity and the backup traffic stream is established over the wavelength. Otherwise, the backup path can be established if the free capacity C_F is at least $C_{new} - C_{prev}$. The algorithm for the dynamic case is shown in Algorithm 12.1(a).

It should be noted that to allocate a primary path, say p_{sd} of capacity c_{sd}, one needs to ensure that there is c_{sd} capacity free on the wavelength. When a traffic stream connection leaves the network, the primary and backup traffic stream paths are released. The capacity of the primary channel needs to be released and is added to C_F. For the backup path, the algorithm simply removes the traffic stream from the channel and runs Algorithm 12.1(a) for the link to obtain the new backup capacity. Alternately, one may check whether or not the backup capacity for the released connection must be released. Depending on the scenario under which the algorithm is operating, the two algorithms may result in a similar complexity.

12.2 Grooming traffic streams on the network

Traffic stream paths in the network can be either primary or backup traffic streams. Such traffic streams can be groomed onto the wavelengths in two ways.

(i) Both primary and backup TSPs can be groomed onto the same wavelength. This is referred to as a *mixed primary–backup grooming policy* (MGP).
(ii) The wavelength can consist of either primary or backup traffic streams but not both. This is referred to as a *segregated primary–backup grooming policy* (SGP).

Figure 12.1 depicts an example of MGP and SGP on a link with three wavelengths. The capacity on the wavelength can essentially be grouped into three types: capacity used by the primary TSPs or primary capacity (PC), capacity used by the backup TSPs or backup capacity (BC), and the unused or free capacity (FC).

12.2.1 Sharing backup capacity

In order to use the channels efficiently, the backup multiplexing technique is used. This process of grooming primary and backup TSPs onto the WDM network is illustrated through a simple example. Consider the eight-node network as shown in

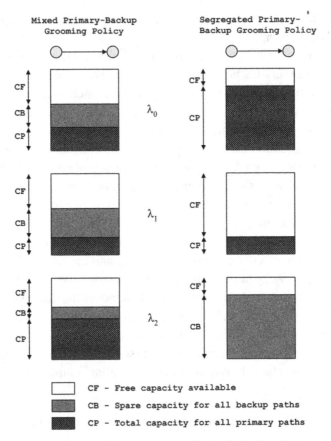

Fig. 12.1. Traffic stream grooming policies in a link.

Fig. 12.2. Each edge consists of bidirectional fiber links with two wavelengths per fiber link in each direction. The example network shown in this figure is represented as a layered graph with two wavelength layers, λ_0 and λ_1. Figures 12.2–12.5 show the allocation of wavelengths and paths to four pairs of primary and backup TSPs, $\langle p_i, b_i \rangle$, $1 \leq i \leq 4$, where b_i is the backup TSP for the primary TSP p_i. Each of the four traffic streams requires a capacity of $0.5C$, arriving in that order and set up sequentially as shown in Figs. 12.2–12.5. The network uses the MGP to assign traffic stream paths to wavelengths.

The first request, for a dependable connection from node 2 to node 4, is established as shown on λ_0. The primary TSP is established on the link 2–4 on λ_0 and the backup TSP is established through links 2–3–5–7–4 on the same wavelength. To establish the second request from node 1 to node 8, the primary TSP is established on λ_1 through links 1–3–5–7–8 and the backup TSP on λ_0 through links 1–2–4–6–8. The primary TSP cannot be established on λ_0 on the same path even if there

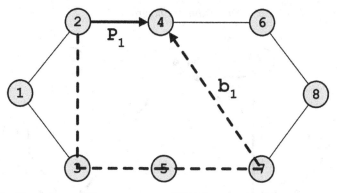

Fig. 12.2. Grooming traffic streams on a WDM network with two wavelengths.

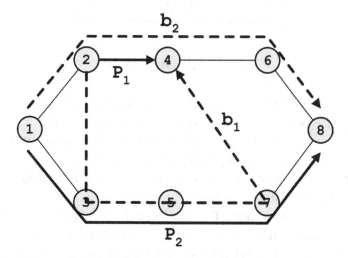

Fig. 12.3. Primary backup multiplexing.

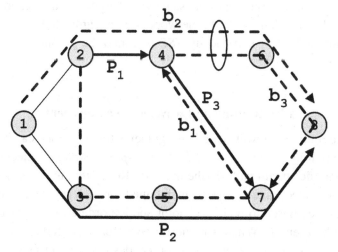

Fig. 12.4. Capacity shared on links 4–6 and 6–8 between b_2 and b_3, on λ_0.

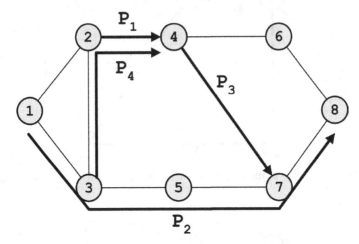

Fig. 12.5. Primary routing.

is capacity available. This is due to the cross-connect constraints at nodes 3 and 7. Specifically, if p_2 were to be established on λ_0 through links 1–3–5–7–8, then the backup TSP b_1 would be disturbed. However, notice that the wavelength channel on λ_0 on link 2–4 carries both a primary and backup TSP.

When the third request for a dependable connection from node 4 to node 7 arrives, the primary TSP is established on link 4–7 on λ_1. However, this primary TSP is link-disjoint with primary TSP p_2. Hence, the backup TSP b_3 can be backup-multiplexed with b_2, that is, they can share the same capacity on links 4–6 and 6–8. In contrast, the backup TSP b_4 between nodes 3 and 4 cannot be backup-multiplexed with b_1 since the primary TSP p_1 is not link-disjoint with primary TSP p_4. Hence, even if they can be groomed onto the same wavelength, λ_0, they cannot share the capacity on the wavelength. If the network uses a segregated grooming policy, then the primary TSP p_1 and backup TSP b_2 cannot be groomed on to the same wavelength λ_0.

12.3 Routing and wavelength assignment

For dynamic routing and wavelength assignment strategies, one can basically use the adaptive routing approach of *fixed alternate-path routing* to select the primary and backup traffic stream paths in the network. Recall that in this scheme, a fixed set of predetermined routes, say k, is maintained for each source–destination pair. These routes are chosen to be link-disjoint and are ordered in non-decreasing order of their hop length. When a request arrives for a dependable traffic stream connection for a capacity c between a source–destination pair (s, d), one selects the two shortest paths out of the k disjoint paths on which the primary and the backup TSP can be established. The primary path can be established if the free capacity on at least one wavelength on each of the links of the path is greater than or equal to the required capacity. The backup path can be established if the sum of the free capacity and the shareable backup capacity available through backup multiplexing on at least one wavelength on the links of the path is greater than or equal to the required capacity. In addition, it should be ensured that the traffic streams can be groomed onto the wavelengths without disturbing the existing traffic streams. The traffic request is assumed to be blocked if it is not possible to establish either the primary or the backup on any of the k predetermined paths. Another approach to adaptive routing is fully-adaptive routing, in which the route is determined dynamically at the time of the connection request. Although this can give better performance, it is slow and the gain may not be worth the trouble and delay. Therefore, one may restrict oneself to the fixed alternate-routing strategy.

Note that it is possible to establish a backup TSP on a path on which the corresponding primary TSP might be blocked. This is due to the fact that the backup TSP can be established by sharing its capacity with the existing backup capacity on the wavelengths even if free capacity is not available. On the other hand, the primary path does need to have enough free capacity on the wavelengths of the links to establish the connection.

12.3.1 Wavelength assignment

The wavelength assignment problem also needs attention for both the primary and backup traffic streams depending on whether the grooming policy used in the network is either *mixed* or *segregated*. If the SGP is used in the network, each wavelength is classified as being either a *primary*, *backup*, or *free* wavelength. A *free* wavelength is one that has no traffic streams multiplexed on it. A *primary* (*backup*) wavelength is one that is composed of one or more primary (backup) traffic streams multiplexed onto it. On the other hand, such a classification is not necessary when the MGP is used in the network, as both primary and backup traffic streams are multiplexed onto the wavelengths. Before applying the wavelength assignment algorithm, the set of available wavelengths on the primary and backup paths needs to be obtained. The procedures described in Section 12.1 are used for each wavelength on each link, to obtain the set of wavelengths on the primary (or backup) path which can accommodate the primary (or backup) traffic stream. In addition, if the SGP is used in the network, the set of wavelengths for the primary path is restricted to those wavelengths which are either *primary* or *free*. A similar restriction applies to the set of wavelengths for the backup path. There is, however, no such restriction if the MGP is used in the network.

If no wavelengths are available for either the primary or backup path, then an alternate route is selected according to the procedure described earlier. Typically one can use one of the following three simple wavelength assignment heuristics to route the primary and backup traffic streams.

(i) First fit (FF): in this case, the first-fit wavelength assignment strategy is used for both primary and backup paths and the wavelength chosen for the traffic stream connection has the smallest index among the set of available wavelengths along the path.

(ii) First fit–last fit (FF–LF): in this case, first-fit wavelength assignment strategy is used for the primary path. However, to assign the wavelengths for the backup traffic stream path, choose the wavelength with the largest index among the set of available wavelengths along the path. The basic idea here is to reduce the conflict between the primary and backup traffic streams.

(iii) Best fit (BF): use the best-fit wavelength assignment for each of the primary and backup traffic stream paths. In the case of the primary traffic stream path, among the available wavelengths in the set, the traffic stream is assigned to that wavelength which has the least free capacity remaining after the traffic stream has been accommodated. For wavelength assignment of the backup traffic stream, select the wavelength that minimizes the total *additional* backup capacity that is needed to set up the backup traffic stream on the links of the path.

More complex algorithms such as max-sum and RCL may be applied and may yield better performance. Since grooming here considers one-to-one path protection, it is possible that the wavelengths of the primary and the backup can be different and their assignments are made independent of each other. In the case of a failure, the source and destination tune to the backup wavelength and switch to the backup path. In addition, the cross-connects at the intermediate nodes of the backup path need to be appropriately configured to activate the backup path.

If the SGP is used, the cross-connects at the intermediate nodes of the backup path need to be appropriately configured to set up the lightpaths that activate the backup path. However, if the MGP is used in the network, the lightpaths are already configured and the backup traffic stream path is ready for use.

12.4 Effect of traffic grooming

The performance of the two grooming policies with the wavelength assignment algorithms is evaluated using a simulation study in [280]. The results are based on the simulations of 10^6 calls each for the three topologies: a 16-node mesh-torus, an eight-node ring and the 14-node NSFnet. The dynamic traffic conditions, the network use, and the traffic assumptions are similar to those given in Section 10.1. A fixed alternate-path routing strategy is used and the overall network blocking performance is used as the metric for comparing the effect of wavelength assignment,

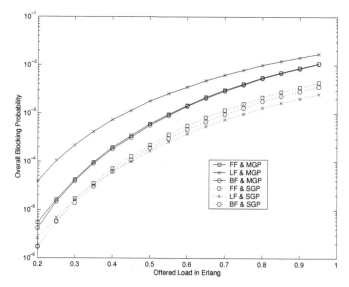

Fig. 12.6. Overall blocking probability vs. offered load per station in Erlangs for a 4 × 4 mesh-torus network with $W = 5$ and $g = 4$ for different wavelength assignment and grooming policies (FF: first fit, LF: last fit, BF: best fit, MGP: mixed grooming policy, SGP: segregated grooming policy).

topology, alternate-path routing, and granularity on the grooming policy used in the network.

12.4.1 Effect of wavelength assignment and topology

Figures 12.6–12.8 depict the blocking probability of various wavelength assignment algorithms for the MGP and the SGP against the offered load for a 4 × 4 mesh-torus, NSFnet, and an eight-node ring, respectively. All the figures illustrate the important role played by the grooming policy in determining the blocking performance. The following observations are made.

(i) In the case of the mesh-torus network and the NSFnet, for high loads, the last-fit algorithm offers the best performance when SGP is used as the grooming policy, but offers the worst performance in the case of MGP.

(ii) In the case of the ring, the first-fit algorithm offers the best performance when the grooming policy is MGP and the worst performance in the case of SGP.

(iii) Overall, for both the mesh-torus and the NSFnet, SGP offers better performance (more than an order of magnitude decrease in blocking probability for the NSFnet). However, in the case of the eight-node ring, it is interesting to note that MGP performs better.

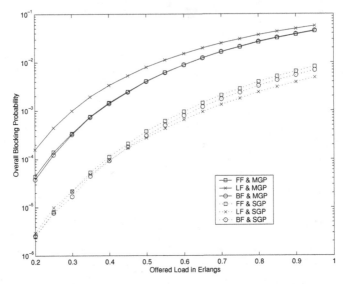

Fig. 12.7. Overall blocking probability vs. offered load per station in Erlangs for a 14-node NSFnet network with $W = 5$ and $g = 4$ for different wavelength assignment and grooming policies (FF: first fit, LF: last fit, BF: best fit, MGP: mixed grooming policy, SGP: segregated grooming policy).

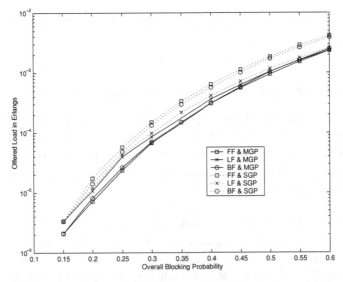

Fig. 12.8. Overall blocking probability vs. offered load per station in Erlangs for an eight-node ring network with $W = 5$ and $g = 4$ for different wavelength assignment and grooming policies (FF: first fit, LF: last fit, BF: best fit, MGP: mixed grooming policy, SGP: segregated grooming policy).

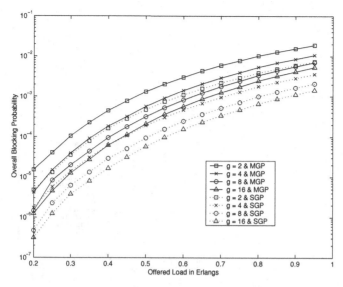

Fig. 12.9. Overall blocking probability vs. offered load per station in Erlangs for a 4×4 mesh-torus network with $W = 5$ and for different granularities (g) and grooming policies (g = granularity, MGP: mixed grooming policy, SGP: segregated grooming policy).

The reason for this reversal in performance is as follows. As explained earlier, using SGP allows one to run the backup traffic streams over the wavelengths without configuring the cross-connects. Mixed grooming, on the other hand, requires one to configure the cross-connects if the wavelengths carry primary traffic along with the backup traffic. Therefore, although MGP offers the inherent advantage of being able to mix primary and backup streams in the wavelengths, it restricts the amount of sharing of backup traffic stream paths that can potentially be done in the network.

(iv) The mesh-torus network has more connectivity than the ring. This helps to generate a good amount of traffic switching and mixing at the nodes. This traffic switching and mixing translates to more sharing of backup streams for the SGP. Therefore, SGP provides better performance for the mesh-torus. On the other hand, in the case of the ring, due to the limited number of paths between the nodes and the high load correlation, most of the traffic is just pass-through and there is very little mixing and traffic switching between wavelengths and the potential for backup traffic stream sharing is low. Hence in this case, MGP provides better performance than SGP.

12.4.2 Effect of the grooming granularity

Figures 12.9, and 12.10 depict the blocking probability for different granularities for both cases of the MGP and the SGP, against the offered load. In the case of both the mesh-torus and the ring, it is observed that as the granularity is increased,

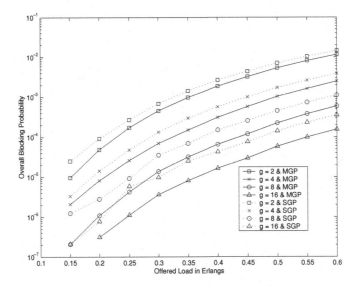

Fig. 12.10. Overall blocking probability vs. offered load per station in Erlangs for an eight-node ring network with $W = 5$ and for different granularities (g) and grooming policies (g = granularity, MGP: mixed grooming policy, SGP: segregated grooming policy).

the blocking probability decreases. In addition, it is observed that one can increase the granularity for a mesh-torus network with the MGP (or SGP in the case of a ring) and make it perform better than a mesh-torus network with an SGP of lower granularity (or correspondingly a ring network with an MGP of lower granularity). A small increase in the difference in blocking probabilities for the MGP and the SGP when the granularity is increased for both the ring and the mesh-torus is also observed.

12.4.3 Effect of the number of alternate paths

Figure 12.11 considers the effect of changing the number of alternate paths for the primary and backup traffic streams in a 4×4 mesh-torus. It is readily observed that a significant decrease in blocking probability can be achieved with even a small increase in the number of alternate paths, although the number of alternate paths which a node-pair can have is limited by the topology. In addition, it is also observed that with an increase in the number of alternate paths, the difference in blocking performance between the MGP and the SGP approaches increases.

The most important observation from the above performance evaluation is that using the SGP provides better performance for the mesh-torus and the NSFnet, while using the MGP provides better performance for the ring. Thus the SGP is

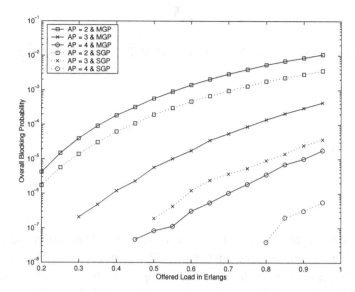

Fig. 12.11. Overall blocking probability vs. offered load per station in Erlangs for a 4 × 4 mesh-torus network with $W = 5$ and $g = 4$ for different numbers of alternate paths and grooming policies (AP = number of alternate paths available for an s-d pair, MGP: mixed grooming policy, SGP: segregated grooming policy).

useful for topologies with good connectivity and a good amount of traffic switching and mixing at the nodes. On the other hand, the MGP is useful for topologies such as rings that have high load correlation and low connectivity. It is also interesting to note that the performance improvement of the SGP over the MGP in the mesh-torus increases as the number of alternate paths for the primary and backup traffic streams is increased.

13

Static survivable grooming network design

Various lightpath protection schemes for a survivable WDM grooming network with dynamic traffic were investigated in Chapter 12. The nodes in the WDM grooming network are assumed to include *ADM (add–drop multiplexer)-constrained grooming nodes*. This chapter deals with the static survivable WDM grooming network design with *wavelength continuity constrained grooming nodes*. For static traffic the problem of *grooming* subwavelength level requests in mesh-restorable WDM networks, the corresponding path selection and wavelength assignment problems are formulated as ILP optimization problems.

13.1 Design problem

To address the survivable grooming network design problem, a network with W wavelengths per fiber and K disjoint alternate paths for each s-d pair can be viewed as $W \times K$ networks, with each of them representing a single wavelength network. For $K = 2$, the first W networks contain the first alternate path for each s-d pair on each wavelength. We number the networks from 1 to W, according to the wavelengths associated with them. The second set of W networks contain the second alternate path for each s-d pair on each wavelength. These networks are numbered from $W + 1$ to $2W$, where the $(W + i)$th network represents the same wavelength as the ith network, $i = 1, 2, \ldots, W$. Figure 13.1 illustrates this layered model for a six-node network with three wavelengths and two link-disjoint alternate paths. For each node-pair, it also depicts routing of two alternate paths for two connections in the network. Notice from Fig. 13.1, for example, when a primary path is selected in networks 1 to W for a request, its corresponding backup paths can only be selected in networks $W + 1$ to $2W$, to guarantee that the primary and backup paths are link-disjoint. Then the problem is to find two alternate paths for each connection request while optimizing the resource utilization.

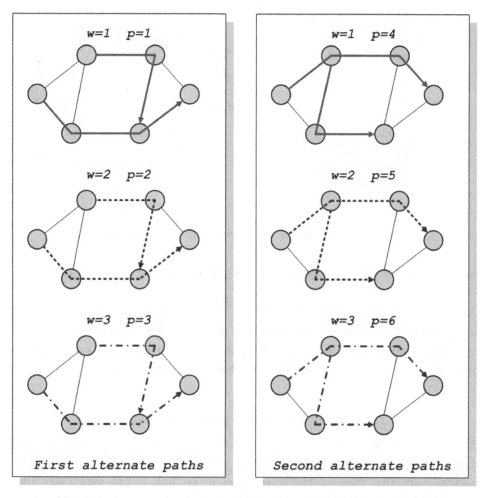

Fig. 13.1. An example of a layered network model with $W = 3$, $K = 2$.

In this formulation, a 100% restoration guarantee for any single-link failure is considered. This implies that the primary (working) paths and the restoration (backup) paths are link-disjoint and are assigned the same capacity, assuming that it is possible in the given network topology.

13.1.1 Dedicated backup reservation

One simple and effective way of assigning backup capacities is to reserve dedicated capacity for each backup path; while choosing primary paths, instead of simply choosing the shortest path, to *minimize* the total *link-primary-sharing* (MLPS)

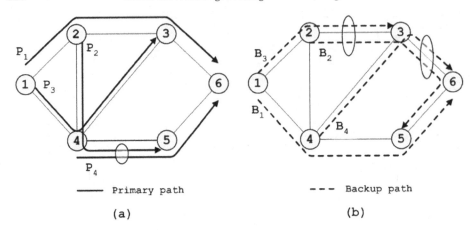

Fig. 13.2. An example of a survivable grooming network.

would be another preferred choice. The link-primary-sharing is defined as follows:

$$s_l = \max(0, P_l - 1) \tag{13.1}$$

where s_l denote the link-primary-sharing of link l and P_l denotes the total number of primary paths that utilize link l. Sharing on link l, s_l can be viewed as the penalty assigned to link l when it is used by more than one primary path.

13.1.2 Capacity reservation with backup multiplexing

As noted earlier in WDM grooming networks, the capacity reserved for restoration paths is more complicated. Let $B = \{b_1, b_2, \ldots, b_k\}$ denote the set of backup paths that traverse the wavelength w on link l. Let their respective capacities be $D = \{d_1, d_2, \ldots, d_k\}$, and their respective primary paths be $P = \{p_1, p_2, \ldots, p_k\}$. If none of the p_i have common links, the capacity required on w is $\max(d_1, d_2, \ldots, d_k)$. If some of the p_i have common links, their backup paths can still be groomed on wavelength w. However, the capacity to be reserved must be up to the summation of their capacities. The primary paths can be grouped according to their common links. Let $P^l = \{p_1^l, p_2^l, \ldots, p_a^l\}$ denote the group of primary paths that have link l as their common link. The capacity required by this group for backup of link l is then given by $D^l = (d_1^l + d_2^l + \cdots + d_a^l)$ when backup capacity sharing is allowed. It is possible that one primary path belongs to more than one group. The reserved capacity on wavelength w on link l is therefore the maximum value of the capacities required by all the groups, that is $D = \max(D^l)$. This is illustrated in Fig. 13.2. In this figure, link $2 \rightarrow 3$ is shared by backup B_2 and B_3. Since their primary paths are link-disjoint, the backup capacity that needs to be reserved is $\max(C(B_2), C(B_3))$. A similar situation exists on link $3 \rightarrow 6$

that is shared by backups B_2 and B_4. But their primaries are not link-disjoint, therefore the backup capacity that needs to be reserved is $C(B_2) + C(B_4)$ on link $3 \rightarrow 6$.

Backup multiplexing and dedicated backup reservation schemes with MLPS are formulated as ILP optimization problems in the following section. The following assumptions are made.

(i) The network is a single-fiber general mesh network.
(ii) A connection request cannot be divided into several lower-speed connection requests and routed separately from the source to the destination. The data traffic on a connection request should always follow the same route.
(iii) The transceivers in a network node are fixed, hence the wavelength continuity constraint applies.
(iv) Each grooming node has unlimited multiplexing and demultiplexing capability. This means that the network node can multiplex/demultiplex as many low-speed traffic streams to a lightpath as needed, as long as the aggregated traffic does not exceed the lightpath capacity.

The following notation is used in the formulations of the network.

- The physical topology of a WDM network is represented as a weighted directed graph $G_p = (V, E)$ with V as the set of network nodes and E as the set of physical links (edges). $|V| = N$ and $|E| = L$. Nodes correspond to network nodes and links correspond to the fibers between nodes. The weights of the links are their costs.
- W: maximum number of wavelengths in each direction in a bidirectional fiber (technology-dependent data).
- C: maximum capacity of each wavelength. (It is assumed that each wavelength has the same capacity.)
- $m, n, s, t = 1, 2, \ldots, N$: number assigned to each node in the network.
- $l = 1, 2, \ldots, L$: number assigned to each link in the network.
- $w = 1, 2, \ldots, W$: number assigned to each wavelength.
- $i, j = 1, 2, \ldots, N \times (N - 1)$: number assigned to each demand (s-d pair).
- $D_{N \times N} = \{d_i\}$: traffic matrix, where d_i indicates the required capacity of low-speed traffic requests in units of OC-1.
- $K = 2$: number of alternate routes between every s-d pair.
- $p, r = 1, 2, \ldots, KW$: number assigned to a path for each s-d pair. A path has an associated wavelength (lightpath). Each route between every s-d pair has W wavelength continuous paths. The first $1 \leq p, r \leq W$ paths belong to route 1 and $W + 1 \leq p, r \leq 2W$ paths belong to route 2.
- $\bar{p}, \bar{r} = 1, 2, \ldots, KW$: if $1 \leq p, r \leq W$ (route 1), then $W + 1 \leq \bar{p}, \bar{r} \leq 2W$ (route 2) and vice versa.
- $\psi_w^{i,p}$: wavelength indicator that takes a value of one if wavelength w is used by the path (i, p) and zero otherwise (data).

The following cost parameter is employed.

- C_l: cost of using a link.

The following information is given regarding link usage and whether two given paths are link- and node-disjoint.

- $\epsilon_l^{i,p}$: link indicator that takes a value of one if link l is used in path $(i,\,p)$ and zero otherwise (data).
- $I_{(i,p),(j,r)}$: takes a value of one if paths $(i,\,p)$ and $(j,\,r)$ have at least one link in common and zero otherwise. If two routes share a link, then all lightpaths using those routes have the corresponding I value set to one and zero otherwise (data).

The following variables are used for path-related information.

- $\delta^{i,p}$: path indicator that takes a value of one if $(i,\,p)$ is chosen as a primary path and zero otherwise (binary variable).
- $v^{i,r}$: path indicator that takes a value of one if $(m,\,r)$ is chosen as a restoration path and zero otherwise (binary variable).

The following variables are used to present wavelength assignment in this grooming network.

- $p_{l,w}^i$: binary variable, one if wavelength w on link l is used by a primary path of demand i and zero otherwise.
- $r_{l,w}^i$: binary variable, one if wavelength w on link l is used by a backup path of demand i and zero otherwise.
- W_l: a non-negative integer, the total number of wavelengths required on link l.
- $M_{l,w}$: non-negative integer, the total capacity assigned to primary paths on wavelength w on link l.
- $R_{l,w}$: a non-negative integer, the total capacity reserved for backup paths on wavelength w on link l.

13.1.3 ILP I: backup multiplexing

13.1.3.1 Objective: minimize the total wavelength links

Given a network topology and a set of point-to-point demands and their link-disjoint primary and backup routes, assign the primary and backup routes in an optimal way so that the total wavelength links cost is minimized. When $C_l = 1$ the objective is to minimize the total number of *wavelength* $\times$ *links*:

$$\min \sum_{l \in E} C_l \times W_l \tag{13.2}$$

13.1.3.2 Constraints

(i) *On physical route variables:* a lightpath can carry traffic for an s-d pair only if it is in the physical route of this request,

$$p^i_{l,w} = \sum_{p=1}^{KW} \delta^{i,p} \epsilon^{i,p}_l \psi^{i,p}_w \tag{13.3}$$

$$r^m_{ij,w} = \sum_{r=1}^{KW} v^{i,r} \epsilon^{i,r}_l \psi^{i,r}_w \tag{13.4}$$

(ii) *On path indicators:* one and only one path is assigned as a primary (backup) path for each request,

$$\sum_{p=1}^{KW} \delta^{i,p} = 1 \tag{13.5}$$

$$\sum_{r=1}^{KW} v^{i,r} = 1 \tag{13.6}$$

(iii) *On the topology diversity of primary and backup paths:* primary and restoration paths of a given demand should be node- and link-disjoint:

$$\sum_{p=1}^{W} \delta^{i,p} = \sum_{r=W+1}^{KW} v^{i,r} \tag{13.7}$$

$$\sum_{p=W+1}^{KW} \delta^{i,p} = \sum_{r=1}^{W} v^{i,r} \tag{13.8}$$

(iv) *On wavelength capacity variables:* primary capacities are aggregated. For each wavelength, the sum of primary capacities and backup capacities should not exceed the total wavelength capacity,

$$M_{l,w} = \sum_i d_i \times p^i_{l,w} \tag{13.9}$$

$$M_{l,w} + R_{l,w} \leq C \tag{13.10}$$

(v) *On fiber capacity constraints:* the number of wavelengths used on a fiber should not exceed the total number of wavelengths carried by the fiber. Equations (13.12)–(13.14) together set $u_{l,w} = 1$, if $x_{l,w} \geq 1$, and zero otherwise. $x_{l,w}$ counts the number of primary and backup paths that use wavelength w on link l, and W_l counts the number of wavelengths used on link l. Recall that single-fiber networks are assumed here,

$$x_{l,w} = \sum_i \left(r^i_{l,w} + p^i_{l,w} \right) \tag{13.11}$$

$$u_{l,w} \leq x_{l,w} \tag{13.12}$$

$$KN(N-1)u_{l,w} \geq x_{l,w} \tag{13.13}$$

$$u_{l,w} \in \{0, 1\} \tag{13.14}$$

$$W_l \geq \sum_w u_{l,w} \tag{13.15}$$

$$W_l \leq W \tag{13.16}$$

(vi) *On the backup multiplexing constraint:* the capacity reserved for backup paths on a link needs to take the correlations between the corresponding primary paths into account. If the primary paths do not have common links, their backup paths can share the same wavelength on their common links.

The reserved capacity is the maximum requested capacity among them. Otherwise, the capacity for their backups on the same wavelength is also to be aggregated. Recall that $R_{l,w}$ denotes the capacity assigned to backup paths on wavelength w on link l, $R_{l,w}$ is given as

$$
\begin{aligned}
R_{l,w} \geq\ & d_i \times v^{i,p} \epsilon_l^{i,p} \psi_w^{i,p} \\
& + \sum_{j \geq i} d_n \times v^{j,p,i,p} \epsilon_l^{j,p} \psi_w^{j,p} \times I_{(i,\bar{p}),(j,\bar{p})} \\
& + \sum_{j \geq i} d_j \times v^{j,\bar{p},i,p} \epsilon_l^{j,\bar{p}} \psi_w^{j,\bar{p}} \times I_{(i,p),(j,\bar{p})} \\
& + \sum_{j \geq i} d_j \times v^{j,p,i,\bar{p}} \epsilon_l^{j,p} \psi_w^{j,p} \times I_{(i,\bar{p}),(j,p)} \\
& + \sum_{j \geq i} d_j \times v^{j,\bar{p},i,\bar{p}} \epsilon_l^{j,\bar{p}} \psi_w^{j,\bar{p}} \times I_{(i,p),(j,p)}
\end{aligned} \tag{13.17}
$$

where $v^{j,p,i,p}$ is a binary variable which takes a value of one when $v^{j,p} = 1$ and $v^{i,p} = 1$. It is given by Eqs. (13.18)–(13.20),

$$v^{j,p,i,p} \geq v^{j,p} + v^{i,p} - 1 \tag{13.18}$$

$$v^{j,p,i,p} \leq v^{j,p} \tag{13.19}$$

$$v^{j,p,i,p} \leq v^{i,p} \tag{13.20}$$

13.1.4 ILP II: dedicated backup with MLPS

13.1.4.1 Objective

Minimize the total number of wavelength links as well as the total link-primary-sharing. Recall that s_l denotes the link-primary-sharing on link l. Let C_{share}^l be the weight of s_l. The objective function is hence given by

$$\min \left(\sum_{l \in E} C_l \times W_l + C_{\text{share}}^l \times s_l \right) \tag{13.21}$$

13.1.4.2 Constraints

The constraints specified in Eqs. (13.3)–(13.15) are still applicable, only the backup capacities are calculated in a different way.

(vii) *On backup wavelength capacity variables:* backup capacities are aggregated when dedicated backup reservation is applied,

$$R_{l,w} = \sum_i d_i \times r^i_{l,w} \tag{13.22}$$

(viii) *On link-primary-sharing:* recall the definition of s_l in Section 13.1.1, s_l is non-negative and given as follows:

$$s_l \geq \sum_i \sum_w p^i_{l,w} - 1 \tag{13.23}$$

$$s_l \leq \sum_i \sum_w p^i_{l,w} \tag{13.24}$$

13.2 Example

The utility of the above ILP formulations is demonstrated using the physical topologies depicted in Figs. 13.3(a) and (b).

The grooming performance depends on the efficiency of *grooming* fractional wavelength traffic onto a full or almost-full wavelength. Hence, the overall efficiency also depends on the traffic pattern. When most of the traffic is of full-wavelength or almost full-wavelength capacity, grooming does not bring much improvement to the wavelength utilization. In this example traffic is randomly generated with each call request having a capacity of OC-12, which is a quarter of the full wavelength capacity. Two link-disjoint alternate paths for each connection are pre-computed based on the fixed shortest-path routing algorithm.

CPLEX Linear Optimizer 7.0 is used to solve ILP formulations I and II. Tables 13.1 and 13.2 show the path selection and wavelength assignment results of the same set of requests on topology given by Fig. 13.3(a) for the two ILP formulations.

From the tables it is observed that 21 wavelength links are needed to carry all 15 requests. The solution for the same request set in the network without traffic grooming capability is obtained from formulation I, as a special case where each request has full wavelength capacity. The results are shown in Table 13.3. It turns out that a minimum of 52 wavelength links are required in the network without traffic grooming capability. This represents a significant saving.

From precomputed path sets, the maximum number of wavelength links that are needed to establish all the primary and backup paths can be calculated. Notice that without traffic grooming and backup multiplexing, 64 wavelength links are

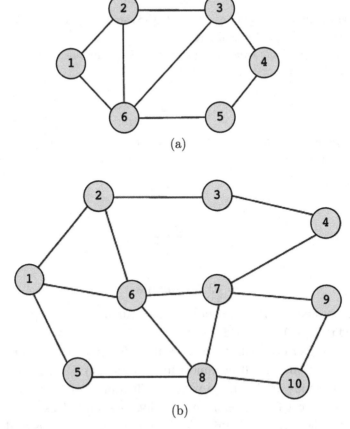

Fig. 13.3. Physical topologies used in experiments.

needed, while backup multiplexing helps to reduce this to 52. The gain in using backup multiplexing is thus 18.75%, and eight wavelength links are saved.

With subwavelength traffic grooming, 21 wavelength links are sufficient. This means that another 31 wavelength links are saved. When taking into account the wavelength capacity granularity, the total required capacity is $64/4 = 16$ OC-12 capacity units. Without grooming, each lightpath uses the full OC-48 capacity, although the requested capacity is OC-12, so in total 52 OC-48 capacity units have been occupied. With traffic grooming, 21 wavelength links have been used. It is still possible to pack other lightpaths onto some wavelengths even without taking backup multiplexing into account, because some wavelengths still have free bandwidth. The total used capacity is exactly 16 OC-12 capacity units. This example clearly demonstrates the improvement of capacity utilization by enabling subwavelength level grooming in the restorable WDM network design.

Table 13.1. *Solution from the backup multiplexing ILP formulation*

s-d pair	Formulation I			
	Primary		Backup	
1–3	1–2–3	w_3	1–6–3	w_2
1–4	1–2–3–4	w_3	1–6–5–4	w_2
1–5	1–2–3–4–5	w_3	1–6–5	w_2
2–4	2–3–4	w_2	2–6–5–4	w_4
2–5	2–6–5	w_4	2–3–4–5	w_3
2–6	2–6	w_4	2–1–6	w_2
3–5	3–4–5	w_3	3–6–5	w_4
3–6	3–2–6	w_4	3–6	w_4
4–2	4–3–2	w_4	4–5–6–2	w_2
4–5	4–3–6–5	w_4	4–5	w_2
5–2	5–6–2	w_2	5–4–3–2	w_4
5–3	5–4–3	w_4	5–6–3	w_2
6–1	6–2–1	w_2	6–1	w_1
6–3	6–3	w_2	6–2–3	w_2
6–4	6–3–4	w_2	6–5–4	w_2

Table 13.2. *Solution from the dedicated backup ILP formulation*

s-d pair	Formulation II			
	Primary		Backup	
1–3	1–2–3	w_3	1–6–3	w_3
1–4	1–6–5–4	w_3	1–2–3–4	w_3
1–5	1–6–5	w_3	1–2–3–4–5	w_3
2–4	2–3–4	w_2	2–6–5–4	w_1
2–5	2–6–5	w_1	2–3–4–5	w_2
2–6	2–6	w_1	2–1–6	w_3
3–5	3–4–5	w_2	3–6–5	w_1
3–6	3–6	w_1	3–2–6	w_1
4–2	4–3–2	w_1	4–5–6–2	w_3
4–5	4–5	w_2	4–3–6–5	w_1
5–2	5–6–2	w_3	5–4–3–2	w_1
5–3	5–6–3	w_3	5–4–3	w_1
6–1	6–1	w_1	6–2–1	w_3
6–3	6–3	w_3	6–2–3	w_3
6–4	6–5–4	w_3	6–3–4	w_3

Table 13.3. *Solution without traffic grooming*

s-d pair	No traffic grooming			
	Primary		Backup	
1–3	1–2–3	w_6	1–6–3	w_6
1–4	1–2–3–4	w_3	1–6–5–4	w_1
1–5	1–6–5	w_3	1–2–3–4–5	w_2
2–4	2–3–4	w_4	2–6–5–4	w_7
2–5	2–6–5	w_2	2–3–4–5	w_7
2–6	2–6	w_8	2–1–6	w_2
3–5	3–6–5	w_8	3–4–5	w_5
3–6	3–6	w_6	3–2–6	w_7
4–2	4–3–2	w_6	4–5–6–2	w_2
4–5	4–5	w_4	4–3–6–5	w_7
5–2	5–6–2	w_4	5–4–3–2	w_7
5–3	5–6–3	w_5	5–4–3	w_1
6–1	6–1	w_3	6–2–1	w_2
6–3	6–3	w_4	6–2–3	w_2
6–4	6–5–4	w_5	6–3–4	w_6

Table 13.4. *Requests matrix for the 10-node 14-link network*

	1	2	3	4	5	6	7	8	9	10
1	0	0	0	12	1	0	0	0	0	0
2	1	0	0	0	0	0	0	0	0	12
3	0	3	0	0	0	0	0	0	0	0
4	0	0	0	0	3	1	0	3	12	0
5	0	0	0	0	0	0	0	0	1	0
6	0	0	3	0	0	0	0	0	0	0
7	0	0	0	0	0	0	0	0	3+1	0
8	1	0	12+12	0	0	0	1	0	0	0
9	0	3	0	0	12	3+3	0	0	0	0
10	3	0	0	0	0	0	0	0	0	0

Although, in the above example, backup multiplexing and dedicated backup with MLPS perform the same in terms of wavelength links, this does not always happen. In this specific scenario MLPS is preferred because fewer working paths are touched by single-link failures. For example, the failure of link $(2, 3)$ would affect four working paths as shown in Table 13.1 and two working paths as shown in Table 13.2. Additionally, with the objective of minimizing the total number of wavelength links, backup multiplexing stops when the objective value no longer decreases. It is still possible to reallocate some primary paths so that there could be more chance of multiplexing backup paths onto some wavelength, and result in more spare capacity on the utilized wavelengths. But the value of the objective function will stay the same.

Different path selections can be observed in Tables 13.1 and 13.2. In order to simply minimize the total wavelength links, grooming tends to exhaust one wavelength before using another wavelength. While the link-primary-share is taken as a link penalty, in formulation II, the preference would be to have a more balanced load for primary paths.

Experiments are performed on the topology in Fig. 13.3(b), which is a 10-node network with 14 bidirectional links. The randomly generated request matrix is shown in Table 13.4.

The solution from formulation I shows that by employing the backup multiplexing technique 28 wavelength links are needed, while formulation II yields a solution that requires 33 wavelength links. In general, formulation II requires more wavelength links in comparison to formulation I. However, this becomes affordable in networks with subwavelength grooming capability, where the wavelength utilization is significantly improved by traffic grooming. Moreover, with respect to the IL formulation, formulation II is less complex than formulation I in terms of the number of constraints and variables, which makes formulation II less computationally expensive and hence more practical.

14

Trunk-switched networks

Previous networks used electronics for both the medium of transmission and the processing technology. Hence, the transmission and processing bandwidths at nodes were approximately of the same order. Electronic technology advanced simultaneously on the transmission and processing sides, leading to a matched growth in the evolution of the networks. With the shift to optical technology, the transmission capacity has taken a quantum leap while the processing capacity has seen only modest improvements in electronics. Optical processing is currently in its infancy and therefore the backbone networks are likely to remain circuit-switched with the possibility of having optical switching at intermediate nodes.

The increase in the transmission capacity in terms of multiple wavelengths each operating at a few tens of gigabits per second with multiple time slots within a wavelength requires an equivalent increase in the electronic processing for efficient operation of the networks. However, it is impractical to match the power of the optical technology with that of electronics if the nodes were to process all the information that is received from different links they are connected to. Hence, the switching trends depend on having multiple simple processing devices that work independently on parts of the information that is received at a node. Such a network model is referred to as a *trunk-switched network* (TSN). A TSN is a two-level network model in which a link is considered as multiple channels and channels are combined together to form groups called *trunks*. This conceptual architecture is capable of grooming subwavelength level traffic over a link.

14.1 Channels and trunks

A trunk-switched network consists of nodes interconnected by links. Note that in an optical network, a link may consist of multiple fibers, multiple wavelengths per fiber, and multiple time slots (or CDMA codes) per wavelength, as shown in Fig. 14.1.

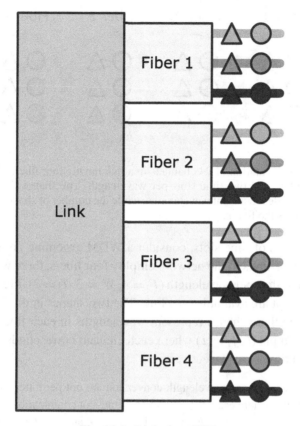

Fig. 14.1. Links in a TSN.

In a TSN, each link ℓ has a set of channels denoted by C_ℓ. A node views a link connected to it as groups of channels called a *trunk*. A *trunk* at a node is defined as a unique non-null set of channels in a link. The number of trunks as viewed by a node i is denoted by K_i. The set of channels of a link ℓ that fall within a trunk x at node i is denoted by $\chi^i_{\ell,x}$. Let $S^i_{\ell,x} = |\chi^i_{\ell,x}|$ denote the number of channels in the set $\chi^i_{\ell,x}$. With the above definition of a trunk, we have $\chi^i_{\ell,x} \cap \chi^i_{\ell,y} = \emptyset$.

14.2 Modeling a WDM grooming network as a TSN

A single-fiber wavelength-routed WDM network employing W wavelengths can be modeled as W trunks with one channel per trunk. A multi-fiber multi-wavelength network with F fibers and W wavelengths and no wavelength conversion can be viewed as W trunks with F channels per trunk each consisting of time slots on a wavelength on all fibers. If full-wavelength conversion is available, then a link can be viewed as a single trunk with FW channels. However, networks that employ limited-wavelength conversion [235, 287] cannot be modeled easily or effectively as a TSN, as full-permutation wavelength conversion is not employed.

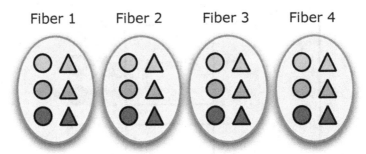

Fig. 14.2. Representation of 24 channels in a link having four fibers, three wavelengths per fiber, and two time slots per wavelength. The shapes represent time slots and the shades represent wavelengths, while the number of shapes of a certain shade represents the fibers.

To see the effects of time slots, consider a WDM grooming network as shown in Fig. 14.1. Let the links in the network employ four fibers, three wavelengths per fiber and two time slots per wavelength ($F = 4, W = 3, T = 2$). Figure 14.2 shows the 24 channels that are available on a link. The two shapes in the figure represent time slots and the three shades represent wavelengths in each fiber. A channel is represented by a tuple (l, f, w, t) where each element corresponds to a link, fiber, wavelength, and time slot, respectively.

- If time-slot interchange and wavelength conversion are not permitted, a node i views a link ℓ as WT trunks where each wavelength and time-slot combination forms a trunk, i.e. $\chi^i_{\ell,(w,t)} = \{(l, f, w, t) \mid 1 \le f \le F\}$, where $1 \le w \le W$ and $1 \le t \le T$. Every trunk has F channels as shown in Figure 14.3(a).
- If time-slot interchange is permitted, but not wavelength conversion, a node i views a link ℓ as W trunks where each wavelength forms a trunk, i.e. $\chi^i_{\ell,w} = \{(l, f, w, t) \mid 1 \le t \le T \text{ and } 1 \le f \le F\}$, where $1 \le w \le W$. Every trunk has FT channels as shown in Figure 14.3(b).
- If full-wavelength conversion is permitted, but not time-slot interchange, then for a given link l, a time slot on all the wavelengths can be grouped to form a trunk, i.e. $\chi^i_{\ell,t} = \{(l, f, w, t) \mid 1 \le w \le W \text{ and } 1 \le f \le F\}$, where $1 \le t \le T$. Every trunk has FW channels as shown in Figure 14.3(c).
- If both full-wavelength conversion and time-slot interchange are permitted, then the entire link is treated as one trunk with FWT channels, as shown in Figure 14.3(d).

14.3 Node architecture

Nodes in a TSN view the links as a set of trunks. Besides being a source or a destination of a connection in a network, a node can also act as an intermediate switching node for other connections that pass through it. Hence, the functionality of a node includes switching of channels from one link to another in order to

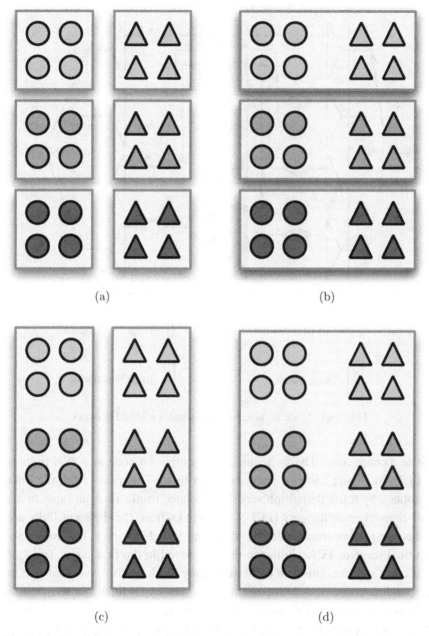

Fig. 14.3. Possible grouping of the channels in a link as trunks. (a) Wavelength–time-slot trunk; (b) wavelength trunk; (c) time-slot trunk; and (d) link is a trunk.

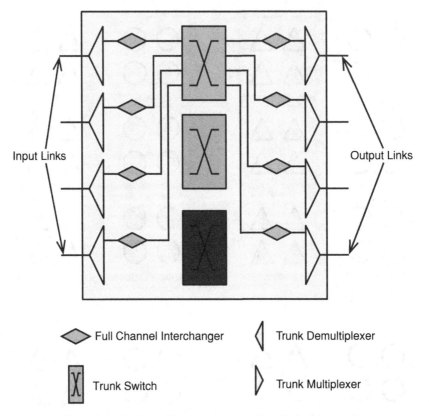

Fig. 14.4. Node architecture in a trunk-switched network.

facilitate a connection. The switching architecture of a node in a TSN is shown in Fig. 14.4. The figure shows a node with four links. The trunks from the links are first isolated by trunk demultiplexers. The isolated trunks form an input to a stage of full-channel interchangers (FCIs). The trunks from the different links are fed into their respective trunk switches. After the switching stage, the trunks are fed into a final stage of FCIs similar to that employed in the first stage. Trunks from the different switches are then combined using trunk multiplexers and sent out to corresponding output links. One of the input and output links at a switching node is dedicated to sourcing and sinking its own traffic.

The switching architecture employed at each node in a TSN obeys the following two conditions.

- A full-channel interchanger is employed for every trunk at the input and output stages of the node, as shown in Fig. 14.4.
- Switching at a node obeys the trunk-continuity constraint, i.e. the channels cannot be switched across trunks. Therefore, the trunk switches at a node function independently.

A full-channel interchanger has the ability to convert any channel within the trunk on a link to any other channel within the same trunk on that link. Note that an FCI switches channels within the same trunk on a link and does not have the capability to switch channels across links. This functionality allows a node to operate in an autonomous manner. For example, it is possible to rearrange the channel assignments at a node without involving co-ordination with neighboring nodes in order to optimize the network operation as long as the channel assignments on the links are kept the same. Another use of such a functionality is to employ specialized monitoring functions on a specific channel on the input link. The ability to rearrange the channels inside the node allows the node to monitor different channels at different instants of time without co-ordinating with the neighboring nodes. The second constraint is employed due to the limited switching resources at a node. It is also assumed that every trunk switch at a node has the same architecture.

The definition of a trunk could be different at different nodes. A TSN is said to be *homogeneous* if the collection of channels that constitute a trunk at a node is the same for all the nodes in the network. Otherwise, it is said to be *heterogeneous*. It should be noted that the above definition does not specify any constraints on the switch architecture employed for a trunk at the node. The architecture of the trunk switches can be different at different nodes, although all the nodes in the network view the links in the same manner.

14.4 Free and busy trunks

A channel on a link is said to be *busy* if it is allocated for a connection. Otherwise, it is said to be free. A trunk on a link at a node is said to be busy if all the channels on the trunk are busy. Otherwise, it is said to be free. The number of channels that are busy on a trunk at the input of a node is the same as the number of channels busy on the trunk at the input to the switch at the node. However, the distribution of busy channels on the trunk at the input of the node may be different from that at the input of the switch. The number of trunks busy at the input of a node is the same as the number of trunks busy at the input of the switch at that node.

Consider a specific trunk at a node that acts as an intermediate switching node for a connection. Figure 14.5 shows the channel occupancy status of the trunk under consideration at the input and the output of the trunk switch at the intermediate node. The node has four incoming links. Note that the channel occupancy described in the figure denotes the channel status after the input FCI stage and before the output FCI stage.

This trunk under consideration is said to be *available* on a two-link path involving a certain link at the input of the switch and a certain link at the output of the switch

Trunk-switchednetworks

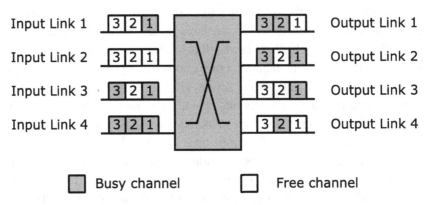

Fig. 14.5. Channel occupancy at the input and output of a trunk switch at an intermediate node on a path. The channel occupancy shown here is after the FCI at the input stage and before the FCI at the output stage.

if any free channel on the input link can be switched to any free channel on the specific output link. For example, if the trunk switch has full-permutation switching capability, then any free channel at any input link can be switched to any free channel at the output link but within the same trunk. In this case, if any of the first three links at the input and any of the output links are chosen, then the trunk is available on the two-link path. However, if link 4 at the input is chosen, then it has no free channels, hence the trunk is not available for the two-link path. When a trunk switch has full-permutation switching capability, it has the freedom to switch the channels in the space and channel domains.

If the trunk switch implements only a space switch, then a free channel on a link at the input must be switched to the same channel at any of the output links. Such a switch does not have the flexibility to switch in the channel domain. Assume that link 1 at the input and link 1 at the output are on two links of a two-link path. At the input of the switch channels 2 and 3 are free while at the output they are not. As the channels can be switched only in the space domain (across links) but not in the channel domain, the trunk cannot be used for establishing a connection on a path using these two links.

A trunk may be free at an input and output link at a node; however, it may not be available on a path including these two links as the trunk switch does not have the capability to switch the free channel at the input to a free channel at the output. On the other hand, if the switching capability is available then a path can be set. For example, if link 1 at the input and link 2 at the output are part of a path, then channel 2 on the trunk is free, hence the trunk is said to be available on the two-link path. A node employing space switches for every trunk is said to employ *channel-space switching* (CS) due to the FCI at the input stage of the node.

14.5 Connection establishment

The primary service provided by a circuit-switched network is to establish a communication path between two nodes in the network. *Requests* or *calls* arrive at nodes in the network that require a connection of a certain bandwidth to another node. Setting up such a communication path between the nodes is referred to as a *connection establishment*.

The two steps in connection establishment are referred to as *routing* and *wavelength assignment* (RWA) in the context of wavelength-routed networks and are solved using integer linear programming (ILP). While such an approach produces an optimal solution for static traffic demands, applying these techniques to dynamic traffic is not practical due to their prohibitively large computation time.

A framework to support a methodology for information collection and routing in optical networks (MICRON) is required for connection establishment in wavelength grooming optical networks. Connection establishment in a circuit-switched network consists of two steps: *path selection* and *resource assignment*. Path selection refers to selecting a path from a source to a destination based on certain criteria. Resource assignment refers to assigning one or more channels depending on the requirement of the request on every link of the chosen path.

14.5.1 Path selection

The first step of a connection establishment process, namely path selection, is carried out using one of the following mechanisms.

(i) If a source–destination pair has one preselected path, then it is referred to as a *fixed-path* approach.
(ii) If a path is selected depending on the network status from a preselected set of candidate paths, then it is referred to as a *fixed alternate-path routing* (FAPR) approach. The set of candidate paths remains the same at all times and does not change with the network status.
(iii) Fixed-path least-congestion routing (FPLCR) is an approach in which a path from a source to a destination is selected from a set of precomputed paths. While FAPR attempts the paths in a specified order, FPLCR selects the least-loaded path.
(iv) If the candidate paths are chosen based on the network status and the available paths, then the path selection process is referred to as a *dynamic routing* approach. Dynamic path selection and dynamic routing approaches require use of network state information.

14.5.2 Network state information

Up-to-date network state information is collected in the network either though *link-state* or *distance-vector* protocols. In the link-state protocol, every node in the network transmits to every other node its own state information and information from the links that it is connected to. Hence, every node in the network has precise knowledge of the topology and the current status of the network. The main drawback of this approach is that it is not scalable. As the network size increases, the amount of information that needs to be collected and maintained at a node also increases. Hence, it is impractical to adopt this approach for large networks. Wide area optical networks are expected to employ only a few nodes. Therefore in the context of optical networks, employing link-state protocols is still a preferred method of information collection.

Distance-vector protocols maintain up-to-date network information by exchanging the node and link information with neighbors. Nodes in a network employing such an approach for information collection do not have knowledge of the network topology. Every node maintains a routing table that would indicate one or more preferred neighbors to reach a destination node. The main drawback of this approach is also scalability; however, for a different reason compared to the link-state approach. As the changes in the network information are propagated through a series of neighborhood information exchanges, it takes a significant amount of time for the changes in one part of the network to be reflected in the other regions. The time required for a change in a network to propagate increases with an increase in network size. Also, such an approach does not always necessarily result in convergence of the network state as viewed by all nodes.

14.5.3 Path-selection algorithms

A preferred approach to path selection in large networks is to employ alternate-path selection. Selecting a path from a fixed set of candidate paths requires information to be collected on those candidate pairs. This information can be collected as a part of the connection establishment procedure by sending requests along all the candidate node-pairs. Note that such an approach can be employed after a request arrival at a node; hence the information that is collected can be tailored to the requirements of the request.

Path-selection algorithms are also classified based on how the path-selection decisions are made. If the source node selects a path to the destination, then it is referred to as *source routing*. Note that such a routing in the network requires the source to have complete knowledge about the network to make a decision. Hence, networks that employ source routing also employ the link-state protocol for information collection.

If the path selection is done independently at different nodes, it is referred to as *distributed routing*. In networks that employ distance-vector protocols for information collection, nodes do not have knowledge of the entire network topology.

Hence, requests for a connection to a certain node are forwarded to a preferred neighbor node. The neighboring node then makes a decision on the next hop of the connection. In such an approach, the source node does not have control over the path that is established in the network.

Both of the above-mentioned approaches have their own advantages and disadvantages. In source routing, the source node has complete control over the established path. Such a control is an advantage when the node attempts to include or avoid certain nodes in the path or implement quality of service requirements. However, a major drawback of source routing is the requirement of knowledge of the entire network to select a path. Distributed routing has its advantage in that an intermediate route could reroute the connection request depending on the current network status, thus adapting to the network changes more easily compared to source routing. However, a disadvantage is that the source node does not have control over the path that is chosen.

Path selection can be made either before or after a request arrival. Algorithms that select a path before a request arrival typically select the path such that most of the requests can be accommodated. Such an approach to path selection is referred to as destination-specific, where a path is selected based only on the destination node. Algorithms that select paths after the arrival of a request can prune the set of candidate paths based on the request characteristics. Such algorithms are referred to as request-specific algorithms. The paths that are selected from a specific source to destination could be different for requests with different requirements. The former approach requires continuous monitoring of the network status. Although a path is readily available when a request arrives when a destination-specific approach is employed, a connection may still not be established as changes in the network need not be reflected immediately. If the network has slowly varying dynamics, then such situations are rare. Hence, a destination-specific approach would have a lower connection establishment time compared to a request-specific approach.

14.5.4 Resource assignment

The second step of the connection establishment process, namely resource assignment, can have many choices as well. On a chosen path, one or more channels can be assigned on every link that is either based on some global metric or just based on available resources on the path. For example, in a wavelength-routed network, resource assignment refers to wavelength assignment. Wavelength assignment algorithms such as random wavelength assignment or first-fit wavelength assignment assign resources based on the status of the path. Wavelength assignment algorithms such as most-used or least-used wavelength assignment policies assign a wavelength based on the wavelength usage across the entire network.

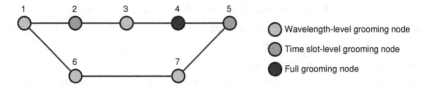

Fig. 14.6. An example network showing two paths from node 1 to node 5.

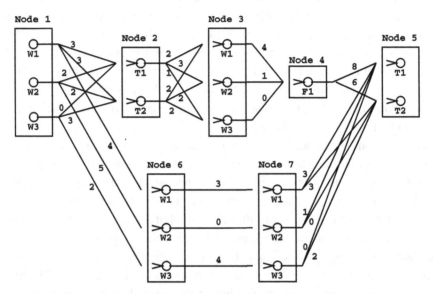

Fig. 14.7. Expanded view of the network with channel occupancy information.

14.6 Grooming network model

Consider a seven-node network as shown in Fig. 14.6. Let the nodes be connected using links employing three fibers each carrying three wavelengths and two time slots per wavelength. Also assume that nodes 1, 3, 6, and 7 are wavelength-level grooming nodes; nodes 2 and 5 are time-slot-level grooming nodes; and node 4 is a full-grooming node. Wavelength-level grooming nodes view the link as three wavelength trunks (denoted by W_1, W_2, and W_3) with six channels in each, time-slot-level grooming nodes view a link as two time-slot trunks (denoted by T_1 and T_2) with nine channels in each, and a full-grooming node views a link as one trunk (denoted by F_1) with 18 channels.

Figure 14.7 shows the expanded view of the network, indicating the different trunks at the nodes. For example, consider trunk W_1 of node 1 and trunk T_1 of node 2. Let ℓ_{12} denote the link that connects node 1 to node 2. The number of channels in the link ℓ_{12} that belongs to both the trunk W_1 at node 1 and T_1 at node 2 is three.

The corresponding three channels are $(\ell_{12}, 1, 1, 1)$, $(\ell_{12}, 2, 1, 1)$, and $(\ell_{12}, 3, 1, 1)$, with each channel belonging to a distinct fiber. The arrow connecting trunk W_1 of node 1 to trunk T_1 of node 2 indicates the number of free channels that belong to both the trunk definitions. A value of 3 indicates that all the channels belonging to both the trunk definitions are free. An arrow connecting a trunk at a node to a trunk at its neighboring node refers to a subtrunk. Note that, if there are no channels present in a subtrunk, then the arrows are not shown. For example, consider nodes 6 and 7. Both the nodes are wavelength-level grooming nodes. Therefore, there are no channels that would belong to a subtrunk Θ_{xy}^{67}, where $x, y \in \{W_1, W_2, W_3\}$ and $x \neq y$. Hence, the corresponding arrows are not shown in the figure.

Assume that the network is observed at an instant of time during its operation and the channel occupancy in the links is known. Let $\psi(l, f, w, t)$ denote the status of the channel: denoted by 0 if occupied by a connection and by 1 if the channel is free. Let $(l, f, w, t).Status$ denote if the channel is assigned for a primary (or working) connection (denoted by PRIMARY) or a backup connection (denoted by BACKUP). Let $(l, f, w, t).BackupList$ denote the set of requests that share the channel (l, f, w, t) for their backup paths. This implies that the primary paths of these requests are link-disjoint.

Consider a request, say r, with a primary path already assigned. Let $(l, f, w, t).LinkDisjoint(r)$ indicate whether or not the request r has its primary path link-disjoint with the primary paths of all the requests that have their backup assigned on the channel (l, f, w, t) (denoted by 1 if it is link-disjoint and 0 otherwise). If the channel is not assigned to either a primary or a backup connection, the value is set to 0. Such information strictly identifies whether the channel can be "shared" with already allocated backup connections.

14.7 MICRON framework

A connection establishment framework includes representation of link information, combining them to obtain the path information, selection of a path, and subtrunk assignment [242, 243, 248]. The steps are described in detail in the following subsections.

The notation employed in the framework is listed in Table 14.1.

14.7.1 Link information

A link in a TSN connects two nodes that could view the channels in a link as different groups of trunks. The information about the channels representing each subtrunk on a link is organized as a matrix. A link connecting node i and j is

Table 14.1. *MICRON framework notation*

Notation	Description
K_i	Number of trunks as viewed by node i
S_i	Number of channels per trunk at node i
$\chi_{\ell,x}^i$	Set of channels on link ℓ that fall within trunk x at node i
$\Theta_{x,y}^{ij}$	Set of channels on link ℓ connecting node i to j that fall within trunk x at node i and trunk y at node j ($= \chi_{\ell,x}^i \cap \chi_{\ell,y}^j$)
L_{ij}	Information matrix for link i–j (dimension $= K_i \times K_j$)
P_{sd}	Path information matrix for a specific path from node s to node d having one or more links (dimension $= K_i \times K_j$)
U_i	Unit row vector at node i (dimension $= 1 \times K_i$); all entries are 1
I_i	Identity matrix at node i (dimension $= K_i \times K_i$)
V_{sd}	Path information vector for a specific path from node s to node d having one or more links (dimension $= 1 \times K_d$); $V_{sd} = U_s P_{sd}$

represented by a matrix L_{ij}:

$$
L_{ij} = \begin{bmatrix}
l_{11} & l_{12} & \cdots & l_{1K_j} \\
l_{21} & l_{22} & \cdots & l_{2K_j} \\
\cdot & & & \\
\cdot & & & \\
l_{K_i 1} & l_{K_i 2} & \cdots & l_{K_i K_j}
\end{bmatrix} \tag{14.1}
$$

where each element l_{xy} denotes a certain property about the channels in the link that belong to the subtrunk Θ_{xy}^{ij}. For example, consider the link 1–2 in the example network shown in Fig. 14.6. Node 1 views each wavelength as a trunk, hence it has three trunks. Node two views each time slot as a trunk, hence it has two trunks. Hence, L_{12} is a 2×3 matrix.

The matrix can denote different properties of the channels that belong to a certain subtrunk. Specific examples effectively illustrate the use of the framework.

Case 1: connectivity. Assume that a connection request that arrives at a node requires a bandwidth of B. Each channel has a capacity of 1. In order to determine if a link has enough capacity in each of its subtrunks, every element l_{xy} of the matrix L_{ij} is denoted by 1 if the number of free channels that belong to a subtrunk Θ_{xy}^{ij} has a capacity of at least B. The matrix L_{ij} is defined as

$$
l_{xy} = \begin{cases} 1 & \text{if } \left[\displaystyle\sum_{(l,f,w,t) \in \Theta_{xy}^{ij}} (l, f, w, t).Availability \right] \geq B \\ 0 & \text{otherwise} \end{cases} \tag{14.2}
$$

$$L_{12} = \begin{bmatrix} 1 & 1 \\ 1 & 1 \\ 0 & 1 \end{bmatrix} \quad L_{23} = \begin{bmatrix} 1 & 1 & 1 \\ 1 & 1 & 1 \end{bmatrix} \quad L_{34} = \begin{bmatrix} 1 \\ 1 \\ 0 \end{bmatrix}$$

$$L_{45} = \begin{bmatrix} 1 & 1 \end{bmatrix} \quad L_{16} = \begin{bmatrix} 1 & 0 & 0 \\ 0 & 1 & 0 \\ 0 & 0 & 1 \end{bmatrix} \quad L_{67} = \begin{bmatrix} 1 & 0 & 0 \\ 0 & 0 & 0 \\ 0 & 0 & 1 \end{bmatrix}$$

$$L_{75} = \begin{bmatrix} 1 & 1 \\ 1 & 0 \\ 0 & 1 \end{bmatrix}$$

Fig. 14.8. Link information matrices indicating whether there is at least one free channel in a subtrunk.

$$L_{12} = \begin{bmatrix} 3 & 3 \\ 2 & 2 \\ 0 & 3 \end{bmatrix} \quad L_{23} = \begin{bmatrix} 2 & 3 & 1 \\ 2 & 2 & 2 \end{bmatrix} \quad L_{34} = \begin{bmatrix} 4 \\ 1 \\ 0 \end{bmatrix}$$

$$L_{45} = \begin{bmatrix} 8 & 6 \end{bmatrix} \quad L_{16} = \begin{bmatrix} 4 & 0 & 0 \\ 0 & 5 & 0 \\ 0 & 0 & 2 \end{bmatrix} \quad L_{67} = \begin{bmatrix} 3 & 0 & 0 \\ 0 & 0 & 0 \\ 0 & 0 & 4 \end{bmatrix}$$

$$L_{75} = \begin{bmatrix} 3 & 3 \\ 1 & 0 \\ 0 & 2 \end{bmatrix}$$

Fig. 14.9. Link information matrices indicating the number of free channels in a subtrunk.

where $1 \le x \le K_i$ and $1 \le y \le K_j$. It is observed that the matrix gives the connectivity information to route a call that requires a capacity of B without splitting the connection across subtrunks. For the example network considered in Fig. 14.6, the link information matrices for different links for connection requests of one channel capacity are shown in Fig. 14.8.

Case 2: available capacity. In order to represent available capacity on each subtrunk of a link, every element l_{xy} of the matrix L_{ij} is defined as the number of free channels that belong to a subtrunk Θ_{xy}^{ij} as

$$l_{xy} = \sum_{(l,f,w,t) \in \Theta_{xy}^{ij}} (l, f, w, t).Availability \tag{14.3}$$

where $1 \le x \le K_i$ and $1 \le y \le K_j$. The matrix representation for different links in the example network in Fig. 14.7 is shown in Fig. 14.9.

Note that the matrices obtained using the available subtrunk capacity also contain the information for the matrices representing connectivity information. Depending on the level of information that is required in the network, different matrix representations can be employed.

Case 3: backup sharing. In this case, it is assumed that a path for establishing a primary connection has been selected. A backup path for the request has to be computed, hence the matrix is filled with information related to the backup capacity that is freely available. This implies that the capacity has already been allocated as backup to some other connection in the network and the current request can be overloaded on that capacity as their primary paths are link-disjoint. The matrix L_{ij} is defined as

$$l_{xy} = \sum_{(l,f,w,t)\in\Theta_{xy}^{ij}} (l,\,f,\,w,\,t).LinkDisjoint(r) \tag{14.4}$$

For example, assume that the link information depicted in Fig. 14.7 is the capacity that has already been allocated as backup to other connections in the network. This could also be shared as the backup for the request under consideration. The matrix representation in this case would be the same as that described for the case of available information in Fig.14.9.

14.7.2 Path information

Information concerning a certain path from nodes i to k that are not physically connected by a fiber is obtained by combining the link information in the path. The matrix representation for a path is defined in a manner similar to that of a link. A path matrix from node i to node k through j is obtained as a matrix multiplication of individual path segments P_{ij} and P_{jk} as

$$P_{ik} = P_{ij}P_{jk} \tag{14.5}$$

A generalized version of matrix multiplication is employed to compute the path metric. An element p_{xy}^{ik} (the superscript ik denotes the matrix to which the element belongs) is obtained as

$$p_{xy}^{ik} = \left(p_{x1}^{ij} \otimes p_{1y}^{jk}\right) \oplus \left(p_{x2}^{ij} \otimes p_{2y}^{jk}\right) \oplus \cdots \oplus \left(p_{xK_j}^{ij} \otimes p_{K_jy}^{jk}\right) \tag{14.6}$$

The operators $\otimes$ and $\oplus$, denoted as a tuple $(\otimes, \oplus)$, can be defined in different combinations so that several meaningful results are obtained. It is observed that when $\otimes$ is an integer or real-number multiplication operation and $\oplus$ is an integer or real-number addition operation, the above equation denotes the traditional matrix multiplication.

To illustrate the significance of different operators, consider the following two examples of matrix representation of links, namely the connectivity and available capacity matrix representations, and apply two different sets of operators to obtain different information from the network.

Case 1: arithmetic operators. In this case, integer multiplication ($\times$) and integer addition ($+$) are used as the operators for $\otimes$ and $\oplus$, respectively. Consider the matrix representation shown in Fig. 14.8 for a request that requires connection of one channel capacity. Applying the operator on the path 1–2–3–4–5, we obtain the path information matrix as

$$P_{1-2-3-4-5} = \begin{bmatrix} 4 & 4 \\ 4 & 4 \\ 2 & 2 \end{bmatrix} \tag{14.7}$$

An element p_{xy} of the above matrix denotes the number of distinct subtrunk selections available from trunk x of node 1 to trunk y of node 5. For example, there are four paths that start at trunk W_1 of node 1 and end at trunk T_1 of node 5. These four trunk assignments on the path are represented as a set of tuples containing node numbers and the trunk number on that node through which the connection passes. The four possible trunk assignments on paths, denoted by P_1 through P_4, are:

$$
\begin{aligned}
P_1: & \quad \{(1, W_1), (2, T_1), (3, W_1), (4, F_1), (5, T_1)\} \\
P_2: & \quad \{(1, W_1), (2, T_1), (3, W_2), (4, F_1), (5, T_1)\} \\
P_3: & \quad \{(1, W_1), (2, T_2), (3, W_1), (4, F_1), (5, T_1)\} \\
P_4: & \quad \{(1, W_1), (2, T_2), (3, W_2), (4, F_1), (5, T_1)\}
\end{aligned}
$$

The existence of trunk assignment for other trunk pairs can be easily verified from Fig. 14.6.

Consider the link information as shown in Fig. 14.9. Applying the operator ($\times, +$) on these matrices results in the path information matrix for path 1–2–3–4–5 as

$$P_{1-2-3-4-5} = \begin{bmatrix} 504 & 378 \\ 336 & 252 \\ 240 & 180 \end{bmatrix} \tag{14.8}$$

An element p_{xy} of the above matrix denotes the number of possible channel assignment combinations on the path that start at a certain trunk x on node 1 and end at a trunk y on node 5. For example, consider the possible trunk assignments that start at trunk W_1 at node 1 and end at trunk T_1 at node 5. On every subtrunk on the path the number of ways of assigning a channel is the same as the number of channels in the subtrunk. Hence, the number of possible channel assignments on a specific trunk assignment on the path is the product of the number of channels on the assigned subtrunk on every link. The number of possible channel assignments on four possible trunk assignments P_1 through P_4 that start the connection at trunk

W_1 at node 1 and end at trunk T_1 at node 2 are 192, 72, 192, and 48, respectively, adding up to 504 possible ways of channel assignment.

Case 2: selection operators. In this case, the operator $\otimes$ indicates the minimum of the two operands while the operator $\oplus$ indicates the maximum of the two operands. Applying this set of operations to the matrix representation in Fig. 14.8 for connectivity yields the matrix representation for the path 1–2–3–4–5 as

$$P_{1-2-3-4-5} = \begin{bmatrix} 1 & 1 \\ 1 & 1 \\ 1 & 1 \end{bmatrix} \tag{14.9}$$

The elements of the matrix indicate the existence of a channel allocation scheme for a one-channel capacity call that would start at trunk x at node 1 and end at trunk y at node 5. Note that the matrix in Eq. (14.9) can be obtained from the matrix in Eqs. (14.7) or (14.8) by replacing every element in the matrix by 1 if it is non-zero.

Applying this set of operations to the matrix representation in Fig. 14.9, the maximum capacity that can be routed from node 1 to node 5 without splitting the connection is obtained. The matrix representation for the path is obtained as

$$P_{1-2-3-4-5} = \begin{bmatrix} 2 & 2 \\ 2 & 2 \\ 2 & 2 \end{bmatrix} \tag{14.10}$$

Consider the possible trunk assignments that start the connection at trunk W_1 at node 1 and end at trunk T_1 at node 2. Consider the four possible trunk assignments P_1 through P_4. It is observed from Fig. 14.6 that the trunk assignments P_1 and P_3 have the link connecting node 2 to node 3 as a bottleneck with a channel capacity of two. The trunk assignments P_2 and P_4 have the link connecting node 3 to node 4 as a bottleneck with a channel capacity of one. Hence, a maximum of a two-channel capacity connection can be routed from node 1 to node 5 starting at trunk W_1 at node 1 and ending at trunk T_1 at node 5.

The link information, depicted in Fig. 14.7, is assumed to indicate the capacity available on each subtrunk that has already been assigned to other connections as backup and could be shared with the request that is under consideration, employing the operator set (min, max) would give a path matrix as shown in Eq. (14.10). In this scenario, the elements of the matrix denote the maximum capacity that could be shared on the path by the request.

14.8 A two-pass approach

When a call arrives at a node, a request for connection establishment is sent along a set of candidate paths. The connection establishment is carried out in two passes:

forward pass and reverse pass [233]. During the forward pass, the connection request is forwarded to the nodes along the path along with a vector, called the path information vector (PIV). The path information vector at a node k for a path p with source i and destination k, denoted by V_{ik}, is of dimension $1 \times K_k$. V_{ik} is obtained as a product of the path information vector at the source node and the information matrix of the path connecting nodes i and k:

$$V_{ik} = U_i P_{ik} \qquad (14.11)$$

where U_i denotes the path information matrix at the source node which is always set as a unit row vector.

Assume that the path from node i to node k passes through node j. Rewriting the above equation yields the relationship between the PIV vectors at node j and node k,

$$V_{ik} = U_i P_{ik} = U_i P_{ij} P_{jk} = V_{ij} P_{jk} \qquad (14.12)$$

The matrix-vector multiplication employed above is similar to the generalized matrix multiplication given in Section 14.7.2 with the operator tuple $(\otimes, \oplus)$. The elements of PIV at a node indicate specific properties about paths that end at a certain trunk. For example, if the link information matrix represented in Fig. 14.8 and operator $(\times, +)$ are employed, then the resulting PIV at each node indicates the number of possible trunk assignments on a path that would terminate the connection on a certain trunk at that node.

During the forward pass of the connection establishment, a node j on path p with source i can forward either the path information matrix L_{ij} to its neighboring node or the path information vector P_{ij}. Forwarding the latter has the advantage of minimizing the amount of information forwarded. Note that the reduction in information exchange is significant when the number of trunks at a node is large. Hence, the path information vector at a node is known to the successive node in the path and will be used to assign a subtrunk on the reverse pass.

14.8.1 Path selection

The path information vector can be used to select a suitable path from a given source–destination pair.

For example, consider the two paths from node 1 to node 5: 1–2–3–4–5 and 1–6–7–5. Employing the matrix information represented in Fig. 14.8 and the operator $(\times, +)$, the path information vectors for the two paths are given below:

$$V_{1-2-3-4-5} = [10 \quad 10]$$
$$V_{1-6-7-5} = [1 \quad 2]$$

With these matrices known at the destination, one could employ different comparison algorithms to select a path. For example, the total number of trunk assignments possible on a path is obtained by summing all the elements of the matrix. A path that has the maximum value for this metric can be chosen for establishing the connection in order to distribute the traffic in the network.

If the matrix representation in Fig. 14.9 and the operator (min, max) are employed, the path information vectors for the two paths are obtained as

$$V_{1-2-3-4-5} = [2 \quad 2]$$
$$V_{1-6-7-5} = [3 \quad 3]$$

It is observed that the path 1–6–7–5 can route a call for a three-channel capacity request without splitting, while the other path cannot. Hence, if traffic requirements in the network are diverse and destination-based path selection is employed, then the path 1–6–7–5 could be chosen so as to minimize the blocking at that time instance.

The selection of the path need not be based only on the path information vector. Different metrics such as the hop length, the delay, the abstract cost, etc. that could be included for a link-state vector and possible path selection schemes for WDM grooming networks [247]. The path information vector can also be extended to include multiple metrics in order to select a path.

14.8.2 Subtrunk assignment

At the end of the forward pass, the destination node has the path information vector for the different probed paths and selects a path using a path selection algorithm. Once a path is chosen, a subtrunk has to be selected on every link of the path in order to complete the connection establishment. The subtrunk assignment is carried out in two steps.

(i) The destination node first selects the trunk at its node where the connection would terminate.
(ii) Every node in the network selects the output trunk at its previous node. If a link connects nodes i and j, then node j selects the output trunk at node i, hence the subtrunk assignment on the link $i-j$.

Consider the information matrix represented in Fig. 14.8 and the operator $(\times, +)$. The path information vectors obtained at different nodes are shown in Fig. 14.10.

The trunk assignment to end the connection at the destination node can be made using the path information vector. Several trunk selection schemes such as first-fit, best-fit, random, etc. could be employed. Random subtrunk assignment is used in

$$V_{11} = \begin{bmatrix} 1 & 1 & 1 \end{bmatrix}$$

$$V_{12} = \begin{bmatrix} 1 & 1 & 1 \end{bmatrix} \begin{bmatrix} 1 & 1 \\ 1 & 1 \\ 0 & 1 \end{bmatrix} = \begin{bmatrix} 2 & 3 \end{bmatrix}$$

$$V_{13} = \begin{bmatrix} 2 & 3 \end{bmatrix} \begin{bmatrix} 1 & 1 & 1 \\ 1 & 1 & 1 \end{bmatrix} = \begin{bmatrix} 5 & 5 & 5 \end{bmatrix}$$

$$V_{14} = \begin{bmatrix} 5 & 5 & 5 \end{bmatrix} \begin{bmatrix} 1 \\ 1 \\ 0 \end{bmatrix} = \begin{bmatrix} 10 \end{bmatrix}$$

$$V_{15} = \begin{bmatrix} 10 \end{bmatrix} \begin{bmatrix} 1 & 1 \end{bmatrix} = \begin{bmatrix} 10 & 10 \end{bmatrix}$$

Fig. 14.10. Path information vector computed at the nodes along the path 1–2–3–4–5.

the following. Let x_k denote the trunk that is chosen to accommodate the connection at node k.

In order to select a subtrunk on the link j–k, a *ratio vector* is computed at node k. The vector, denoted by R_{jk}, is obtained as the product of the vector at the previous node of the path information vector V_{ij} and the column vector of the link information matrix L_{jk} corresponding to x_k:

$$\begin{aligned} R_{jk} &= V_{ij} \times L_{jk}^{\mathrm{T}}(x_k) \\ &= [v_1 \cdots v_{K_j}] \circ [l_{1x_k} \cdots l_{K_j x_k}] \\ &= [v_1 \circ l_{1x_k} \cdots v_{K_j} \circ l_{K_j x_k}] \end{aligned} \tag{14.13}$$

where $L_{jk}^{\mathrm{T}}(x_k)$ denotes the transpose of the column vector corresponding to the column x_k of the matrix L_{jk} and the operator $\circ$ denotes the elementwise operation on the row vectors. Again, one could define different operators depending on the construction of the information matrix. Since the input trunk at node k is decided, the choice of output trunk at node j is also dictated by the channel occupancy of the channels that fall within $\Theta_{yx_k}^{jk}$. The output channel at node j can be selected in various ways using the ratio vector.

During the reverse pass, a subtrunk is allocated on the path. As node 5 is the destination, it selects a trunk for the connection to terminate. A random subtrunk assignment is used here. The operator $\circ$ denotes an integer multiplication. The operator $\circ$ is the same as the operator $\otimes$ since the elementwise operation that is evaluated here is similar to the matrix-vector multiplication employed for computing the path information vector.

The PIV at node 5 indicates that there are 20 possible subtrunk assignments with each trunk being able to carry 10 each. Hence, one of the two is chosen with equal probability. In general, if p_x subtrunk assignments are possible on the path that would terminate the connection at the destination node at trunk x, then the trunk x is chosen with a probability $p_x / \sum_{y=1}^{K_d} p_y$, where K_d denotes the number of trunks at the destination node d. In the example considered here, one of the two trunks is selected with equal probability. Assume that the trunk chosen is T_2. The node also selects the output trunk at its previous node. In order to select this, the ratio vector is computed as

$$R_{45} = [\,10\,] \circ [\,1\,] = [\,10\,] \tag{14.14}$$

In this case, one output trunk is available and that is selected. Hence, on link 4–5, a channel that belongs to trunk F_1 of node 4 and trunk T_2 of node 5 is selected. Node 5 confirms the selection of output trunk F_1 to node 4 during the reverse pass in the network.

As the trunk assignment for the connection at node 4 is decided by node 5, node 4 chooses the output trunk at node 3 by computing a similar ratio vector as

$$R_{34} = [5 \quad 5 \quad 5] \circ [1 \quad 1 \quad 0] = [5 \quad 5 \quad 0] \tag{14.15}$$

The R_{34} vector denotes the selection ratio for the three output trunks at node 3. Note that although there are five possible paths that could end at trunk W_3 at node 3, there are no free channels on link 3–4 that are within trunk W_3 of node 3 and trunk F_1 of node 4. This information is reflected in the selection ratio vector SR_{34} as a zero entry corresponding to the ratio for trunk W_3. Hence, trunk W_1 or W_2 is selected with equal probability. Assume that trunk W_2 is selected in this case.

At node 3, the vector R_{23} is computed as

$$R_{23} = [2 \quad 3] \circ [1 \quad 1] = [2 \quad 3] \tag{14.16}$$

and node 3 selects the output trunk at node 2 in the ratio of 2:3, i.e. trunk T_1 is selected with a probability of 0.4 while trunk T_2 is selected with a probability of 0.6. Assume that trunk T_1 is chosen.

At node 2, the vector R_{12} is computed as

$$S_{12} = [1 \quad 1 \quad 1] \circ [1 \quad 1 \quad 0] = [1 \quad 1 \quad 0] \tag{14.17}$$

One of the trunks W_1 or W_2 is chosen with equal probability. Assume that W_1 is chosen. This selection is sent to node 1 completing the subtrunk assignment. Now, the path established for the connection can be written as a set of node–trunk pairs assigned at each node on the path $\{(1, W_1), (2, T_1), (3, W_2), (4, F_1), (5, T_2)\}$ or equivalently as a set of link and subtrunk pairs on the path $\{(\ell_{12}, \Theta_{W_1 T_1}), (\ell_{23}, \Theta_{T_1 W_2}), (\ell_{34}, \Theta_{W_2 F_1}), (\ell_{45}, \Theta_{F_1 T_2})\}$. Any channel belonging to the subtrunk assigned at a link can be chosen for establishing the channel as every node has full-permutation switching capability within a trunk.

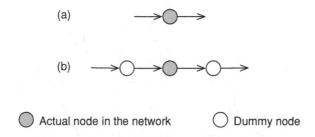

Fig. 14.11. (a) A node in a network employing channel-space switching. (b) The node is modeled with two dummy nodes and two additional links.

It can be observed that such a trunk selection strategy selects with uniform probability a possible subtrunk assignment on the path. In the above trunk selection process, if with the information matrix representation shown in Fig. 14.9, operator $(\times, +)$, the operator $\circ$ is set to multiplication, and there is random channel assignment within a chosen trunk on each link, then the resulting channel assignment algorithm selects a channel uniformly from the set of all possible channel assignments possible on the path.

Several other channel trunk selection approaches can be developed based on the proposed connection establishment framework. Selecting the trunk that has the minimum or maximum value in the selection ratio vector would have an effect similar to packing and spreading the connections across the available subtrunk assignments in the path. In order to implement a first-fit subtrunk assignment the first available trunk is chosen from the ratio vector. It can be easily observed that several routing strategies proposed in the literature for wavelength-routed networks can be easily derived from this framework.

14.9 Modeling a channel-space switch in MICRON

The connection establishment framework can be employed to simulate a network where the nodes employ channel-space switches. Modeling such a switching architecture is achieved by incorporating dummy nodes in the network.

Figure 14.11(a) shows a node in a network with an incoming and outgoing link. Assume that this node implements channel-space switching. This node is modeled with two dummy nodes as shown in Fig. 14.11(b).

Example. Consider a 3×3 unidirectional mesh-torus network as shown in Fig.14.12. Assume that every link has F fibers, W wavelengths per fiber, and T time slots per wavelength. Let C denote the total number of channels in a link. For the sake of simplicity, assume that all the links have the same number of channels. Let a node i view the links attached to it as K_i trunks with S_i channels in each.

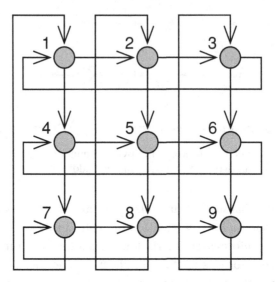

Fig. 14.12. A 3×3 unidirectional mesh-torus network.

Figure 14.13 shows modeling of the 3×3 unidirectional mesh-torus network with dummy nodes. The shaded nodes represent the actual nodes in the network while the unshaded ones represent the dummy nodes. Let the dummy nodes around actual nodes be labeled based on the direction (with the right-hand side node referring to east). Let iE, iW, iN, and iS denote the dummy node connected to the right, left, top, and bottom of a node i.

The actual nodes in the network are made to view links as FWT trunks and the dummy nodes that are immediate neighbors of a node i view the link as K_i trunks. A link connecting a dummy node to an actual node in the network is represented by a $K_i \times C$ matrix while a link connecting an actual node to a dummy node is represented as a $C \times K_i$ matrix. A link connecting two dummy nodes is represented as a $K_i \times K_j$ matrix, where K_i and K_j denote the number of trunks as viewed by the source and destination dummy nodes.

For example, assume that node 1 is a wavelength grooming (WG) node and node 2 is a trunk grooming (TG) node, viewing the links as three and two trunks, respectively. The transpose of the information matrix of the link connecting node 1 to 1E is given by (the information matrix is represented in its transposed form)

$$
L^{\mathrm{T}}_{1,1E} =
\begin{bmatrix}
1 & 1 & 1 & 1 & 1 & 1 & 0 & 0 & 0 & 0 & 0 & 0 & 0 & 0 & 0 & 0 & 0 & 0 \\
0 & 0 & 0 & 0 & 0 & 0 & 1 & 1 & 1 & 1 & 1 & 1 & 0 & 0 & 0 & 0 & 0 & 0 \\
0 & 0 & 0 & 0 & 0 & 0 & 0 & 0 & 0 & 0 & 0 & 0 & 1 & 1 & 1 & 1 & 1 & 1
\end{bmatrix}
$$

$$(14.18)$$

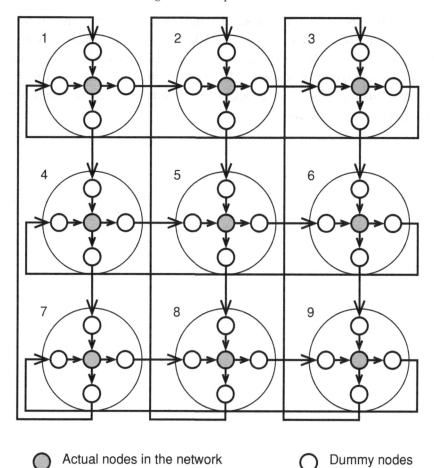

● Actual nodes in the network   ○ Dummy nodes

Fig. 14.13. A 3×3 unidirectional mesh-torus network.

The information matrix of the link connecting node $2W$ to 2 is given by

$$L_{2W,2} = \begin{bmatrix} 1 & 1 & 1 & 1 & 1 & 1 & 1 & 1 & 1 & 0 & 0 & 0 & 0 & 0 & 0 & 0 & 0 & 0 \\ 0 & 0 & 0 & 0 & 0 & 0 & 0 & 0 & 0 & 1 & 1 & 1 & 1 & 1 & 1 & 1 & 1 & 1 \end{bmatrix}$$

(14.19)

The information matrix of the link connecting node $1E$ to node $2W$ is given by

$$L_{1E,2W} = \begin{bmatrix} 3 & 3 \\ 3 & 3 \\ 3 & 3 \end{bmatrix}$$

(14.20)

Note that the dummy nodes represent the full-channel interchangers that are present at the input and output stages at a node. It can be easily seen that if two nodes view the links in exactly the same manner, the link connecting the dummy node of the first link to that of the second link would be a diagonal matrix. In such a case, one of the dummy nodes can be removed from the network. If such a modified network is employed for simulation, the traffic in the network is generated and terminated only at the actual nodes in the network, hence dummy nodes do not contribute to the network traffic.

15

Blocking in TSN

One of the important performance metrics by which a wide area network is evaluated is based on the success ratio of the number of requests that are accepted in the network. This metric is usually posed in its alternate form as the *blocking probability*, which refers to the rejection ratio of the requests in the network. The smaller the rejection ratio is, the better the network performance. Although other performance metrics exist, such as the effective traffic carried in the network, the fairness of request rejections with respect to requests requiring different capacity requirements or different path lengths, the most meaningful way to measure the performance of a wide-area network is through the blocking performance. To some extent the other performance metrics described above can be obtained as functions of the blocking performance.

Analytical models that evaluate the blocking performance of wide-area circuit-switched networks are employed during the design phase of a network. In the design phase these models are typically employed as an *elimination test*, rather than as an acceptance test. In other words, the analytical models are employed as back of the envelope calculations to evaluate a network design, rejecting those designs that are below a certain threshold.

15.1 Blocking model

The following assumptions are made to develop an analytical model for evaluating the blocking performance of a TSN.

- The network has N nodes.
- The call arrival at every node follows a Poisson process with rate λ_n. The choice of Poisson traffic is to keep the analysis tractable.
- The calls can be either unicast connections with one destination or multicast connections with more than one destination. The traffic due to multicast connections is negligible.

261

Hence, for the derivation of path and tree blocking probabilities, the network load is assumed to be entirely due to unicast connections.

- The probability that a call requires capacity b and is destined to a node that has a distance of z hop lengths is $p_{z,b}$. Note that the distance refers to the length of the connection that needs to be established.
- The maximum bandwidth requirement of a call is B and the maximum path length in the network is $N - 1$.
- The path selection is predetermined (a fixed-path routing), such as the shortest-path algorithm.
- A request cannot be split among multiple trunks at a node.
- The holding time of every call follows an exponential distribution with mean $1/\mu$. The Erlang load offered by a node is $\rho_n = \lambda_n/\mu$.
- Blocked calls are not reattempted.
- A call is assigned a channel randomly from a set of available channels in a subtrunk on a link.

15.2 Estimation of call arrival rates on a link

Typically, the network traffic is specified in terms of the offered load between node-pairs. The call arrival rates at the nodes have to be translated into arrival rates at individual links in the network. The computation of the blocking probability depends on the link arrival rates, and the link arrival rates, in turn, depend on the network blocking probability. However, if the blocking probability in the network is small, then its effect on the link arrival rates can be ignored.

For a network with N nodes and L links, the average path length of a connection in the network is given by

$$Z_{\text{av}} = \sum_{z=1}^{N-1} z \, p_z \tag{15.1}$$

where p_z is the path-length distribution. The path-length distribution is obtained from the joint probability distribution of the path length and the call capacity as follows:

$$p_z = \sum_{b=1}^{B} p_{z,b} \tag{15.2}$$

Similarly, the average capacity requirement of a connection B_{av} is given by

$$B_{\text{av}} = \sum_{b=1}^{B} b \, p_b \tag{15.3}$$

where p_b is the probability that a call requires a bandwidth of b channels and is computed as

$$p_b = \sum_{z=1}^{N-1} p_{z,b} \tag{15.4}$$

The average resource required by a call is computed as

$$R_{av} = \sum_{z=1}^{N-1} \sum_{b=1}^{B} z b p_{z,b} \tag{15.5}$$

Recall that λ_n denotes the call arrival rate at a node. Let λ denote the average link arrival rate. It is computed using the following:

$$\lambda = \frac{N\lambda_n \sum_{z=1}^{N-1} \sum_{b=1}^{B} z b p_{z,b}}{L} \tag{15.6}$$

The fraction of traffic that is not destined for a node is obtained as the ratio of the number of links to a path that are not the last hop to the total number of links in the path. For a path with z links, there are $(z - 1)$ intermediate links. Let δ_c denote the fraction of traffic on a link that continues on any neighboring links at a node. δ_c is computed as given below:

$$\delta_c = \frac{\sum_{z=1}^{N-1} \sum_{b=1}^{B} b(z-1) p_{z,b}}{\sum_{z=1}^{N-1} \sum_{b=1}^{B} z b p_{z,b}} \tag{15.7}$$

$$= 1 - \frac{B_{av}}{R_{av}} \tag{15.8}$$

It should be noted that the above expression gives the fraction of the traffic that is not destined for a node. Such traffic could continue on any of the output links at the node. The link load correlation is defined as the probability that a call on a link would continue to a successive link on a chosen path and is given by

$$\gamma_c = \left(1 - \frac{B_{av}}{R_{av}}\right) \frac{1}{E} \tag{15.9}$$

where E denotes the number of links at the node that do not connect the node to any of the previous nodes in the path, referred to as *exit links*. Hence, the arrival rate of traffic on a link that would continue to a successive link on a path is given by $\lambda_c = \gamma_c \lambda$.

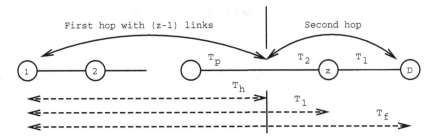

Fig. 15.1. A z-link path © IEEE. Source: R. Srinivasan and A. K. Somani, Analysis of multi-rate traffic in WDM grooming networks, in *Proceedings of ICCCN 2002* [245].

If the capacity requirement of every call in the network is one channel, then the link load correlation reduces to

$$\gamma_c = \left(1 - \frac{1}{Z_{av}}\right)\frac{1}{E} \tag{15.10}$$

15.3 Path blocking performance

The network blocking probability is computed as the average blocking probability experienced over different path lengths. Consider a z-link path model as shown in Fig. 15.1. The analysis that is developed in this section assumes that the capacity requirement of a call is r channels.

Let $P_z(T_f)$ denote the probability of T_f trunks being available on a z-link path as viewed by the last node on the path. It should be noted that the destination is not considered as the last node in the path (node z). The definition of the trunk is as viewed by the node denoted by the subscript on P. $P_z(T_f = 0)$ denotes the blocking probability over the z-link path.

Let $P_z(T_f, T_l)$ denote the probability of T_f trunks being available on a z-link path with T_l trunks free on the last link. Notice that the last link should have at least T_f trunks free, therefore $T_l \geq T_f$. $P_z(T_f)$ can then be written as follows:

$$P_z(T_f) = \sum_{T_l=T_f}^{K_z} P_z(T_f, T_l) \tag{15.11}$$

where K_z denotes the number of trunks in the link as viewed by node z.

A z-link path is analyzed as a two-hop path by considering the first $z - 1$ links as the first hop and the last two links as the second hop, as shown in Fig. 15.1. Let T_h and T_p denote the number of trunks available on the first hop and that which are free on the last link of the first hop (link $z - 1$), respectively, as viewed by the last node on the first hop (node $z - 1$). Let T_1 and T_2 denote the number of trunks free on the first hop and number of trunks free on the last link of the first hop as seen by the node in the second hop (node z). $P_z(T_f, T_l)$ can then be recursively computed as in Eq. (15.12):

$$
P_z(T_f, T_l) = \sum_{T_h=0}^{K_{z-1}} \sum_{T_p=T_h}^{K_{z-1}} \sum_{T_1=T_f}^{K_z} \sum_{T_2=T_1}^{K_z} P_{z-1}(T_1, T_2)
$$
$$
\times P_{z,z-1}(T_h, T_p \mid T_1, T_2) \, P_z(T_f, T_l \mid T_h, T_p) \qquad (15.12)
$$

where $P_z(T_f, T_l \mid T_h, T_p)$ denotes the probability of T_f trunks being available on the second hop with T_l trunks free on the last link of the second hop given that T_g trunks are available on the first hop with T_p trunks free at the input to the node on the second hop.

Let $P_{z,z-1}(T_h, T_p \mid T_1, T_2)$ denote the probability that the number of trunks available on the first hop and the number of trunks free on the last link of the first hop as viewed by the node in the second hop (node z) are T_h and T_p, respectively, given that the trunk availability as viewed by the last node (node $z - 1$) on the first hop is T_1 and T_2. For homogeneous TSNs, for any two successive nodes $z - 1$ and z on a path $P_{z,z-1}(T_h, T_p \mid T_1, T_2)$ is defined as in Eq. (15.12):

$$
P_{z,z-1}(T_h, T_p \mid T_1, T_2) = \begin{cases} 1 & \text{if } T_h = T_1 \text{ and } T_p = T_2 \\ 0 & \text{otherwise} \end{cases} \qquad (15.13)
$$

For a homogeneous TSN, Eq. (15.12), therefore, reduces to

$$
P_z(T_f, T_l) = \sum_{T_h=T_f}^{K} \sum_{T_p=T_h}^{K} P_{z-1}(T_h, T_p) \, P(T_f, T_l \mid T_h, T_p) \qquad (15.14)
$$

where K denotes the number of trunks in a link as viewed by the nodes in the network.

The starting point of the recursion, for $z = 1$, is defined as

$$
P_1(T_f, T_l) = \begin{cases} P(T_l) & \text{if } T_f = T_l \\ 0 & \text{otherwise} \end{cases} \qquad (15.15)
$$

where $P(T_l)$ denotes the probability of T_l trunks being free on a link. The computation of $P(T_l)$ is derived in Section 15.4.

$P_z(T_f, T_l \mid T_h, T_p)$ is computed by conditioning on the number of trunks free on the last link as viewed by node z. From this point on, the second hop, which is a two-link path, needs to be analyzed. The definition of a trunk is assumed to be as that viewed by the intermediate node in the two-link path.

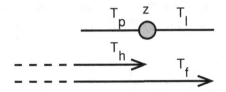

Fig. 15.2. Two-link path model with the free and available trunks as viewed by the intermediate node.

$P_z(T_f, T_l \mid T_h, T_p)$ is computed as

$$P_z(T_f, T_l \mid T_h, T_p) = \begin{cases} P_z(T_f \mid T_h, T_p, T_l)\, P_z(T_l \mid T_h, T_p) & \text{if } T_h \geq T_f \\ 0 & \text{otherwise} \end{cases} \quad (15.16)$$

where $P_z(T_l \mid T_h, T_p)$ denotes the probability of T_l trunks being free on the last link given that T_h trunks are available on the first hop with T_p trunks free on the last link of the first hop as viewed by node z. The number of trunks free on the last link depends on the number of trunks free on the previous links.

If the correlation of traffic on a link is assumed to be only due to its previous link, then it is referred to as the *Markovian correlation*. With the assumption of Markovian correlation, $P_z(T_l \mid T_h, T_p)$ can be reduced to $P_z(T_l \mid T_p)$. Hence, Eq. (15.16) can be written as

$$P_z(T_f, T_l \mid T_h, T_p) = \begin{cases} P_z(T_f \mid T_h, T_p, T_l)\, P_z(T_l \mid T_p) & \text{if } T_h \geq T_f \\ 0 & \text{otherwise} \end{cases} \quad (15.17)$$

$P_z(T_f \mid T_h, T_p, T_l)$ denotes the probability that T_f trunks are available on the two-hop path given that T_l trunks are free on the last link and T_h trunks are available on the first hop with T_p trunks free on the last link of the first hop. $P_z(T_f \mid T_h, T_p, T_l)$ is computed by considering a two-link path as shown in Fig. 15.2. The number of trunks free on the first link and that which are free on the second link are denoted by T_p and T_l, respectively.

The trunk on a two-link path can be in any one of the following four states, as shown Fig. 15.3.

- **Case 1:** the trunk is busy on both the links. The trunk can be either partially or fully occupied by continuing calls. Let V_c denote the number of trunks busy on both the links.
- **Case 2:** the trunk is busy on the first link but not on the second.
- **Case 3:** the trunk is busy on the second link but not on the first.
- **Case 4:** the trunk is free on both the links. Let T_b denote the number of trunks free on both links. However, this does not imply that these trunks are available on the two-link

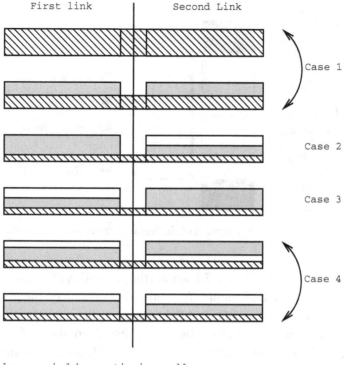

Fig. 15.3. Different possible states for trunk occupancy on a two-link path.

path. Let T_a ($T_a \leq T_b$) denote the number of trunks available on the two-link path. When the node connecting the two links employs a full-permutation switch, a trunk is available on the two-link path if it is free on both the links. Hence, $T_a = T_b$.

Let $P_z(T_a, T_b \mid T_p, T_l)$ denote the probability that T_b trunks are free on both the first and second link with T_a among them being available on the two-link path given that T_p and T_l trunks are free on the first and second links, respectively. $P_z(T_f \mid T_h, T_p, T_l)$ can then be written as

$$P_z(T_f \mid T_h, T_p, T_l) = \sum_{T_a=T_f}^{\min(T_p,T_l)} \sum_{T_b=T_a}^{\min(T_p,T_l)} P_z(T_f \mid T_a, T_b, T_h, T_p, T_l)$$
$$\times P_z(T_a, T_b \mid T_p, T_l) \qquad (15.18)$$

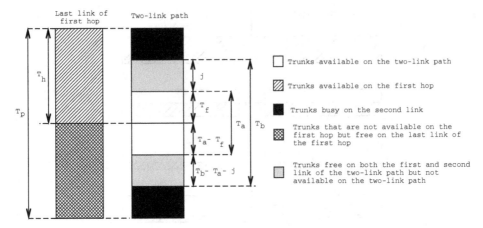

Fig. 15.4. Arrangement of the trunk distribution on a two-link path.

where $P_z(T_f \mid T_a, T_b, T_h, T_p, T_l)$ denotes the probability of T_f trunks being available on the two-hop path given that T_h trunks are available on the first hop, T_p trunks are free on the first link (the first link in the two-link model is the last link on the first hop), T_l trunks are free on the second link, T_b trunks are free on both the first and second links, and T_a among them are available on the two-link path.

Figure 15.4 shows the pictorial view of the distribution of free trunks on the last link of the first hop and that of the two-link path model. From this figure, $P_z(T_f \mid T_a, T_b, T_h, T_p, T_l)$ can be computed using the following line of thinking.

Assume that T_p trunks are free on the last link of the first hop with T_h among them available on the first hop. Also, assume that T_b trunks are free on both links of the two-link path with T_a among them being available on the two-link path. Among the T_a available trunks on the two-link path exactly T_f trunks overlap with the T_h trunks available on the first hop. The remaining trunks that are available on the first hop, $T_h - T_f$, are not available on the two-link path. This could occur in two cases: (i) the corresponding trunk is busy on the second link of the two-link path or (ii) the trunk is free on the second link, but is not available on the two-link path (due to switching constraints). The number of trunks that satisfy the latter case is $T_b - T_a$. The required probability is computed by assuming that j of the trunks that are free on both the first and second link, but not available on the two-link path, overlap with the remaining $T_h - T_f$ available trunks of the first hop. The trunks on the second link corresponding to the remaining $T_h - T_f - j$ available trunks on the first hop are busy. Thus, $P_z(T_f \mid T_a, T_b, T_h, T_p, T_l)$ can be written as

$$P_z(T_f \mid T_a, T_b, T_h, T_p, T_l) = \frac{\binom{T_h}{T_f}\binom{T_p-T_h}{T_a-T_f}\sum_j \binom{T_h-T_f}{j}\binom{T_p-T_h-T_a+T_f}{T_b-T_a-j}}{\binom{T_p}{T_b}\binom{T_b}{T_a}} \tag{15.19}$$

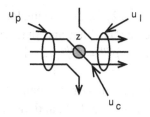

Fig. 15.5. Two-link model with channel distribution.

where $\max(0, T_b + T_h - T_p - T_f) \leq j \leq \min(T_h - T_f, T_b - T_a)$. For the special case, when the switch at a node has full-permutation switching capability, the above equation reduces to the following:

$$
P_z(T_f \mid T_a, T_b, T_h, T_p, T_l) = \begin{cases} \dfrac{\dbinom{T_h}{T_f}\dbinom{T_p - T_h}{T_a - T_f}}{\dbinom{T_p}{T_a}} & \text{if } T_a - T_f \leq T_p - T_h \\ 0 & \text{otherwise.} \end{cases}
$$

$$(15.20)$$

Also, for this case, $P_z(T_a, T_b \mid T_p, T_l) = 0$ if $T_a \neq T_b$. Hence, Eq. (15.18) can be written as

$$
P_z(T_f \mid T_h, T_p, T_l) = \sum_{T_a=T_f}^{\min(T_p, T_l)} P(T_f \mid T_a, T_a, T_h, T_p, T_l)\, P(T_a, T_a \mid T_p, T_l)
$$

$$(15.21)$$

The probability values, $P_z(T_a, T_b \mid T_p, T_l)$, $P_z(T_l \mid T_p)$, and $P_z(T_l)$ are computed by considering a switch model as explained in the following.

15.4 Free trunk distribution

Consider a two-link path model as shown in Fig. 15.5. Let u_p, u_l, and u_c denote the number of channels busy on the first link, the number of channels busy on the second link, and the number of channels occupied by calls that continue from the first link to the second, respectively. Note that $u_c \leq \min(u_p, u_l)$.

The number of channels busy on a trunk at the input of node z is the same as the number of channels busy at the input of the switch (after the FCI) at the node, while the distribution of the busy channels at the input of the switch is independent of the distribution at the input of the node. Also, the state of a trunk (busy or free) at the input of the node is the same as that at the input to the switch. Therefore, the first link as referred to here corresponds to the link viewed at the input of the switch.

Let $\lambda_p^{(b)}$, $\lambda_l^{(b)}$, and $\lambda_c^{(b)}$ denote the arrival rate for calls that request capacity b to the first link, the second link, and those that continue from the first link to the second. Note that $\lambda_c^{(b)} \leq \min(\lambda_p^{(b)}, \lambda_l^{(b)})$. The Erlang loads corresponding to the calls that occupy the first link, the second link, and that which continue from the first to the second can be written as, $\rho_p^{(b)} = \lambda_p^{(b)}/\mu$, $\rho_l^{(b)} = \lambda_l^{(b)}/\mu$, and $\rho_c^{(b)} = \lambda_c^{(b)}/\mu$, respectively.

Let u_p, u_l, and u_c denote the number of channels busy on the first link, the number of channels busy on the second link, and the number of channels occupied by calls that continue from the first link to the second, respectively. Note that $u_c \leq \min(u_p, u_l)$.

The channel distribution on a two-link path can be characterized as a three-dimensional Markov chain. The state-space is denoted by the 3-tuple (u_p, u_l, u_c). Note that such a representation of the state of the two links does not take into account the number of calls that require a certain specified bandwidth. The steady-state probability for the states is computed recursively by considering the number of calls that require a certain capacity, with B being the maximum capacity requirement of a connection, as

$$\Pi(u_p, u_l, u_c) = \frac{1}{G} \xi_B(u_p, u_l, u_c) \tag{15.22}$$

where $0 \leq u_p \leq KS$, $0 \leq u_l \leq KS$, and $0 \leq u_c \leq \min(u_p, u_l)$. G is the normalization constant and is defined as follows:

$$G = \sum_{u_c=0}^{KS} \sum_{u_p=0}^{KS} \sum_{u_l=0}^{KS} \xi_B(u_p, u_l, u_c) \tag{15.23}$$

$\xi_B(u_p, u_l, u_c)$ is computed recursively as given below:

$\xi_b(u_p, u_l, u_c)$

$$= \begin{cases} \sum_{z=0}^{\lfloor u_c/b \rfloor} \sum_{x=z}^{\lfloor u_p/b \rfloor} \sum_{y=z}^{\lfloor u_l/b \rfloor} \dfrac{\left(\rho_p^{(b)} - \rho_c^{(b)}\right)^{x-z} \left(\rho_c^{(b)}\right)^z \left(\rho_p^{(b)} - \rho_c^{(b)}\right)^{y-z}}{(x-z)! \quad z! \quad (y-z)!} \\ \quad \times \xi_{b-1}(u_p - bx, u_l - by, u_c - bz) & \text{if } b > 1 \\[4pt] \dfrac{\left(\rho_p^{(b)} - \rho_c^{(b)}\right)^{u_p - u_c} \left(\rho_c^{(b)}\right)^{u_c} \left(\rho_p^{(b)} - \rho_c^{(b)}\right)^{u_l - u_c}}{(u_p - u_c)! \quad u_c! \quad (u_l - u_c)!} \\ & \text{if } b = 1 \\[4pt] 0 & \text{otherwise} \end{cases}$$

$$\tag{15.24}$$

Let V_p, V_l, and V_c denote the number of trunks busy on the first link, the number of trunks busy on the second link, and number of trunks that are busy on both the first and second links, respectively. It can be observed that $V_c \leq \min(V_p, V_l)$. The number of trunks free on both the links is given by $T_b = K_z - (V_p + V_l - V_c)$.

The number of trunks available on the two-link path is denoted by T_a. The state-space of the trunk distribution is captured by the 4-tuple (V_p, V_l, V_c, T_a). The steady-state probability of the states can be computed by conditioning on the channel distribution, (u_p, u_l, u_c), as

$$\psi(V_p, V_l, V_c, T_a) = \sum_{u_c=0}^{KS} \sum_{u_p=u_c}^{KS} \sum_{u_l=u_c}^{KS} P(V_p, V_l, V_c, T_a \mid u_p, u_l, u_c) \Pi(u_p, u_l, u_c)$$

(15.25)

where $P_z(V_p, V_l, V_c, T_a \mid u_p, u_l, u_c)$ denotes the probability that the trunk distribution is in state (V_p, V_l, V_c, T_a) given that the channel distribution is (u_p, u_l, u_c). The following probability values that are required to complete the analytical model described in Section 15.3 can then be derived from the above steady-state probability:

$P_z(T_a, T_b \mid T_p, T_l)$

$$= \frac{\psi(K_z - T_p, K - T_l, K_z + T_b - T_p - T_l, T_a)}{\sum_{t_a=0}^{\min(T_p, T_l)} \sum_{t_b=t_a}^{\min(T_p, T_l)} \psi(K_z - T_p, K - T_l, K_z + t_b - T_p - T_l, t_a)}$$

(15.26)

$P_z(T_l \mid T_p)$

$$= \frac{\sum_{t_a=0}^{\min(T_p, T_l)} \sum_{t_b=0}^{\min(T_p, T_l)} \psi(K_z - T_p, K - T_l, K_z + t_b - T_p - T_l, t_a)}{\sum_{t_l=0}^{K_z} \sum_{t_a=0}^{\min(T_p, t_l)} \sum_{t_b=0}^{\min(T_p, t_l)} \psi(K_z - T_p, K - t_l, K_z + t_b - T_p - t_l, t_a)}$$

(15.27)

$$P_z(T_l) = \sum_{t_p=0}^{K_z} \sum_{t_a=0}^{\min(t_p, T_l)} \sum_{t_b=t_a}^{\min(t_p, T_l)} \psi(K_z - t_p, K - T_l, K_z + t_b - t_p - T_l, t_a)$$

(15.28)

The trunk occupancy probability for a given channel distribution is computed as

$$P_z(V_p, V_l, V_c, T_a \mid u_p, u_c, u_l) = \frac{N_{k=K_z}(V_p, V_l, V_c, T_a \mid u_p, u_l, u_c)}{A_{k=K_z}(u_p, u_l, u_c)}$$

(15.29)

where $N_k(V_p, V_l, V_c, T_a \mid u_p, u_l, u_c)$ denotes the number of ways of arranging across k trunks, u_p busy channels on the first link, u_l busy channels on the second link, with u_c channels among them being occupied by calls that continue from the first link to the second, such that V_p trunks are busy on the first link, V_l trunks are busy on the second link with V_c among them being busy on both links, and T_a trunks being available on the two-link path.

Let $A_k(u_p, u_l, u_c)$ denote all possible ways of arranging across k trunks, u_p busy channels on the first link, u_l busy channels on the second link with u_c channels among them being occupied by calls that continue from the first link to the second.

$A_k(u_p, u_l, u_c)$ is recursively computed as

$$A_k(u_p, u_l, u_c) = \sum_{z=0}^{\min(S,u_c)} \sum_{x=z}^{\min(S,u_p)} \sum_{y=z}^{\min(S,u_l)} A_1(x, y, z) A_{k-1}(u_p - x, u_l - y, u_c - z)$$

(15.30)

where $0 \le u_p \le kS$, $0 \le u_l \le kS$, and $0 \le u_c \le \min(u_p, u_l)$. The definition of $A_1(x, y, z)$ depends on the type of switch.

$N_k(V_p, V_l, V_c, T_a \mid u_p, u_l, u_c)$ (written as $N_k(\cdot)$ for short) is assigned 0 if any of the following conditions hold true:

- $\{V_p, V_l, V_c, T_a, u_p, u_l, u_c\} < 0$;
- $\{V_p, V_l, V_c, T_a\} > k$;
- $\{u_p, u_l\} > kS$ or $u_c > \min(u_p, u_l)$;
- $u_p < V_p S$ or $u_l < V_l S$.

Otherwise, it is computed recursively under one of the following four cases, as described in Fig. 15.3:

Case 1: if $V_c > 0$. The required probability is obtained by conditioning on a trunk being busy on both the links.

$$N_k(\cdot) = \frac{k}{V_c} \sum_{z=0}^{S} A_1(S, S, z) N_{k-1}$$
$$\times (V_p - 1, V_l - 1, V_c - 1, T_a \mid u_p - S, u_l - S, u_c - z) \quad (15.31)$$

Case 2: if $V_c = 0$, $V_p > 0$. The required probability is obtained by conditioning on a trunk being busy on the first link but free on the second link.

$$N_k(\cdot) = \frac{k}{V_p} \sum_{z=0}^{\min(S-1,u_c)} \sum_{y=z}^{\min(S-1,u_l)} A_1(S, y, z)$$
$$\times N_{k-1}(V_p - 1, V_l, V_c, T_a \mid u_p - S, u_l - y, u_c - z) \quad (15.32)$$

Case 3: if $V_c = 0$, $V_p = 0$, $V_l > 0$. The required probability is obtained by conditioning on a trunk being free on the first link but busy on the second link:

$$N_k(\cdot) = \frac{k}{V_l} \sum_{z=0}^{\min(S-1,u_c)} \sum_{x=z}^{\min(S-1,u_p)} A_1(x, S, z)$$
$$\times N_{k-1}(V_p, V_l - 1, V_c, T_a \mid u_p - x, u_l - S, u_c - z) \quad (15.33)$$

Case 4: if $V_c = 0$, $V_p = 0$, $V_l = 0$. The required probability is obtained on the condition that a trunk is free on both the links. Two possible cases need to be considered: (i) the trunk is available on the two-link path or (ii) the trunk is not available on the two-link path.

Let $B_1(x, y, z)$ denote the number of ways of arranging on a trunk, x busy channels on the first link, y busy channels on the second link, with z channels among them being occupied by calls that continue from the first link to the second, such that the trunk is not available on the two-link path.

Similarly, let $F_1(x, y, z)$ denote the arrangement of the busy channels on a trunk such that the trunk is available on the two-link path. It can be observed that $F_1(x, y, z) + B_1(x, y, z) = A_1(x, y, z)$. $N_k(\cdot)$ can then be computed as

$$N_k(\cdot) = \sum_{z=0}^{\min(S-1,u_c)} \sum_{x=z}^{\min(S-1,u_p)} \sum_{y=z}^{\min(S-1,u_l)}$$
$$\times \left[F_1(x, y, z) N_{k-1}(V_p, V_l, V_c, T_a - 1 \mid u_p - x, u_l - y, u_c - z) \right.$$
$$\left. + B_1(x, y, z) N_{k-1}(V_p, V_l, V_c, T_a \mid u_p - x, u_l - y, u_c - z) \right] \quad (15.34)$$

The starting point, denoted by $N_1(V_p, V_l, V_c, T_a \mid u_p, u_l, u_c)$, of the recursion (for $k = 1$) is assigned 0 if any of the following conditions hold true.

(i) $V_p = 0$ and $u_p = S$.
(ii) $V_l = 0$ and $u_l = S$.
(iii) $V_c = 0$ and $\min(u_p, u_c) = S$.

Otherwise, it is defined in terms of $B_1(u_p, u_l, u_c)$ and $F_1(u_p, u_l, u_c)$ as follows:

$$N_1(\cdot) = \begin{cases} F_1(u_p, u_l, u_c) & \text{if } T_a = 1 \text{ and } V_p = V_l = V_c = 0 \\ B_1(u_p, u_l, u_c) & \text{if } T_a = 0 \\ 0 & \text{otherwise} \end{cases} \quad (15.35)$$

The definitions of $A_1(u_p, u_l, u_c)$, $B_1(u_p, u_l, u_c)$, and $F_1(u_p, u_l, u_c)$ depend on the switch architecture.

15.5 Modeling switches

Two kinds of switches are modeled in this section: space-only and full-permutation switches. For a space-only switch, the channel continuity constraint is enforced by the switch. Hence, a call continuing from the first link to the second one occupies the same channel at the input and output of the switch. Note that although the switch is space-only, the switching provided by the node is channel-space due to the full-channel interchanger at the input of the node. A full-permutation switch, on the other hand, can switch any free channel at the input to any free channel at the output. Consider a call that requires a capacity of b channels, for a trunk with S channels, $A_1(u_p, u_l, u_c)$ and $B_1(u_p, u_l, u_c)$ are defined as below in the two cases.

Space-only switch:

$$A_1(u_p, u_l, u_c) = \begin{cases} \binom{S}{u_p}\binom{u_p}{u_c}\binom{S - u_c}{u_l - u_c} & \text{if } 0 \leq u_p, u_l \leq S \\ & \text{and } u_c \leq \min(u_p, u_l) \\ 0 & \text{otherwise} \end{cases} \quad (15.36)$$

$$B_1(u_p, u_l, u_c) = \begin{cases} \sum_{r=0}^{b-1} \binom{S - r}{u_p}\binom{u_p}{u_c}\binom{u_p - u_c}{S - r - u_l} & \text{if } 0 \leq u_p, u_l \leq S \\ & u_c \leq \min(u_p, u_l), \text{ and} \\ & u_p + u_l - u_c \geq S - b \\ 0 & \text{otherwise} \end{cases}$$

$$(15.37)$$

Full-permutation switch:

$$A_1(u_p, u_l, u_c) = \begin{cases} \binom{S}{u_p}\binom{S}{u_l} & \text{if } 0 \leq u_p, u_l \leq S \\ & \text{and } u_c \leq \min(u_p, u_l) \\ 0 & \text{otherwise} \end{cases} \quad (15.38)$$

$$B_1(u_p, u_l, u_c) = \begin{cases} \binom{S}{u_p}\binom{S}{u_l} & \text{if } \min(S - u_p, S - ul) < b \\ & \text{and } u_c \leq \min(u_p, u_l) \\ 0 & \text{otherwise} \end{cases} \quad (15.39)$$

Several analytical models proposed previously in the literature can be derived from this generalized model by setting the following values:

- model for wavelength routing from [179, 180] by setting $K = W$, $S = 1$, $\gamma_c = 0$;
- model for wavelength routing from [268] by setting $K = W$, $S = 1$;
- model for wavelength routing with a full-permutation switch [143] by setting $K = W$, $S = T$, $\gamma_c = 0$;
- model for wavelength routing with a full-permutation switch by setting $K = W$, $S = F$.

15.6 Heterogeneous switch architectures

In order to analyze networks with heterogeneous node architectures, the mapping of the trunk distributions from one node architecture to the other has to be computed.

Consider an intermediate link of a two-hop path connected by two nodes with different switching architectures as shown in Fig. 15.6.

The first node views the link as K_1 trunks with S_1 channels per trunk while the second node views the link as K_2 trunks with S_2 channels per trunk.

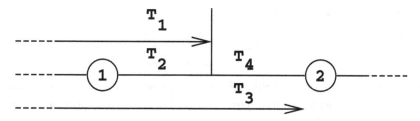

Fig. 15.6. Trunk distribution of a link as viewed by two nodes with different switching architectures.

Let T_1 denote the number of available trunks on the first hop and T_2 denote the number of free trunks on the last link of the first hop as viewed by the first node in the second hop. Let f_2 denote the number of available channels. Available channels are those free channels in the set of available trunks and f_1 denotes the number of free channels on the link.

The required mapping, $P(T_3, T_4 \mid T_1, T_2)$ is then computed as

$$P(T_3, T_4 \mid T_1, T_2) = \sum_{f_1=0}^{KS} \sum_{f_2=f_1}^{KS} P(f_1, f_2 \mid T_1, T_2) \, P(T_3, T_4 \mid f_1, f_2) \qquad (15.40)$$

where $P(f_1, f_2 \mid T_1, T_2)$ denotes the probability of f_2 channels being free on the link with f_1 among them being available given that T_2 trunks are free on the link with T_1 among them being available for a path up to that link. $P(T_3, T_4 \mid f_1, f_2)$ denotes the probability of having T_4 free trunks with T_3 among them available as viewed by another node given that f_2 channels are free on the link with f_1 among them being available.

$P(f_1, f_2 \mid T_1, T_2)$ is computed by conditioning on the number of free channels available in the link as

$$P(f_1, f_2 \mid T_1, T_2) = P(f_2 \mid T_1, T_2) \, P(f_1 \mid T_1, T_2, f_1) \qquad (15.41)$$
$$= P(f_2 \mid T_2) \, P(f_1 \mid T_1, T_2, f_1) \qquad (15.42)$$

where $P(f_2 \mid T_1, T_2)$ denotes the probability of having f_2 channels free in the link given that T_2 trunks are free with T_1 among them being available. This is reduced to $P(f_2 \mid T_2)$ as the number of free channels on the link does not depend on the available trunks in the path. This probability is computed using the two-link model described earlier.

$P(f_1 \mid T_1, T_2, f_2)$ denotes the probability of f_1 channels being available on the link given that T_2 trunks are free with T_2 of them being available and f_2 channels free. It is computed as follows:

$$P(f_1 \mid T_1, T_2, f_2) = \frac{\xi_{T_1}(f_1)\xi_{T_2-T_1}(f_2 - f_1)}{\sum_{f=0}^{f_2} \xi_{T_1}(f)\xi_{T_2-T_1}(f_2 - f)} \tag{15.43}$$

In the above $\xi_t(f)$ denotes the number of ways of arranging f free (or available) channels across t free (or available) trunks such that each trunk has at least one free (or available) channel in it. $\xi_t(f)$ is computed as follows:

$$\xi_T(f) = \sum_{x=1}^{\min(S_1, f)} \xi_1(x)\xi_{t-1}(f - x) \tag{15.44}$$

Let $\xi_1(x)$ denote the number of ways of arranging x free channels over a trunk. With S_1 channels per trunk, $\xi_1(x)$ is computed as follows:

$$\xi_1(x) = \begin{cases} \binom{S_1}{x} & \text{if } 1 \leq x \leq S_1 \\ 0 & \text{otherwise} \end{cases} \tag{15.45}$$

The probability of finding the trunk distribution as viewed by the second node given the trunk distribution and the channel distribution as viewed by the first node depends on how the channels are distributed across the trunks at the two nodes. While there could be several possible choices, there are two cases that are of interest: (i) only the number of trunks that a link is viewed as is known for both the nodes and the exact architectures are not known and (ii) the precise grooming architecture at the two nodes is known.

Case 1: architecture-independent mapping. In this case, the exact architecture of the two nodes connected to a link is not known. The only information that is known is the number of trunks that each node views the link as having. The precise mapping of channels from a trunk as viewed by one node to that viewed by the other is not known. Due to the lack of knowledge of the exact architecture, knowledge of the channel distribution alone is employed for mapping the trunk distribution. The required probability, $P(T_3, T_4 \mid T_1, T_2, f_1, f_2)$, is computed as

$$P(T_3, T_4 \mid T_1, T_2, f_1, f_2) = \begin{cases} \dfrac{\zeta_{T_4}(T_3, f_1, f_2)}{\sum_{t_3=0}^{K_2} \sum_{t_4=t_3}^{K_2} \zeta_{t_4}(t_3, f_1, f_2)} & \text{if } T_3 \leq T_4 \\ 0 & \text{otherwise} \end{cases} \tag{15.46}$$

where $\zeta_t(t', f_1, f_2)$ denotes the number of ways of arranging f_2 calls across t trunks such that a trunk having a call belonging to f_1 available trunks would result

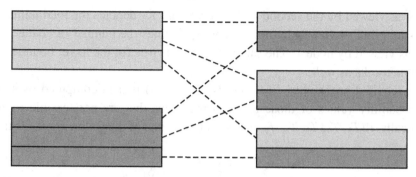

Fig. 15.7. Channel distribution across trunks as viewed by two nodes employing different switching architectures.

in exactly t' trunks being available. This value is computed using the following:

$$\zeta_t(t', f_1, f_2) = \begin{cases} \frac{t}{t'} \sum\limits_{x=1}^{\min(f_1, S_2)} \sum\limits_{y=x}^{\min(f_2, S_2)} \zeta_{t-1}(t'-1, u_1 - x, u_2 - y)\zeta_1(1, x, y) \\ \quad \text{if } t' > 0 \text{ and } u_1 > 0 \\ \sum\limits_{y=1}^{\min(f_2, S_2)} \zeta_{t-1}(t, f_1, f_2 - y)\zeta_1(1, x, y) \\ \quad \text{if } t' = 0 \text{ and } u_1 = 0 \end{cases} \quad (15.47)$$

It has to be noted that the computation of the probability $P(T_3, T_4 \mid T_1, T_2, f_1, f_2)$ does not depend on T_1 and T_2.

Case 2: architecture-dependent mapping. This choice arises from an architectural viewpoint. Note that when a link has multiple fibers, wavelengths, and time slots, the alternatives for trunk switching are limited to treating either the wavelength or the time slot as a trunk, when only limited switching is allowed. In such a case, two nodes that view the link differently would have the channel distribution as considered here. For example, consider a link with two wavelengths and three time slots per wavelength. Let node 1 view the link as wavelength trunks, i.e. two trunks with three channels in each. Let node 2 view the link as time-slot trunks, i.e. three trunks with two channels in each. This scenario is depicted in Fig. 15.7 showing the distribution of channels across trunks as seen by the two nodes.

In this case, due to the regularity in the channel distribution, the knowledge of the trunk and channel distribution as seen by node 1 could be used to derive the lower bound on the trunk distribution as seen by node 2. For example, if three channels are free with one trunk being free as viewed by node 1, then a minimum of three trunks need to be free as viewed by node 2. In general, if f_1 channels and T_1 ($f_1 > 0$ and $T_1 > 0$) trunks are free, then a minimum of $f_1 K_2 / T_1 S_1$ trunks must be free as viewed by the second node. Recall that K_2 denotes the total number of trunks in a link as viewed by node 2 and S_1 denotes the number of channels per trunk as viewed by node 1. The same reasoning is true for the lower bound on the available trunks as well.

The required probability $P(T_3, T_4 \mid T_1, T_2, f_1, f_2)$ is then computed by setting the probability values of those trunk distributions that are not feasible to zero, specifically $P(T_3, T_4 \mid T_1, T_2, f_1, f_2)$ is set to 0 if one of the following holds true:

$$T_1 > 0 \quad \text{and} \quad T_3 < \frac{f_1 K_2}{T_1 S_1} \tag{15.48}$$

$$T_2 > 0 \quad \text{and} \quad T_4 < \frac{f_2 K_2}{T_2 S_1} \tag{15.49}$$

The probabilities are then normalized to set the sum of all the conditional probabilities to 1. This pruning of state-space depends entirely on the architecture, hence it will be different for different architectures.

15.7 Improving the accuracy of the analytical model

It can be observed that the analytical model is developed based on a two-level approach. First, the channel distributions are considered to evaluate trunk distributions at nodes and mapping probabilities. Employing these trunk distribution and mapping probabilities, the blocking performance on a path is obtained. The computation of the blocking performance on a path does not explicitly include the channel distribution on the link. Hence, the analytical model developed in this case is not an exact computation of the blocking performance. However, in the next chapter it is shown that sufficient accuracy is obtained in estimating the path and tree blocking performance.

Some of the interesting features that are exhibited by networks cannot be observed if the analytical model is evaluated only once. For example, if a network rejects a larger number of calls that travel longer distances compared to another network, then it would accept a larger number of calls that travel shorter distances. In order to obtain the finer behavior, the analytical model needs to be evaluated more than once by adjusting the parameters based on the results obtained in the earlier runs.

The two main input parameters to the analytical model are the link load and link load correlation. The computation of these two parameters is described in detail in the next chapter. When calls are rejected by the network, these two parameters are affected. The link load and link load correlation experienced by the network are only due to the calls that are accepted in the network. Hence, blocked calls do not have any effect on these parameters.

Let $P_{b,z}$ denote the probability of a call that requires a connection of length z hops and a bandwidth of b channels. The adjusted link load seen in the network is computed as follows:

$$\lambda' = \frac{N\lambda_n \sum_{z=1}^{N-1} \sum_{b=1}^{B} zbp_{z,b}[1 - P_{b,z}]}{L} \tag{15.50}$$

The average path length and average capacity requirement of a call are adjusted using the following relations:

$$Z'_{av} = \sum_{z=1}^{N-1} \sum_{b=1}^{B} z \, p_{z,b} \, [1 - P_{z,b}] \tag{15.51}$$

$$B'_{av} = \sum_{z=1}^{N-1} \sum_{b=1}^{B} b \, p_{z,b} \, [1 - P_{z,b}] \tag{15.52}$$

The average resource required by a call is computed as follows:

$$R'_{av} = \sum_{z=1}^{N-1} \sum_{b=1}^{B} zbp_{z,b} \, [1 - P_{z,b}] \tag{15.53}$$

The adjusted link load correlation is obtained using the following relation:

$$\gamma'_c = \left(1 - \frac{B'_{av}}{R'_{av}}\right) \frac{1}{E} \tag{15.54}$$

where E denotes the number of exit links in the network.

The adjusted values for link load and link load correlation can be employed to evaluate the blocking performance iteratively. Such an iterative procedure could provide significant insights into the working of the network when the blocking probabilities are beyond a certain threshold. However, if the blocking probability values are very low, then the reduction in the link load and correlation values is not significant. Hence, such iterative procedures do not improve the accuracy of the model at low loads.

16

Validation of the TSN model

In order to evaluate the accuracy of the proposed analytical framework, a set of different network topologies as listed below are used. These topologies are evaluated using a simulation setup described in Section 16.1 and the performances of analytical and simulation models are compared.

(i) Unidirectional ring network with N nodes (N odd)

$$P(z) = \begin{cases} \dfrac{1}{N-1} & 1 \leq z \leq N-1 \\ 0 & \text{otherwise} \end{cases} \tag{16.1}$$

$$E = 1 \tag{16.2}$$

(ii) Bidirectional ring network with N nodes (N odd)

$$P(z) = \begin{cases} \dfrac{2}{N-1} & 1 \leq z \leq \dfrac{N-1}{2} \\ 0 & \text{otherwise} \end{cases} \tag{16.3}$$

$$E = 1 \tag{16.4}$$

(iii) Bidirectional $M \times M$ mesh-torus network (M odd)

$$P(z) = \begin{cases} \dfrac{4z}{M^2-1} & \text{if } 1 \leq z \leq \dfrac{M-1}{2} \\ \dfrac{4(M-z)}{M^2-1} & \dfrac{M-1}{2} < z \leq M-1 \\ 0 & \text{otherwise} \end{cases} \tag{16.5}$$

$$E = 3 \tag{16.6}$$

Table 16.1. *Networks with their average shortest path length and link load correlation*

Network	Average shortest path length	Correlation
25-node bidirectional ring	6.5	0.8462
25-node unidirectional ring	12.5	0.92
11-node unidirectional ring	5.5	0.8182
5 × 5 bidirectional mesh-torus	2.5	0.2
7 × 7 bidirectional mesh-torus	3.5	0.2381
3 × 3 unidirectional mesh-torus	2.25	0.2778
3 × 5 unidirectional mesh-torus	3.214	0.3444
3 × 6 unidirectional mesh-torus	3.706	0.3651

(iv) Unidirectional $R \times C$ mesh-torus network

$$P(z) = \begin{cases} \dfrac{z+1}{RC-1} & 1 \le z < \min(R, C) \\[2mm] \dfrac{\min(R, C)}{RC-1} & \min(R, C) \le z < \max(R, C) \\[2mm] \dfrac{R + C - z - 1}{RC-1} & \max(R, C) \le z < R + C - 1 \\[2mm] 0 & \text{otherwise} \end{cases} \tag{16.7}$$

$$E = 2 \tag{16.8}$$

The above selection of networks represents a variety of average lengths for the shortest path between node-pairs, and hence various link load correlations. Ring networks have longer path lengths and hence have a higher link load correlation. A mesh-torus has the same number of nodes as a ring network, but has a smaller average shortest path length due to the increased connectivity. Hence the mesh-torus has low values of correlation. Table 16.1 lists the specific incarnations of the network topologies and their respective average shortest path length and link load correlation.

16.1 Simulation setup

To validate the analytical model a network simulation framework is used. A network is composed of source–destination nodes that can generate requests and act as a destination for requests generated at other source–destination nodes. A network also comprises dummy nodes that switch traffic across and are not capable of generating requests or being the destination of requests generated at other nodes. With the use of the above two kinds of nodes it has been shown that a node channel-space switch can be modeled. Such a network specification allows one to model networks with a heterogeneous switching architecture.

The requests are generated according to a Poisson process with a rate $N_{(s)}\lambda$, where $N_{(s)}$ denotes the number of source–destination nodes. Any source–destination node in the network is equally likely to be the source of the request. The destination of a request is uniformly selected from the remaining source–destination nodes. A request that is generated is fed to multiple networks that are derived from a physical network, but with differing network architectures. The requests are considered by different networks independently. The networks are assumed to employ a shortest-path routing strategy. If more than one path with the minimum path length is available, then one of them is selected at random. A connection is attempted along the path. If sufficient resources are available, then the call is accepted. Otherwise, the call is rejected. It is to be noted that not all of the paths with minimum length are attempted to establish the call. It is possible that among the paths that have the minimum length, there is one path that could accommodate the call, but is not chosen when selected randomly.

The results presented here are based on the following simulations. The experiments are run for a total of 500 000 requests with performance metrics obtained after every 100 000 requests resulting in five sets of values. The blocking probability is obtained as the fraction of the number of calls blocked to the total number of calls received in the network.

16.2 Homogeneous networks performance

The following two networks are considered: (i) a 25-node bidirectional ring network and (ii) a 5×5 bidirectional mesh-torus network. Three different trunk–channel combinations are considered:

(i) one trunk with 20 channels;
(ii) two trunks with 10 channels each; and
(iii) four trunks with five channels each.

Two different switch architectures are considered: (i) full-permutation switching per trunk (FP) and (ii) channel-space switching (CS).

Full-permutation switching. For full-permutation switching per trunk employed at every node in the network, Fig. 16.1 shows the blocking performance with respect to different path lengths obtained through an analytical model and simulation for the total network load of 60 Erlangs. The simulation results are shown as points,

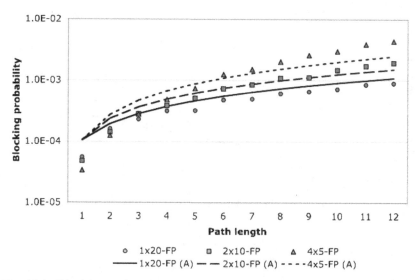

Fig. 16.1. Blocking performance of a 25-node bidirectional ring network with full-permutation switching per trunk for calls of varying path lengths for a network load of 60 Erlangs (link load of 7.8 Erlangs).

while the lines show the values predicted by the analytical model. It is observed that the accuracy of the analytical model improves with an increase in path length. For calls with a path length of one hop, the analytical model predicts a blocking probability that is approximately a factor of 4 higher than the simulation values. It is also observed that the accuracy of the analytical model improves with the decrease in the number of trunks.

One of the effects that is observed with the different trunk and channel combinations is that a network that blocks more calls with longer path lengths tends to accept more calls with shorter path lengths. These are revealed in the simulation. However, the trends are not reflected in the analytical model. If the nodes in a network view a link as having more trunks with fewer channels in each, then it reflects a lower switching capability in the network. In such networks, calls that require longer paths will experience more blocking compared to a network that has more switching capability. Because of this, a network with a lower switching capability would accept more calls traveling over a fewer number of hops compared to networks with a higher switching capability. Analytical models do not predict such a trend when they are employed once. Analytical models directly predict the capability of the switch for a specific load in the network. In order to obtain the trend as seen in the simulation, the link load and the correlation have to be computed analytically based on the blocking probabilities obtained. The computed average link load and the correlation can be used to recompute the blocking probabilities.

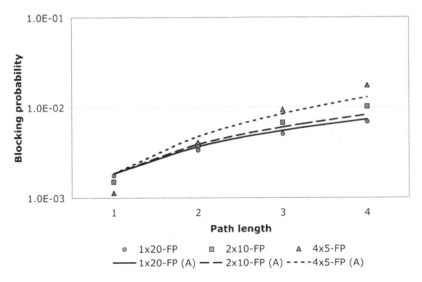

Fig. 16.2. Blocking performance of a 5 × 5 bidirectional mesh-torus network for calls of varying path lengths for a network load of 400 Erlangs (link load of 10 Erlangs).

Figure 16.2 shows the blocking performance on a 5 × 5 bidirectional mesh-torus network for calls of different path lengths for offered network loads of 400 Erlangs (corresponding to link loads of 10 Erlangs). It is observed that the results of the analytical model closely match the simulation results. As observed in the ring network, the analytical model predicts the same blocking performance for different trunk and channel combinations for calls that travel shorter lengths.

Channel-space switching. A channel-space switch is modeled by converting every node in the physical network as one source–destination node with dummy nodes as described in Section 15.5. It is to be noted that despite the full-channel interchanger at the input of every trunk, channel continuity has to be satisfied at the switch. Hence, these switches are expected to block more calls compared to the full-permutation switches.

Figure 16.3 shows the blocking performance of a 25-node bidirectional ring network with a channel-space switch for calls of different path lengths. It is again observed that the analytical model provides an accurate prediction at low loads and the accuracy drops as the load increases.

Comparing the performance of the channel-space switch with that of the full-permutation switching (refer to Fig. 16.1 for the performance of full-permutation switching), it is observed that the blocking performance when employing the channel-space method do not vary significantly. Hence, in networks with a higher link load correlation, channel-space switches can be employed with minimal impact

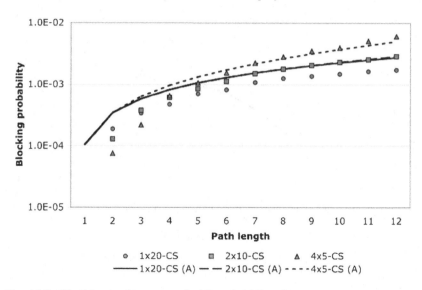

Fig. 16.3. Blocking performance of a 25-node bidirectional ring network, employing channel-space switching, for calls of varying path lengths for a network load of 60 Erlangs (link load of 7.8 Erlangs).

on the blocking performance. Note that a channel-space switch has a lower implementation complexity compared to a full-permutation switch. The increase in the complexity of the switches would necessitate power compensation and clock synchronization at different stages in the switch, thus increasing the cost of switching.

Instead of simulating a more complex 5 × 5 bidirectional mesh network, a 3 × 3 unidirectional network is used for simulation. These two networks are similar in many ways. Both networks have the same maximum hop length. The distribution of the hop lengths are different, hence the link load correlation for the two networks differ, but not significantly. For a 3 × 3 network, the channel-space switch has limited switching capability, the blocking probability increases at a higher rate with increasing path length compared to a full-permutation switch. Hence, as the correlation in the network decreases, the individual link loads become more independent, hence the blocking on longer paths increases.

Figure 16.4 shows the blocking performance of a 3 × 3 unidirectional mesh-torus network with nodes employing channel-space switching for an offered network load of 72 Erlangs. It is observed that the analytical model closely approximates the simulation results. For connections with a path length of one hop, the blocking probability predicted by the analytical model is approximately a factor of 4 higher than the simulation values. This is again due to the effect of networks with a lower switching capability having a tendency to accept more connections of shorter path length compared to networks with a higher switching capability. For path lengths

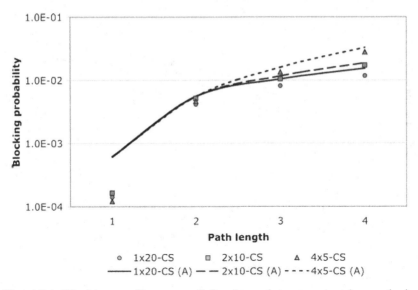

Fig. 16.4. Blocking performance of 3×3 mesh-torus network, employing channel-space switching, for calls of varying path lengths for a network load of 72 Erlangs (link load of 9 Erlangs).

of two through four, the analytical model gives a good estimate of the blocking performance.

For a network with a lower link load correlation, the rate of increase of the blocking performance with increasing path length is higher. One of the trends in the blocking observed with the channel-space switch is that the blocking performance varies by up to two orders of magnitude between path lengths of one and four. Comparing this performance with the performance of full-permutation switching in a 5×5 bidirectional mesh-torus network (Fig. 16.2), the rate of increase due to the full-permutation switch is lower. Also, note that the link load correlation of a 5×5 bidirectional mesh-torus network is lower than that of a 3×3 unidirectional mesh-torus network. While this observation favors employing full-permutation switching, channel-space switching with other trunk assignment algorithms such as first-fit, best-fit, etc. or rearranging the connections would help to reduce the blocking performance even when channel-space switching is employed.

An important observation made from the performance of homogeneous networks is that the blocking performance of establishing unicast connections under different grooming capabilities does not vary significantly. The difference in blocking performance is well within an order of magnitude. Therefore, in networks that have predominantly unicast connections requiring one time-slot capacity, improving the grooming capability of the network may not result in a significant improvement in the blocking performance. For example, it is observed that a significant gain in performance is not achieved by employing one wavelength with 20 time slots (1×20) as compared to four wavelengths with five time slots (4×5). The advantage of employing four wavelengths is that the switching speed in the network can be four times as slow as that employed in a single-wavelength network.

16.3 Heterogeneous networks performance

The blocking performance of a nine-node unidirectional ring network and 3×3 unidirectional mesh-torus networks employing heterogeneous switching and grooming architectures is evaluated in this section. A link with 20 channels organized as two fibers, five wavelengths per fiber, and two time slots per wavelength is considered.

A node can be classified into any one of the following categories based on the level of grooming:

  (i) a time-slot level grooming node;
 (ii) a wavelength-level grooming node; or
(iii) a full-grooming node.

A time-slot-level grooming node would view the link as two trunks with ten channels each. A wavelength-level grooming node views a link as five trunks with four channels each, while a full-grooming network views a link as one trunk. Two different switching architectures are employed at a node for a trunk:

 (i) full-permutation switch and
(ii) channel-space switch.

All of the nodes employ a similar switching architecture within a trunk, while the trunk definition could vary. The nine-node unidirectional ring network and 3×3 unidirectional mesh-torus network with heterogeneous node architectures are organized as shown in Figs. 16.5 and 16.6 with nodes of the same architecture being equally spaced.

It can be seen that any path with a certain path length can be classified into three categories depending on the source. A path may originate from a time-slot-level grooming node, a wavelength-level grooming node, or a full-grooming node. These paths are referred to as path-1, path-2, and path-3, respectively.

Each call needs one time-slot capacity. The results of the analytical model shown in the graphs are obtained without employing knowledge concerning the trunk distribution for mapping the trunk distribution between adjacent nodes. The difference in blocking performance obtained with and without using the exact trunk information is found to be less than 2%. Hence, these are not plotted in the graphs for the sake of clarity.

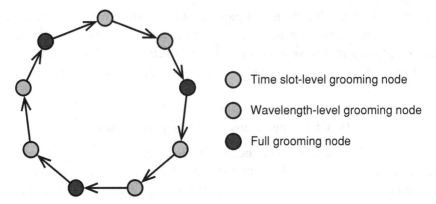

Fig. 16.5. A nine-node heterogeneous unidirectional ring network with three different switch architectures.

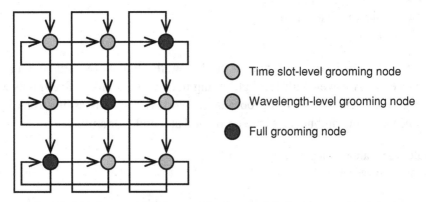

Fig. 16.6. A nine-node heterogeneous bidirectional ring network with three different switch architectures.

Figure 16.7 shows the blocking performance of a nine-node unidirectional ring network with nodes employing full-permutation switching in each trunk for three different path types with varying path lengths for an offered network load of 15 Erlangs (link loads of 7.5 Erlangs).

It is observed that the performance trend observed with the simulation for the different path lengths is also observed through the analysis. The blocking performance as estimated by the analysis is the same for all the paths with a length of one hop. This is due to the reason that a single-hop blocking performance remains the same for a given link load and correlation factor for any switch architecture. For two-hop paths, the blocking performance depends on the switching capability of the intermediate node. The following observations are made.

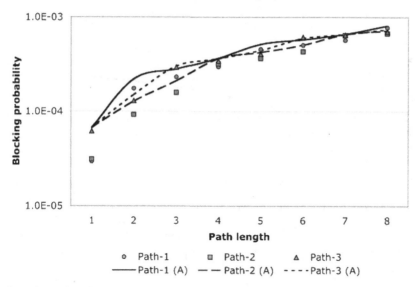

Fig. 16.7. Blocking performance of a nine-node heterogeneous unidirectional ring network with nodes employing full-permutation switching in each trunk for an offered network load of 15 Erlangs.

(i) Calls that would have the intermediate node as a full-grooming node (path-2) would experience the least blocking.

(ii) Calls with the intermediate node as a wavelength-level grooming node (path-1) have the highest blocking among calls that require two-hop connections.

(iii) Similarly, for three-hop connections, calls with intermediate nodes as full-grooming (FG) and TG nodes (path-2) experience the level of lowest blocking.

(iv) Calls with TG and WG nodes as intermediate nodes (path-3) would experience maximum blocking. Now, note that more calls requiring connections with two and three hops are rejected at a wavelength-level grooming node due to insufficient switching capacity at the immediate neighboring node.

(v) Calls requiring a one-hop connection originating at the WG node experience a lower blocking performance.

It is also observed that these performance trends remain the same with increasing load, and thus only depend on where the nodes are positioned.

It is to be noted that although the analytical model shows these trends, the difference in blocking performance between calls of different categories is not exactly the same as that seen in the simulation results. Hence, minor differences in the blocking probabilities seen through simulations may not be observed through the analytical model.

Figure 16.8 shows the blocking performance of a nine-node heterogeneous ring network with nodes employing channel-space switching. The configuration of the

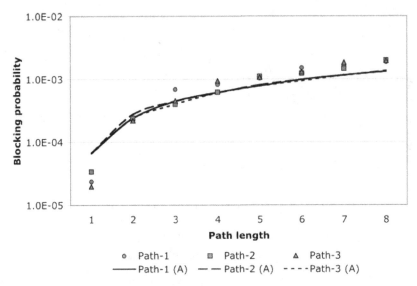

Fig. 16.8. Blocking performance of a nine-node heterogeneous unidirectional ring network with nodes employing channel-space switching in each trunk for an offered network load of 15 Erlangs.

nodes in the ring is similar to that considered earlier (shown in Fig. 16.5). It is observed that the difference in blocking performance observed through simulation for calls originating in different nodes, but having the same path length, are as pronounced as observed when full-permutation switching is employed at the nodes. Hence, the analytical model predicts almost the same blocking performance for paths originating at different nodes but having the same length.

Figure 16.9 shows the blocking performance for paths originating at different nodes versus the path length for offered network loads of 72 Erlangs, in a 3 × 3 heterogeneous unidirectional mesh-torus network (as shown in Fig. 16.6) with nodes employing full-permutation switching in every trunk. It is observed that the performance trends exhibited by the simulation are also reflected by the analytical prediction, although the difference in blocking performance as predicted by simulation and analysis are different.

Figure 16.10 shows the blocking performance for paths originating at different nodes versus the path length for the offered network load of 72 Erlangs in a 3 × 3 heterogeneous unidirectional mesh-torus network (as shown in Fig. 16.6) with nodes employing channel-space switching in every trunk. It is observed that the difference in blocking probability among calls originating at different nodes, but having the same path length, is not as pronounced as that exhibited when full-permutation switching is employed.

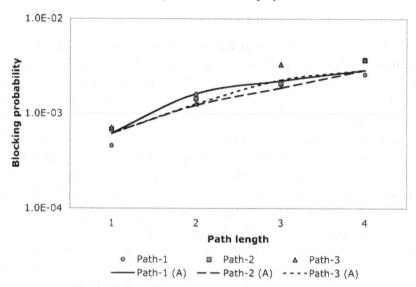

Fig. 16.9. Blocking performance of a 3 × 3 heterogeneous unidirectional mesh-torus network with nodes employing full-permutation switching in each trunk for an offered network load of 72 Erlangs.

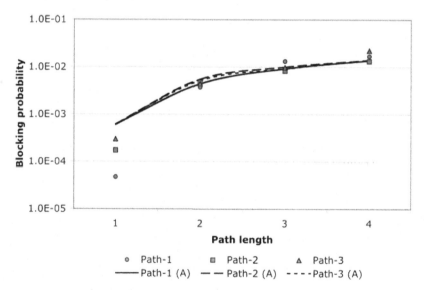

Fig. 16.10. Blocking performance of a 3 × 3 heterogeneous unidirectional mesh-torus network with nodes employing channel-space switching in each trunk for an offered network load of 72 Erlangs.

16.4 Observations

Note that the observations made with regard to the difference in the blocking probabilities of calls originating at different nodes, but having the same path length, exhibiting significant differences in the case of full-permutation switching compared to channel-space switching, cannot be generalized for any arbitrary arrangement of nodes in the ring.

The important observation to be made in the case of heterogeneous networks is that connections with the same hop length could see different blocking probabilities depending on the switching capability of the intermediate nodes. Therefore, it becomes critical to evaluate the importance of a node when the network is upgraded. For example, if only a few wavelength converters are available, then it is critical which nodes in the network are upgraded with this additional flexibility.

Determining the criticality of a node implies evaluating the blocking performance of paths that pass through it under different grooming scenarios. The analytical model allows one to predict the blocking performance of the path with more than one switching architecture. Resource placement algorithms that depend on evaluating path-blocking probabilities to identify an optimal placement of resources in the network employ the analytical model developed in Chapter 15.

17

Performance of dynamic routing in WDM grooming networks

The performance of dynamic routing schemes for WDM grooming networks is explained using the example network shown in Fig. 17.1. Assume that every link carries two wavelengths with four time slots per wavelength. The figure shows the available wavelength capacity (in time slots) on the two wavelengths on each of the links at some point of time during the network operation.

17.1 Information collection

Every node in the network is assumed to maintain the global state information through a link-state protocol. The information collection and path selection are based on two metrics: the available wavelength capacity and the hop length. A path with W wavelengths with C channels per wavelength is denoted by a vector $\{(A_1, A_2, \ldots, A_W); H\}$, where each A_w $(1 \leq w \leq W)$ denotes the number of available channels on a wavelength and H denotes the hop count. For a link vector, the value of H is 1. The available capacity on a wavelength w and the hop length of a path p are denoted by A_p^w and H_p, respectively. It should be noted that the link information is represented here in a vector form for simplicity.

Dijkstra's shortest-path algorithm is extended to the above link-state vector, referred to as extended Dijkstra's shortest-path (EDSP) algorithm, and is employed at every node in the network. The EDSP algorithm uses the link-state vector as defined above instead of the single metric that is traditionally used. The EDSP algorithm has two important operations: (i) combining two path vectors and (ii) selecting the best path vector. Let ψ_{ik} and ψ_{kj} denote the path vectors from node i to node k and from node k to node j, respectively. The path vector from node i to node j through k is obtained by combining the path vectors ψ_{ik} and ψ_{kj}, denoted by $\psi_{ij} = \psi_{ik} \oplus \psi_{kj}$. The vectors are combined in different ways depending on the grooming capability of node k.

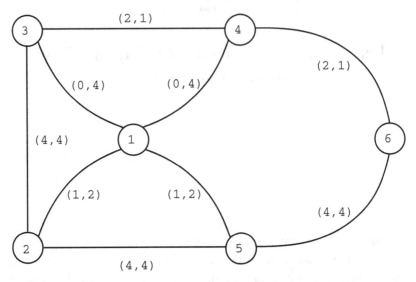

Fig. 17.1. Example network employing two wavelengths per fiber. The tuples denote the available capacity on the two wavelengths, © IEEE. Source: R. Srinivasan and A. K. Somani, Request-specific routing in WDM grooming networks, in *ICC 2002* [187].

The second operation of selecting the best path vector from a given set of path vectors is defined by a specific path selection policy. For example, the traditional shortest-path algorithm selects a path with minimum hop length.

17.1.1 Wavelength-level grooming networks

In wavelength-level grooming networks, connections cannot be switched from one wavelength to another. Hence, a wavelength continuity constraint is obeyed. Two paths vectors ψ_{ik} and ψ_{kj} are combined at a WG node to obtain ψ_{ij} where $A_{ij}^w = \min(A_{ik}^w, A_{kj}^w)$ and $H_{ij} = H_{ik} + H_{kj}$.

Consider the example network shown in Fig. 17.1 and assume that node 4 can perform wavelength-level grooming. The path from node 1 to node 6 through node 4 is described by the vector $\psi_{16} = \{(0, 1); 2\}$.

17.1.2 Sparse full-grooming networks

In sparse full-grooming networks, a few nodes in the network have full-grooming capability. Low-rate traffic streams can be switched across wavelengths at these nodes. Hence, the maximum capacity of a connection that can be switched by an FG node corresponds to the maximum available capacity across different wavelengths on an output link. In such a scenario, the available capacity on a path P_{ij}

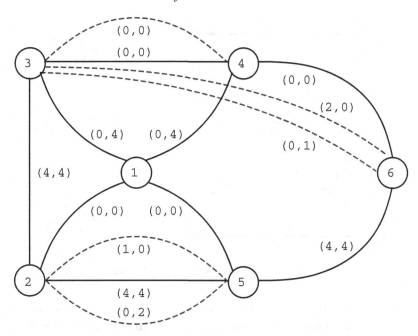

Fig. 17.2. Visualizing a constrained grooming network.

on a wavelength w is obtained by combining two path metrics P_{ik} and P_{kj} as $A_{ij}^w = \min(A_{ik}^w, A_{kj}^{max})$ where A_{kj}^{max} denotes the maximum available capacity across different wavelengths on the path from k to j. The hop length is computed as $H_{ij} = H_{ik} + H_{kj}$.

Again consider the example network in Fig. 17.1 with the assumption that full-grooming is available at node 4. The vector for the path from node 1 to node 6 through node 4 is obtained as $\psi_{16} = \{(0, 2); 2\}$.

17.1.3 Constrained grooming networks

In constrained grooming networks, grooming is accomplished only on the dropped wavelengths. Again, consider the example in Fig. 17.1. Let two connections exist between nodes 3 and 6 through node 4. Assume that the first connection occupies two channels on the first wavelength while the second occupies three on the second wavelength. Although both of the wavelengths have free channels, they cannot be used to reach node 4 as the wavelengths are not dropped at node 3. Hence, when a lightpath is set up between a source and a destination, they can be treated as logical neighbors. The established lightpaths can then be used to route further connections by updating the link-state information.

If the nodes in the network shown in Fig. 17.1 perform constrained grooming, then the network is viewed as shown in Fig. 17.2. The lightpaths that are established

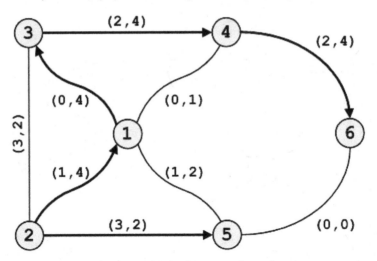

Fig. 17.3. Widest-shortest path routing.

between nodes that are not physical neighbors are shown as dotted lines. Two path vectors in such networks are combined in a manner similar to that of wavelength-level grooming networks.

17.2 Path-selection algorithms

The path-selection algorithms considered are restricted to destination-specific approaches. The different path-selection algorithms specify the rule for selecting the best path vector in the EDSP algorithm. Four examples of path-selection algorithms are listed below.

(i) **Widest-shortest path routing (WSPR).** In this approach, the available wavelength capacity vector on a path is ordered in descending values of the individual wavelength capacities. Thus an available wavelength capacity vector $A'_p = (A'_1, A'_2, \ldots, A'_W)$ is said to be in descending order if $A'_i \geq A'_j$ for $i < j$ and $1 \leq i, j \leq W$.

An ordered vector $A' = (A'_1, A'_2, \ldots, A'_W)$ is said to be smaller than another ordered vector $B' = (B'_1, B'_2, \ldots, B'_W)$ if for some i ($1 \leq i \leq W$), $A'_i < B'_i$ and for all $j < i$, $A'_j = B'_j$. The vectors are said to be equal if $A'_i = B'_i$, for all i, where $1 \leq i \leq W$. Otherwise, A is said to be larger than B. A path with the largest path vector is said to be the *widest* path and is chosen for establishing a connection as illustrated in Fig. 17.3. In the case of a tie, the path with the minimum hop length is chosen.

(ii) **Shortest-widest path routing (SWPR).** This is conventional shortest-path routing based on the hop length. If more than one such path is available, the widest among them is chosen as shown in Fig. 17.4.

(iii) **Available shortest-path routing (ASPR).** In this approach, the shortest path among those that can accommodate the request is chosen. The paths that can accommodate the

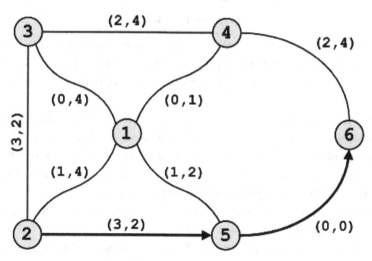

Fig. 17.4. Shortest-widest path routing.

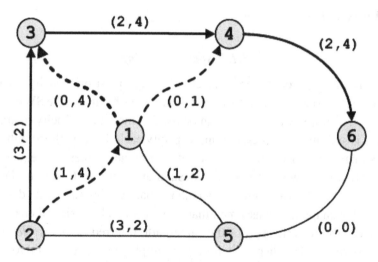

Fig. 17.5. Available shortest-path routing.

request are those that have at least one wavelength that can accommodate the request. If two paths that can accommodate the request have the same hop length, then one of them is chosen at random (Fig. 17.5).

If two path vectors are equal according to any of the above algorithms, one of the path vectors is chosen at random. Note that only the selection of the path vector is based on the ordered available capacity vector. WSPR and SWPR are examples of destination-specific routing schemes while ASPR is an example of request-specific routing. In ASPR, the set of feasible paths is chosen based on the capacity requirement of the request.

17.3 An example

Consider the example network shown in Fig. 17.3. Assume that every link carries two wavelengths and each wavelength is divided into four time slots. The tuples shown in the figure correspond to the available wavelength capacity on each wavelength.

Consider a request that originates from node 2 destined to node 6. The SWPR algorithm selects the path $2 \rightarrow 5 \rightarrow 6$ with path vector $\{(0, 0); 2\}$. The request cannot be accommodated due to a lack of capacity on link $5 \rightarrow 6$. The WSPR algorithm selects the path $2 \rightarrow 1 \rightarrow 3 \rightarrow 4 \rightarrow 6$ with path vector $\{(0, 4); 4\}$. These paths are chosen irrespective of the request requirements.

The ASPR algorithm selects the path based on the request. If the request is for one time slot, then the path $2 \rightarrow 1 \rightarrow 4 \rightarrow 6$ with a path vector $\{(0, 1); 3\}$ or $2 \rightarrow 3 \rightarrow 4 \rightarrow 6$ with path vector $\{(2, 2); 3\}$ is chosen. If the request is for two time slots, the path $2 \rightarrow 3 \rightarrow 4 \rightarrow 6$ with path vector $\{(2, 2); 3\}$ is chosen. If the request is for three or four time slots, then the path $2 \rightarrow 1 \rightarrow 3 \rightarrow 4 \rightarrow 6$ with path vector $\{(0, 4); 4\}$ is chosen.

17.3.1 Dispersity routing

In another routing approach, the connection for a request of capacity b has to be established on just one wavelength, then it is only possible to use the SWP and WSP algorithms. If multiple wavelengths can be used to meet the capacity requirement, the request is split into b requests of unit capacity each. If the path from the source to the destination can accommodate the set of b requests, then the request is said to be accepted. Otherwise, it is blocked. Such an approach to routing larger capacity requests by splitting them into smaller capacity requests is called *dispersity routing*. In this chapter, it is assumed that a request can be assigned channels that are dispersed over wavelengths of the same path, referred to as *wavelength-level dispersity routing*. When dispersity routing is employed, a path is said to be wider if the total available capacity on all the wavelengths in the path is higher.

17.4 Performance of routing algorithms

The performance of four path-selection algorithms described in the previous section is evaluated on the 14-node 22-link NSFnet. The performance results reported in this section are restricted to wavelength-level grooming employed at all nodes in the network.

When a request arrives at a node, the path to the destination is chosen using one of the above-mentioned path-selection schemes. The wavelength allocated to establish the connection is the one that can just accommodate the capacity of the request (best-fit wavelength assignment). SWPR and WSPR algorithms are used in networks that do not allow dispersity routing while SMSPR and MSSPR are used by networks that employ dispersity routing.

17.5 Experimental setup

The experimental setup for the simulation is based on the following assumptions.

- The arrival of requests at a node follows a Poisson process with rate λ and are equally likely to be destined to go to any other node.
- The holding time of the requests follows an exponential distribution with unit mean.
- The capacity requirement of a request is equally likely to take integer values from 1 to 8.
- Every link has a 128-channel capacity divided over W wavelengths.

A network with each link consisting of one fiber with 16 wavelengths in each fiber and eight channels per wavelength is referred to as a 16×8 network. Four different wavelength–channel combinations are considered: (i) 16×8, (ii) 8×16, (iii) 4×32, and (iv) 1×128. The requests are generated independently at a rate of $N\lambda$, where N denotes the number of nodes in the network. The requests are equally likely to have any of the N nodes as their source. The generated requests are fed to the different networks running in parallel and their performances are measured. A total of 6×10^5 requests were generated with performance metrics being measured in batches of 10^5 requests. The average of the performance metrics over six observed sets of values are reported in the results.

17.5.1 Performance metrics

The performance metrics that are measured are:

 (i) the request blocking probability;
 (ii) the average path length of an accepted connection (Z);
(iii) the average shortest path length of an accepted request (Z_m); and
(iv) the network utilization (η).

The blocking probability is computed as the ratio of the number of blocked requests to the number of total requests generated. Z is computed as the average of the length of the paths assigned to the accepted requests by a specific routing algorithm. Z_m is computed as the average of the shortest path length of the requests accepted by the routing algorithm. It can be observed that SWPR would have $Z = Z_m$ while other routing schemes would have $Z \geq Z_m$.

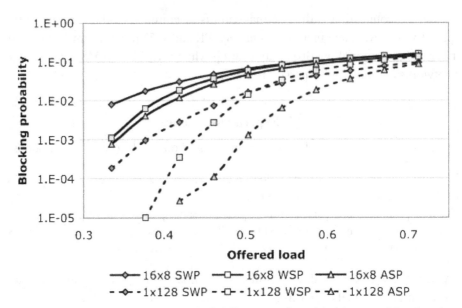

Fig. 17.6. Blocking performance of different routing algorithms on 16×8 and 1×128 NSFnet.

The *network utilization* is computed by assigning an effective network capacity requirement for a request. A request r for capacity b from source s to destination d has an effective capacity requirement of $b \times h_s$, where h_s is the shortest path length from the source to the destination. This effective capacity requirement of a request is the minimum capacity that is required in the network to support the request, irrespective of the routing algorithm. If a routing algorithm selects a path of length h for the connection, $b(h - h_s)$ denotes the additional capacity used by the network to support the connection.

The effective network capacity utilized at an instant of time, denoted by U, is defined as the sum of the effective network capacity requirement of all the connections that are active at that instant. The value of U at any instant of time is bounded by $L \times C$, where L is the total number of links in the network and C is the capacity on each link. The network utilization is then computed as the ratio of the effective used capacity to the maximum capacity of the network as $\eta = U/LC$.

17.5.2 Blocking performance

Figure 17.6 shows the blocking performance of different routing algorithms on 16×8 and 1×128 NSFnet. It is observed that ASPR performs better than SWPR and WSPR. It is also observed that as the network load is increased, the blocking

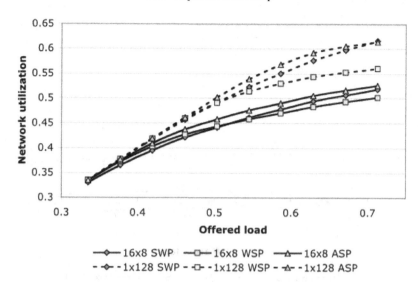

Fig. 17.7. Network utilization of different routing algorithms on 16×8 and 1×128 NSFnet.

performance of WSPR worsens as it routes connections over wider but longer paths resulting in a wastage of bandwidth.

17.5.3 Network utilization

Figure 17.7 shows the network utilization under different routing algorithms on 16×8 and 1×128 NSFnet. It is observed that ASPR achieves the maximum utilization compared to WSPR and SWPR. As the offered load increases, the difference between the network utilization is higher for a 1×128 network over a 16×8 network.

17.5.4 Average path length

More insights into the workings of the algorithms are obtained by observing the average path length of the connections established. Figure 17.8 shows the average path length of connections established in the networks by different routing algorithms. It is observed that WSPR selects longer paths for establishing connections compared to ASPR and SWPR. This difference is significant when the grooming capability in the network is increased. This indicates that increasing the grooming capability helps dynamic routing algorithms to find more paths but at the expense of longer path lengths. SWPR has the least value for this metric as it selects only shortest paths.

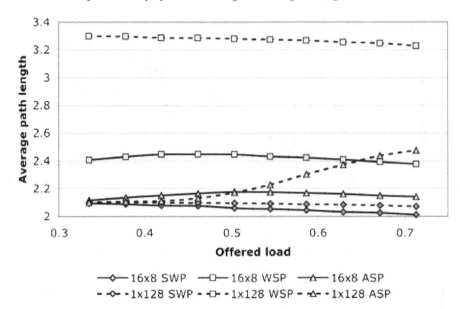

Fig. 17.8. Average length of connections established by different routing algo-rithms on 16×8 and 1×128 NSFnet.

The average path length for a connection established under WSPR remains al-most constant with load as preference is given to distributing the load over the entire network. On the other hand, SWPR only attempts this on the shortest path. As the network load increases, more longer path requests are blocked, hence this results in a decrease in the average path length. ASPR behaves similarly to SWPR under low loads. However, as the offered load to the network is increased, ASPR attempts to route connections on the longer paths, hence the trend of increasing average path length with increasing offered network load.

The average path length for all the routing algorithms for a 1×128 network is higher than that of a 16×8 network because more requests are accepted, but along longer paths in the former network due to the increased grooming capability. Increasing the grooming capability improves the chances of finding a path between nodes, though it would result in a wastage of network resources.

17.5.5 Average shortest path length

Figure 17.9 shows the average shortest path length of accepted requests for different routing schemes. At low loads, very few requests are rejected. Hence, the average shortest path length of accepted requests is the same for different routing schemes. When the offered load to the network is increased, requests with a longer shortest path length experience more blocking, resulting in a bias in favor of requests with a smaller shortest path length. The lower the value of this metric for a routing

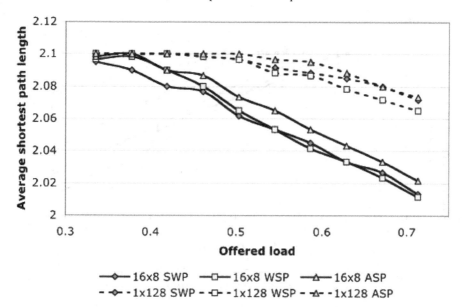

Fig. 17.9. Average shortest path length of connections established by different routing algorithms on 16 × 8 and 1 × 128 NSFnet.

algorithm, the stronger is the bias in favor of requests with a smaller path length. ASPR performs the best with respect to this fairness metric. It is observed that increasing the grooming capability enhances the performance of the routing schemes with respect to this metric.

The routing schemes also exhibit a bias in favor of smaller capacity connections when the offered load to the network is increased. Requests for larger capacity experience more blocking than those for smaller capacity. Such a behavior is pronounced in networks that have less grooming capability. Figure 17.10 shows the average capacity of accepted requests for different routing schemes. It is observed that increasing the grooming capability enhances the fairness of the routing algorithms with respect to requests of different capacity requirement.

The average shortest path length and average capacity of accepted requests quantify the fairness property of the routing algorithms. An ideal routing algorithm would have a constant value for these metrics at all network loads.

It is observed that ASPR offers better performance over SWPR and WSPR algorithms with respect to various performance metrics. Similar performance results are obtained for 8 × 16 and 4 × 32 NSFnet.

17.5.6 Effect of dispersity routing

In order to improve the network performance under different routing algorithms, dispersity routing is also employed. Dispersity routing removes the constraint of

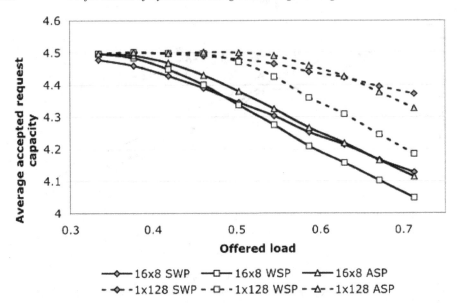

Fig. 17.10. Average capacity of accepted requests for different routing algorithms on 16 × 8 and 1 × 128 NSFnet.

routing a connection entirely on a wavelength, and hence provides greater flexibility in assigning connections. Figures 17.11 and 17.12 show the performance of different routing algorithms on a 16 × 8 NSFnet. It is observed that ASPR performs the best when dispersity routing is employed.

From the results it is clear that employing dispersity routing significantly improves network performance. At an offered load of 0.7, a network employing ASPR with dispersity routing achieves a utilization of 0.62, while the utilization achieved by employing ASPR without dispersity routing is 0.526, a 17.9% improvement. Under dispersity routing, only the end nodes need to maintain information regarding how the connection is split, while the intermediate nodes would route the connection as if they were unit capacity connections.

17.5.7 Effect of dispersity vs. grooming capability

While wavelength-level dispersity routing is one mechanism for achieving increased network performance, alternatives to improve grooming capability can also be considered as a solution. For example, instead of employing 16 wavelengths with eight time slots, one could employ eight wavelengths and 16 time slots. As call requirements still vary only between one and eight time slots, the latter wavelength and time-slot combination would result in reduced blocking. However, such a change increases the transmission speed on a wavelength, requiring faster switches at the

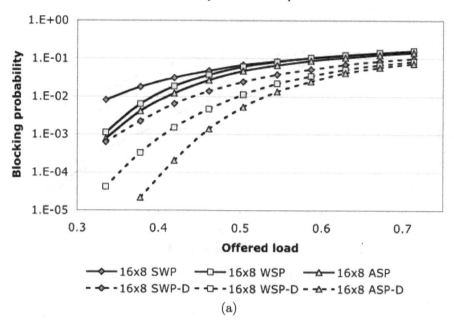

(a)

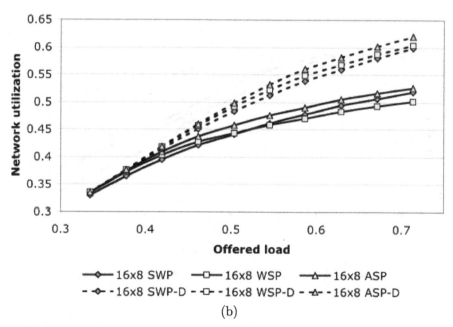

(b)

Fig. 17.11. Performance of different routing algorithms on a 16×8 NSFnet with and without dispersity routing: (a) blocking probability; (b) network utilization.

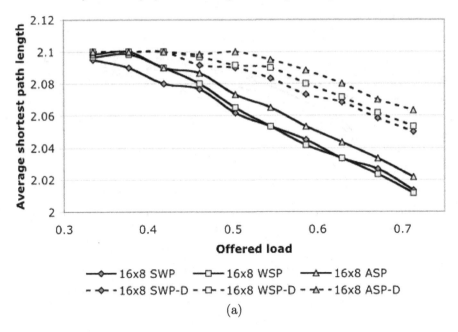

(a)

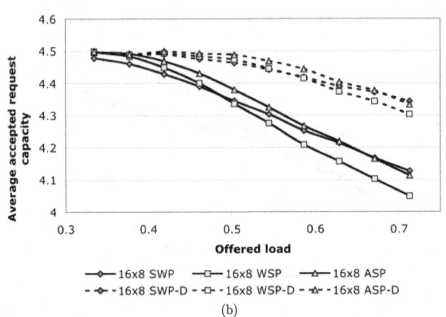

(b)

Fig. 17.12. Performance of different routing algorithms on a 16×8 NSFnet with and without dispersity routing: (a) average shortest path length; (b) average accepted request capacity.

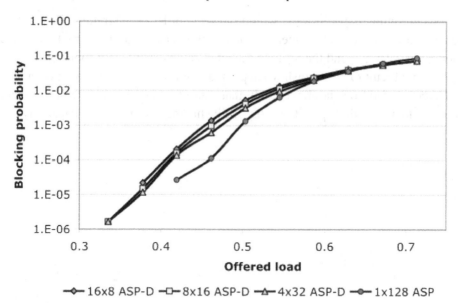

Fig. 17.13. Blocking performance of different routing algorithms on a NSFnet with different levels of grooming capability and dispersity routing.

nodes. In the extreme, one could consider one wavelength with 128 time slots. In this case, the switching speed has to be 16 times faster compared to that in a 16-wavelength eight-time-slot network.

Another approach to achieving improved performance is to employ multiple wavelengths, but to include a wavelength conversion capability. This would eliminate the need for faster switches. For example, consider a network with eight wavelengths and 16 time slots in each wavelength. If a limited wavelength conversion capability is provided at a node where wavelengths W_1 and W_2, W_3 and W_4, W_5 and W_6, and W_7 and W_8 can be interchanged, then this is similar to a network employing four wavelengths and 32 time slots in each wavelength. In such an architecture, the network cost is higher due to the wavelength conversion capability at every node in the network.

The performance of dispersity routing and varying grooming capability is studied to evaluate the tradeoff using four different wavelength and time-slot combinations: 16×8, 8×16, 4×32, and 1×128. In the case of 1×128 there is no distinction between routing a connection with or without dispersity.

Figure 17.13 shows the blocking performance of ASPR with dispersity routing for the four different wavelength and time-slot combinations.

It is observed that the blocking performance is reduced with increasing grooming capability. This performance improvement is observed to be gradual. It is also observed that at offered loads of 0.5 and above, the blocking performance with various grooming capabilities is within the same order of magnitude. The performance with respect to other metrics such as the network utilization, the average shortest path length, and the average accepted call capacity were found to be very close and difficult to distinguish when plotted in graphs, hence they are not shown.

It is concluded from the above results that a significant performance improvement can be achieved with dispersity routing while having less grooming capability.

18

IP over WDM traffic grooming

The popularity of the Internet and internet protocol- (IP-) based internet services is promising enormous growth in data traffic originating from hosts that are IP endpoints. This growth is being fueled by various applications such as those driven by the World Wide Web (WWW) and by the indirect impact of increased computing power and storage capacity at the end systems. The advent of new services with increasing intelligence and the corresponding bandwidth demands are further adding to the traffic growth. New access technologies such as asymmetric digital subscriber line (ADSL), high-bit-rate digital subscriber line (HDSL), and fiber to the home (FTTH) would remove the access bottlenecks and enforce an even faster growth of demand on the backbone network. As noted earlier, these changing trends have led to a fundamental shift in traffic patterns and the traffic is mostly due to data communications.

In the past, the amount of data traffic on carrier networks was small compared with voice-centric traffic. Therefore, the carrier networks were designed to primarily support voice traffic, and the data traffic was transmitted using the voice channels. Now, the core networks are being designed primarily for data traffic with voice support at the edges. Voice traffic can be carried in the core networks using "voice-over-IP" or similar paradigms. To meet these growing demands, the use of WDM will continue to increase in backbone networks. Architectures will be required to satisfy the need for better quality of service (QoS), protection, and availability guarantees in IP networks.

WDM significantly increases the fiber capacity utilization by dividing the available bandwidth into non-overlapping wavelength channels, supporting connections between the two end nodes by establishing an all-optical channel, namely a lightpath, that allows the use of different formats, bit-rates, and protocol transparency. At the same time, most network designers believe that IP is going to be the common traffic convergence layer in communication networks. Consequently, IP over WDM has been envisioned as the winning combination for the network architecture.

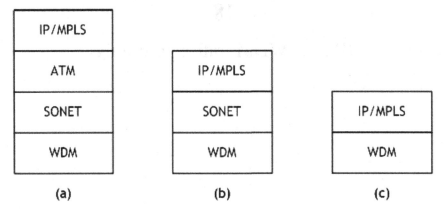

Fig. 18.1. Possible layering architectures.

At present, WDM is mostly deployed in point-to-point networks and the current four-layer architecture is shown in Fig. 18.1(a), in which IP routers are connected to ATM switches. These networks use WDM to send ATM cells over SONET devices that are connected to a WDM transport system. ATM switches are required for multi-service integration (integrating voice and data). In addition, routers are generally limited in speed compared to ATM switches. SONET is required for aggregation (combining 155 Mbit/s ATM streams to OC-48 SONET streams) and to provide protection.

As IP routers become significantly faster and support the quality of service requirement in IP, the need for ATM diminishes. Beginning in 1996, packet over SONET or IP over PPP over SONET started to become a popular approach. The four-layer model depicted in Fig. 18.1(a) hence reduces to a three-layer architecture as shown in Fig. 18.1(b), where IP data traffic is directly transmitted over SONET.

In 1999, several router manufacturers announced fast OC-192 interfaces. That brought the need for traffic aggregation using SONET back under reconsideration. Routers with SONET interfaces that can use the capacity of an entire wavelength are available. Moreover, the protection and restoration function, which is provided by SONET add–drop multiplexers (ADMs), can also be included in the IP and WDM equipment. In 2000, ethernet framing also started gaining a foothold with the evolution of 10 gigabit ethernet. Some researchers believe that SONET would also not be required. The reduced architecture is called "IP over WDM" where IP and WDM are the only two layers that are needed. This two-layer model is shown in Fig. 18.1(c), which aims at a direct integration of IP with WDM optical layers.

Multi-protocol label switching (MPLS) may provide an integration structure between IP and the WDM layer. In an MPLS network, incoming packets are assigned a "label" by a "label edge router" (LER). Packets are forwarded along a "label switch path" (LSP) where each "label switch router" (LSR) makes forwarding decisions based solely on the contents of the label. At each hop, the LSR strips off the existing label and applies a new label which tells the next hop how to forward the packet.

MPLS evolved from numerous prior technologies including Cisco's "tag switching" [308], IBM's "ARIS" [33], and Toshiba's "cell-switched router" [312]. The initial goal of label-based switching was to bring the speed of layer 2 (such as ATM, frame relay or ethernet) switching to layer 3 (such as IP) by replacing the complex IP address-based route lookup with fast label-based switching methods. This initial justification for techniques such as MPLS is no longer perceived as the main benefit, as layer 3 switches are now able to perform route lookups at sufficient speeds to support most interface types. However, MPLS brings many other benefits to IP-based networks such as the following:

- traffic engineering;
- virtual private networks (VPNs); and
- elimination of multiple layers.

Most carrier networks employ an overlay model. In these models SONET/SDH is deployed at layer 1, ATM is used at layer 2, and IP is used at layer 3. Using MPLS carriers can migrate many of the functions of the SONET/SDH and ATM control plane to layer 3, thereby simplifying network management and network complexity. Eventually, carrier networks may migrate away from SONET/SDH and ATM all together. An extension of MPLS, namely multi-protocol lambda switching (MPλS) has also been developed.

18.1 IP traffic grooming in WDM networks

A challenging problem for carrying IP traffic over WDM optical networks is the huge opto-electronic bandwidth mismatch. The bandwidth on a wavelength is 10 Gbit/s or more, while the subrate traffic connections can vary from STS-1 (51.84 Mbit/s) to the full wavelength capacity. Thus, the bandwidth of a full wavelength is becoming too large for a single request. Therefore, the wavelength capacity might be underutilized for IP-centric traffic unless it is filled up by efficiently aggregated traffic.

One approach to provisioning fractional wavelength capacity, as discussed earlier, is to divide a wavelength into multiple subchannels using time-, frequency-, or code-division multiplexing, and then multiplex traffic on the wavelength, i.e. *traffic grooming*. However, optical processing and buffer technologies are still not mature enough to achieve online routing decisions at high speed. With the development of

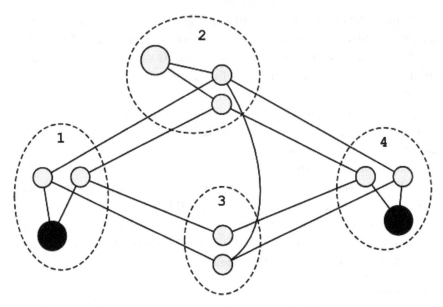

Fig. 18.2. Network representation for integrated routing computation.

MPLS and generalized MPLS (GMPLS) standards, it is possible to aggregate a set of IP packets for transport over a single lightpath. Therefore, traffic grooming in IP over WDM optical networks is performed at two layers, namely *IP traffic grooming* and *WDM traffic grooming*. IP traffic grooming is the aggregation of smaller granularity IP-layer traffic streams. It is performed at MPLS/GMPLS-enabled IP routers by using transmitters and receivers. These aggregated traffic streams are then sent to the optical layer where WDM traffic grooming (or wavelength-level traffic grooming) is performed by utilizing optical add–drop multiplexors (OADMs). The two-layered grooming reduces the workload at both IP and optical layers.

In the IP environment, the network topology is a general mesh and the traffic is typically neither static nor known in advance. Static and dynamic traffic grooming problems have been studied by various researchers. A novel algorithm for integrated dynamic routing of bandwidth guaranteed paths in MPLS networks is developed in [176]. In this work a node is viewed as W subnodes, where W denotes the number of wavelengths. A super-node is created for the node which has wavelength conversion capability. Three different types of nodes, namely routers and OXCs (with or without wavelength conversion capability), are considered. Different logical links are created accordingly so as to create a new network representation. Figure 18.2 gives an example of the network representation for integrated routing computation.

In this example, each link is assumed to have two wavelengths, λ_1 and λ_2. Nodes 1 and 4 are routers, node 2 is an OXC with wavelength conversion, and node 3 is an

OXC without wavelength conversion. Consider a request for 0.1 unit from node 1 to node 4 in Fig. 18.2. If this demand is routed from nodes 1 to 3 to 4 using λ_1, node 3 cannot use λ_1 to route traffic along the path 2–3–4. This is due to the fact that node 3 is an OXC and cannot switch between different wavelengths.

Routing in such a network is therefore decided by taking into account the combined topology and resource usage information at the IP and optical layers, with constraints on the maximum delay or the number of hops. However, the network representation of Fig. 18.2 becomes very complex quickly with the increase in the number of wavelengths. Therefore, it is hard to apply this algorithm in practical DWDM optical networks.

The study in [110] also proposed another auxiliary graph according to the given networking configuration. In this model a node is viewed as $W + 2$ layers with two nodes at each layer, one acting as the input and the other being the output. Apart from W layers with one for each wavelength, two layers called the access layer and the lightpath layer are added. This more general graph model is applicable in heterogeneous WDM mesh networks. An integrated traffic grooming algorithm and an integrated grooming procedure that jointly solve traffic grooming subproblems are developed. Several grooming policies are compared and evaluated through simulations. However, this approach may also face a scalability problem as the number of wavelengths increases. IP over WDM grooming has also been studied in [40, 118, 133, 192, 193, 229].

18.1.1 IP traffic grooming issues

The main cost in IP traffic grooming is due to the transmitters and receivers at the end nodes rather than the number of wavelengths, which was the main cost for grooming in a ring network design. It has been shown that minimizing the required number of transmitters and receivers is equivalent to minimizing the number of lightpaths that are needed, since each lightpath needs one transmitter and one receiver. The problem of minimizing the number of transmitters and receivers for a general topology is studied in [293]. An ILP formulation is developed to solve the transmitter–receiver minimization problem. A heuristic algorithm is presented based on successively deleting lightpaths from an initial topology.

18.2 IP traffic grooming problem formulation

In this section, the design problem for a more general IP traffic grooming network is formulated as an ILP optimization problem. A lower and an upper bound of the transmitter–receiver problem is developed as well as a heuristic algorithm based on traffic matrix transformation.

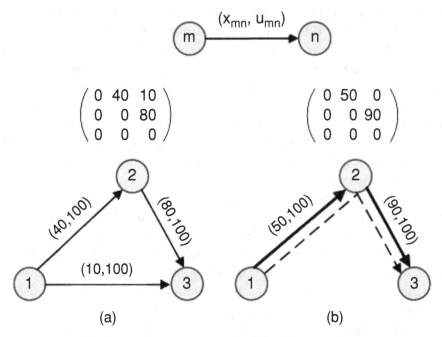

Fig. 18.3. Illustrative example of IP traffic grooming.

Network model. There are two topologies associated with WDM optical networks:

- *physical topology* is represented by a graph $G_p(V, E)$, where V is the set of nodes and E is the set of physical links;
- *virtual topology (logical topology)* is represented by a graph $G_l(V, L)$ with nodes corresponding to the nodes in the physical network and edges corresponding to the lightpaths.

Each lightpath may extend over several physical links (spans). Lightpaths can be viewed as chains of physical channels through which packets are moved from a router to another router toward their destinations. The link flow and link capacity for link (m, n) (from node m to node n) are denoted by x_{mn} and u_{mn}, respectively.

As mentioned earlier, the main cost in IP traffic grooming is due to the transmitters and receivers. The number of transmitters and receivers is equivalent to the number of lightpaths in the network. Figure 18.3 depicts an illustrative example that shows how IP traffic grooming helps to reduce the number of transmitters and receivers in a three-node network.

Assume that each link has a capacity of 100 units. The matrix in Fig. 18.3(a) is the original traffic matrix. It includes the location and capacity of three requests. Figure 18.3(a) depicts one solution in the absence of IP traffic grooming; it simply establishes a lightpath (connection) for each s-d pair. It requires one transmitter and one receiver at each node.

Figure 18.3(b) depicts another solution based on the fact that the capacity requested by s-d pair (1, 3) is relatively smaller. Thus, instead of reserving a separate lightpath for it, the spare capacity along lightpaths $1 \to 2$ and $2 \to 3$ can be reused to accommodate the traffic of the s-d pair (1, 3). That is, the traffic from node 1 to nodes 2 and 3 all take the route from node 1 to node 2. Node 2 receives and analyzes the traffic, drops the traffic that is destined for it and forwards the remaining traffic (from node 1 to node 3) along with its own traffic (from node 2 to node 3) to node 3. This add-and-drop procedure is performed by transmitters and receivers at node 2. In this scenario, the traffic carried by the optical layer is represented by the matrix in Fig. 18.3(b).

The scheme shown in Fig. 18.3(b) results in one fewer transmitters and receivers in comparison to the scheme shown in Fig. 18.3(a). However, the lower size traffic request (1, 3) takes a longer route in the IP layer to avoid reserving an entire wavelength for it. This is the tradeoff that needs to be made in order to alleviate wavelength underutilization in the optical layer. A formal problem statement of the IP traffic grooming problem is given in the next section.

18.3 Solution for an optimal strategy

Let $D_{N \times N} = \{d_{st}\}$ denote the traffic matrix, which represents the capacity requirement of the systems, where d_{st} denotes the traffic capacity required from source node s to destination node t.

The IP traffic grooming problem can be described as follows. *Given a traffic matrix for a network, how does one aggregate the traffic requests, such that the total number of transmitters (and receivers) required in the network is minimized?*

In the virtual topology, each arc corresponds to a lightpath between the node-pair. Hence the problem of minimizing the number of lightpaths is equivalent to minimizing the number of arcs required in the virtual topology.

Notice that if each request is assigned a dedicated lightpath, the virtual topology would be a fully connected network if there is a request for each node-pair. The desired grooming network is the one with a minimum number of transmitters and receivers, which is a solution with a minimum set of arcs in its virtual topology that is sufficient to carry the given traffic.

To simplify the problem, it is assumed that each request has a capacity smaller than or equal to the full-wavelength capacity. Note that for a capacity requirement of more than a full wavelength, there have to be some full-wavelength paths assigned to this request and its remaining capacity needs would be fulfilled using the traffic grooming algorithm. The terms "link" and "arc" are used interchangeably here.

This problem is similar to a capacitated multi-commodity flow design problem [218] with limited link capacities. Therefore, this problem can be formulated as an ILP optimization problem. It is assumed that a request from the same s-d pair will always take the same route. Also, it is assumed that each link has the same capacity that is given by $W \times C$, where W denotes the number of wavelengths carried by a link and C denotes the full-wavelength capacity.

18.3.1 Notation

Parameters.

- W: maximum number of wavelengths in each direction in a bidirectional fiber (technology-dependent data).
- C: maximum capacity of each wavelength. (It is assumed that each wavelength has the same capacity.)
- $s, t = 1, 2, \ldots, N$: number assigned to each node in the network.
- $l = 1, 2, \ldots, L$: number assigned to each link in the network.
- L_{st}^k: (data) for each s-d node-pair, list all possible routes from the source node s to the destination node t, excluding routes that pass through a node more than once, number them using k as an index, e.g., $r_{1,6}^3$ indicates the third route from node 1 to node 6.
- $A_{st}^{l,k}$: (binary data), takes a value of 1 if arc l is on the kth path from node s to node t and is zero otherwise.

Variables.

- γ_{st}^k: binary variable, route usage indicator, takes a value of 1 if route r_{st}^k is taken and zero otherwise.
- u_l: integer variable, logical link usage indicator, keeps an account of the number of lightpaths on arc l in the virtual topology.

18.3.2 Problem formulation

(i) *Objective:* the objective is to minimize the number of arcs in the virtual topology. This reflects the minimum number of lightpaths in the optical layer. Recall that the variable u_l counts the number of lightpaths on arc i in the virtual topology. If the capacity carried by arc i exceeds the full-wavelength capacity, multiple lightpaths between the same node-pair are required. Thus the number of transmitters (and receivers) increases:

$$\min \sum_{l \in L} u_l \qquad (18.1)$$

(ii) *Fiber link capacity constraint:* let TC^l be the total capacity carried by link l, which is given by Eq. (18.2). Constraint (18.3) guarantees that the aggregated capacity on any arc does not exceed the total fiber capacity, which is bounded by $W \times C$:

$$TC^l = \sum_{(s,t),s \neq t} \sum_k \gamma_{st}^k A_{st}^{l,k} d_{st} \tag{18.2}$$

$$TC^l \leq W \times C \tag{18.3}$$

(iii) *Traffic routes constraint:* Eqs. (18.4) and (18.5) ensure that if there is a request from node s to t, one and only one route is assigned to the request. In other words, $d_{st} \geq 0$, set $\sum_k \gamma_{st}^k = 1$. Otherwise, there is no traffic request from node s to node t and none of the routes from node s to node t will be taken, hence, $\sum_k \gamma_{st}^k = 0$.

$$\sum_k \gamma_{st}^k \leq d_{st} \tag{18.4}$$

$$NC \sum_k \gamma_{st}^k \geq d_{st} \tag{18.5}$$

(iv) *Arc usage constraint:* recall that the arc usage indicator u_l counts the number of lightpaths required on arc l (logical link l) in order to carry the aggregated traffic TC^l. $u_l = \lceil TC^l/C \rceil$. This is obtained by using Eqs. (18.6) and (18.7). For example, if $C = 48$ and $TC^i = 62$, $\lceil 62/48 \rceil = 2$ lightpaths are required on logical link i from its start node to its end node,

$$C \times u_l \geq TC^l \tag{18.6}$$

$$C \times u_l \leq TC^l + C \tag{18.7}$$

Notice that from Eqs. (18.3) and (18.6), the total number of lightpaths on a logical link l is bounded by the number of wavelengths on the optical fiber.

Further constraints, such as the limited number of transmitters on each node, can be easily added to this formulation. This helps to capture the cost on each node in the networks.

The limitation of this exact ILP formulation is that it enumerates all the possible routers for each s-d pair and searches for an optimal set of arcs in the virtual topology. In a fully connected network of N nodes, there are up to $\sum_{h=0}^{N-2} P_{N-2}^h$ possible routes for each s-d pair, where P_m^n is the permutation operation. This search requires a large computation time as the network size increases. The formulation can be further simplified by adding a *hop-length constraint* such that the number of possible routes is reduced to a reasonable number; consequently, computation time is saved. However, this network design problem is still a special case of the multi-commodity flow problem, which becomes unmanageable even for moderate-sized networks. Therefore, a heuristic approach would be desired to obtain "good" solutions in a reasonable amount of time that capture all the constraints of the ILP solution.

18.4 Approximate approach

For a network $G(V, E)$, in the absence of IP traffic grooming, the number of transmitters and receivers required at node s, denoted by $Tx_s^{\max}$ and $Rx_s^{\max}$, respectively, can be derived from the matrix $D_{N \times N}$:

$$Tx_s^{\max} = \sum_{t:(s,t)\in E} \left\lceil \frac{d_{st}}{C} \right\rceil \tag{18.8}$$

$$Rx_s^{\max} = \sum_{t:(t,s)\in E} \left\lceil \frac{d_{ts}}{C} \right\rceil \tag{18.9}$$

where C denotes the full wavelength capacity that can be utilized. This is because request d_{st} requires at most $\lceil d_{st}/C \rceil$ transmitters at node s to transmit traffic d_{st}; likewise, it requires at most $\lceil d_{st}/C \rceil$ receivers at node t to receive traffic d_{st} from node s.

From the perspective of network flows, the total amount of outgoing traffic flow seen by node s is $\sum_{t\neq s} d_{st}$, the total amount of incoming flow to node s is $\sum_{t\neq s} d_{ts}$. Hence, the minimum number of transmitters and receivers needed in the network to carry the traffic in $D_{N \times N}$ can be derived using the following two equations:

$$Tx_s^{\min} = \left\lceil \frac{\sum_{t:(s,t)\in E} d_{st}}{C} \right\rceil \tag{18.10}$$

$$Rx_s^{\min} = \left\lceil \frac{\sum_{t:(t,s)\in E} d_{ts}}{C} \right\rceil \tag{18.11}$$

In general, $Tx_s^{\min}$ and $Rx_s^{\min}$ are loose lower bounds. The reason for this is that in order to reduce the number of transmitters (and receivers) some s-d pairs may have to take multiple hops and hence increase the link load in the virtual topology. This overhead load is not captured in Eqs. (18.10) and (18.11), and it is dependent on the traffic pattern.

18.5 Traffic aggregation algorithm

To develop a *traffic aggregation* heuristic approach, the basic idea is to merge the smaller traffic request onto bigger bundles to reduce the number of transmitters and receivers. Although the total number of lightpaths required in the network is reduced, the finer granularity requests may take multiple hops and longer routes. This may introduce a delay for lower-rate requests: it would be affordable in the future slim IP-over-WDM control plane. As a matter of fact, this is a tradeoff which has to be made in order to reduce the overall network cost.

An element in the traffic matrix can be *reallocated* by merging it with other traffic streams. Thus there is no need to establish a direct path for that s-d pair. An element in the traffic matrix can be *aggregated* if it is smaller than the full capacity, i.e. has spare capacity on a wavelength channel and allows other traffic streams to be merged on it. Each element in the traffic matrix can be viewed as being in one of three states:

- state 0: if it can be reallocated or be aggregated;
- state 1: if it cannot be reallocated, but can be aggregated;
- state 2: if it cannot be eliminated or aggregated. For example, if $d_{st} = 0$, there is no traffic to be reallocated, and there is no need to allocate traffic.

The goal of the traffic aggregation algorithm is to choose a traffic stream d_{st} that can be merged with some other traffic streams d_{sn} and d_{nt}, so that d_{st} can be carried using a multiple-hop path and not burden the system to establish a new path for it. After selecting d_{st}, the basic traffic aggregation operation on traffic matrix D consists of the following three steps:

(i) $d_{sn} \leftarrow d_{st} + d_{sn}$;
(ii) $d_{nt} \leftarrow d_{st} + d_{nt}$;
(iii) $d_{st} \leftarrow 0$.

After this operation, the traffic request between s-d pair (s, t) is aggregated on s-d pairs (s, n) and (n, t). Let $TR(T_{s,t,n})$ be the number of transmitters (equal to the number of receivers) needed after merging d_{st} with d_{sn} and d_{nt}. $TR(T^0)$ is called the upper bound, where T^0 is the original traffic matrix.

The key here is to select d_{st} and node n to reduce the value of $TR(T_{s,t,n})$. In experimenting with the ILP formulation, it is observed that the ILP solution uses multi-hop routes for smaller requests, while the bigger requests tend to use direct single-hop paths. This observation is used to develop a heuristic solution. Figure 18.4 gives the *traffic aggregation* algorithm. The resulting new traffic matrix gives the structure of a virtual topology and the required capacity on each physical link. The idea behind this is to integrate a smaller traffic request, say d_{st}, into the larger traffic requests, d_{sn} and d_{nt}, to saturate the existing wavelength paths before establishing a new one. This would force some smaller granularity traffic to take longer routes with multiple hops, while saving some lightpaths.

The algorithm starts by finding the s-d pair with minimum request capacity that is in state 0 (Step (ii) in Fig. 18.4), say d_{st}. Next, it searches for a set of all eligible intermediate nodes, namely K (step (iv)(a) in Fig. 18.4). Define the index value of an item v in set K as $index(v) = \max(d_{sv}, d_{vt})$. The intermediate node n is selected from K to saturate some wavelengths. Hence, if K is not empty, n is chosen as the

---

Input: graph $G(V, E)$ and a traffic matrix $D_{N \times N}$.
Output: rearranged traffic matrix $D_{N \times N}$.
Algorithm:

(i) Initialize s-d pair status:
    if $d_{st} \geq 0$ then $d_{st}.state = 0$,
        else $d_{st}.state = 2$.
(ii) $target = \min(d_{st} : d_{st}.state = 0)$.
(iii) If target = NULL, *terminate*.
(iv) else
    (a) Set K = new stack. Pick node v that satisfies:
        1. $t_{sv}.state \leq 1, t_{vt}.state \leq 1$;
        2. $d_{st} + d_{sv} \leq C$, $d_{st} + d_{vt} \leq C$;
        3. $TR(T_{s,t,v}) < TR(T)$.
        K.push $\{v\}$.
    (b) Define $index(v) = \max(d_{sv}, d_{vt})$, $v \in K$.
    (c) If $K = \Phi$, then $d_{st}.state \leftarrow 1$, go to (ii).
    (d) else $n = \arg \max_{v \in K} \{index(v) : v \in K\}$.
    (e) Update traffic matrix $D_{N \times N}$:
        1. $d_{sn} \leftarrow d_{st} + d_{sn}$;
        2. $d_{nt} \leftarrow d_{st} + d_{nt}$;
        3. $d_{st} \leftarrow 0, d_{st}.state \leftarrow 2$.
(v) Go to (ii).

Fig. 18.4. Approximate approach: traffic aggregation.

node with the maximum *index* value. The algorithm then updates the current traffic matrix after an intermediate node is decided (step (iv)(e) in Fig. 18.4). If K is empty, no eligible intermediate node is found for this s-d pair, $d_{st}.state$ is changed from 0 to 1, which means that request d_{st} cannot be reallocated, but could be aggregated. The algorithm keeps searching for the next s-d candidate for aggregation until no eligible s-d pairs in state 0 can be found.

18.5.1 Complexity analysis

One s-d pair is changed from state 0 to either state 1 or state 2 in each step. Thus the algorithm terminates after at most N^2 passes. The run time for searching *target* in each loop is up to N^2; it takes another N loops to find the set K. Thus, the overall computation complexity of this algorithm is $O(N^5)$. In practice one will never see this complexity and the algorithm terminates much faster than this. One way to improve this is to use effective data structures to make the search more efficient and faster.

18.6 Example of traffic aggregation

Figure 18.5 illustrates an example of how the traffic aggregation algorithm performs. Assume that each wavelength has a capacity of OC-48 (2.5 Gbit/s), and the minimum allocatable unit is OC-1. Thus, $C = 48$. Consider a traffic matrix that is composed of a random combination of OC-1, OC-3, and OC-12. An original traffic matrix includes all possible s-d pairs, shown as the top left-hand matrix in Fig. 18.5.

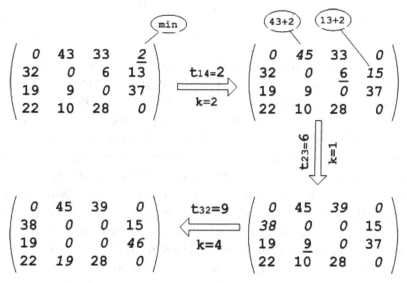

Fig. 18.5. An illustrative example of the traffic aggregation algorithm.

The algorithm starts by finding the minimum eligible s-d pair that can be reallocated, which is (1, 4) with $d_{1,4} = 2$ in this example. Next, it finds the possible intermediate nodes to include into set K. It can be observed that $K = \{2, 3\}$ with $index(2) = 43$ and $index(3) = 37$. Amongst the candidates nodes in K, the one with the highest *index* value is chosen, that is $n = 2$. Next, the current traffic matrix is updated by removing $d_{1,4}$ from the original position and aggregating it with $d_{1,2}$ and $d_{2,4}$. This results in the matrix on the top right-hand side in Fig. 18.5. Next, the algorithm selects $d_{2,3} = 6$ and completes its processing by choosing $n = 1$. The algorithm continues until no more relocatable s-d pairs exist, as shown in Fig. 18.5. The bottom left-hand matrix shows the final results. Application of Eqs. (18.8) and (18.9) indicate that 12 transmitters (and receivers) are required for the original traffic matrix. After traffic aggregation, this number is reduced by three. A more detailed performance study is provided in Section 18.7.

18.6.1 Solutions and results

The ILP formulation of Section 18.3.2 is solved using the CPLEX Linear Optimizer 7.0. The ILP formulation and the traffic aggregation approach are applied to

Table 18.1. *Requests matrix for a six-node network*

	1	2	3	4	5	6
1	0	3	3+1+1	12+12	3+1+1	12+12
2	12+12+12+3	0	3	1+3	0	1+1+12
3	3	1	0	12+12	3+1+1	0
4	3	12	3+12+3+3	0	1	3+1+1+12
5	3	3+12	12	0	0	3+1
6	1+3	12	0	3+12	0	0

solve the IP traffic grooming problem for a six-node network, with $W = 6, C = 48$. Table 18.1 gives a traffic matrix with 50 randomly generated requests. The integer numbers indicate the request capacity in units of OC-1 (51.84 Mbit/s). The objective is to design a network with as few logical links as possible. Notice that there are in total $P_4^0 + P_4^1 + P_4^2 + P_4^3 + P_4^4 = 65$ routes for each s-d pair in a six-node network, and this number increases dramatically as the network size increases. It would be a great burden and might also be unnecessary to obtain optimality by searching among all possible routes. Experiments with different maximum hop lengths (3,4, and 5) are performed on this six-node network. The results show that limiting the hop length to 3 still yields close to an optimal solution while the number of all candidate paths for each s-d pair is effectively reduced from 65 to $P_4^0 + P_4^1 + P_4^2 = 17$. This significantly reduces the size of the feasible region for this ILP formulation, hence it reduces the computation complexity needed in solving the ILP optimization problem.

The results obtained from solving the ILP with a hop length of 3 and traffic aggregation approach are shown in Figs. 18.6(a) and (b), respectively.

According to Eqs. (18.10) and (18.11), at least nine transmitters (receivers) are required. Figure 18.6(a) shows an optimal solution consisting of 11 lightpaths by solving the ILP formulation with a maximum hop-length limit of 3. Figure 18.6(b) shows a solution with 12 transmitters (receivers) using the traffic aggregation approach. Table 18.2 shows the virtual topology routing assignments obtained by solving the ILP formulation and the traffic aggregation heuristic algorithm.

18.6.2 Observations

Figure 18.6 also shows the similarity between the virtual topology design obtained from solving the ILP formulation and the heuristic approach. More specifically, the ILP formulation tends to keep bigger requests on shorter paths in the virtual topology and tries to integrate smaller traffic streams onto bigger bundles. The ILP approach provides an optimal solution by performing an exhaustive search among all possible routes. The traffic aggregation heuristic algorithm also yields a

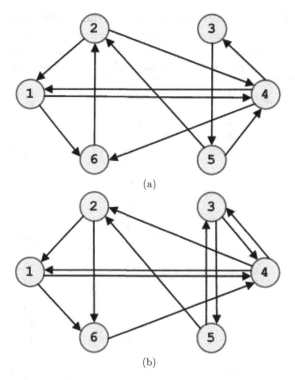

Fig. 18.6. Comparison of the ILP solution and the heuristic approach: an illustrative example. (a) Results obtained by solving the ILP optimization problem with a hop-length limit of 3. (b) Results obtained from the traffic aggregation approach.

very good solution in this example by just performing a local search, which takes much less computation time. However, as an approximate approach, the traffic aggregation heuristic cannot guarantee optimality.

The integration of the traffic helps to reduce the number of transmitters and receivers. On the other hand, it also introduces overhead traffic to the network and impacts on the resource utilization. Besides, it adds potential delays to the requests, which have been reallocated to take multiple hops in the virtual topology. From Table 18.2 it can be observed that the average hop length in the ILP solution is $80/50 = 1.6$. The average hop length in the traffic aggregation heuristic is $77/50 = 1.54$, while without grooming, given enough resources, the minimum average hop length is 1. The more one saves on transmitters and receivers, the longer the average hop length is, and accordingly the longer the average delay. This is an unavoidable tradeoff one would have to face.

The ILP approach becomes unmanageable quickly as the size of the network increases. The reason for this is that the total number of possible arcs in the corresponding fully connected network increases dramatically as the number of nodes

Table 18.2. *Resulting routes in virtual topologies*

Node-pair	Requested capacity	ILP formulation Route on VT	Traffic aggregation Route on VT
1–2	3	1–6–2	1–4–2
1–3	5	1–4–3	1–4–3
1–4	24	1–4	1–4
1–5	5	1–4–3–5	1–4–3–5
1–6	24	1–6	1–6
2–1	39	2–1	2–1
2–3	3	2–4–3	2–1–4–3
2–4	4	2–4	2–6–4
2–6	14	2–4–6	2–6
3–1	3	3–5–2–1	3–4–1
3–2	1	3–5–2	3–4–2
3–4	24	3–5–4	3–4
3–5	5	3–5	3–5
4–1	3	4–1	4–1
4–2	12	4–6–2	4–2
4–3	21	4–3	4–3
4–5	1	4–3–5	4–3–5
4–6	17	4–6	4–2–6
5–1	3	5–4–1	5–2–1
5–2	15	5–2	5–2
5–3	12	5–2–4–3	5–3
5–6	4	5–4–6	5–2–6
6–1	4	6–2–1	6–4–1
6–2	12	6–2	6–4–2
6–4	15	6–2–4	6–4

increases. The performance of the IP traffic aggregation heuristic approach is studied in terms of wavelength utilization in the following section.

18.7 Performance study

A performance study for the above algorithm is carried out using the following performance metrics.

Effective load. With a given traffic matrix $D_{N \times N}$, where d_{st} is the amount of requested wavelength capacity, given the physical topology for a network $G_p(V, E)$ with N nodes, one can apply Dijkstra's shortest-path algorithm to find the shortest path between all s-d pairs. This forms a distance matrix $H_{N \times N} = \{h_{st}\}$, where h_{st} denotes the physical distance from node s to node t. More specifically, here h_{st} represents the shortest hop length from node s to node t. If the number of wavelengths is sufficient, each request would use the corresponding shortest physical path. Thus the effective network load l_{eff} is defined as

$$l_{\text{eff}} = \sum_{(s,t)} d_{st} \times h_{st} \tag{18.12}$$

This gives the minimum network resources in terms of the actual capacity needed for the given traffic requests.

Offered load. In the wavelength-routed optical network without grooming capability, each request is assigned a full wavelength capacity C, even though the actual requested capacity might be only a fraction of C. The minimum offered load of a WDM network in the absence of grooming if denoted as l_{WDM}. It is given by Eq. (18.13) and represents the physical wavelength link product used without the grooming capacity,

$$l_{\text{WDM}} = \sum_{(s,t)} \left\lceil \frac{d_{st}}{C} \right\rceil C h_{st} \tag{18.13}$$

Similarly, let l_{IP} be the offered load by setting up lightpaths based on the new traffic matrix $\bar{D}_{N \times N} = \{\bar{d}_{st}\}$, which is obtained by using the traffic aggregation approach. With sufficient wavelength resources, each s-d pair in $\bar{D}$ would take its corresponding shortest path. Recall the distance matrix $H_{N \times N} = \{h_{st}\}$, l_{IP} can be obtained using Eq. (18.14). More specifically, this provides the lower bound on the actual reserved capacity for the lightpaths after aggregation,

$$l_{\text{IP}} = \sum_{(s,t)} \left\lceil \frac{\bar{d}_{st}}{C} \right\rceil h_{st} C \tag{18.14}$$

Wavelength utilization. The wavelength utilization is defined as the ratio between the effective network load and the actual offered load. Hence, the wavelength utilization in a WDM network without grooming capability and in IP traffic grooming networks are given by Eqs. (18.15) and (18.16), respectively.

$$\eta_{\text{WDM}} = \frac{l_{\text{eff}}}{l_{\text{WDM}}} \tag{18.15}$$

$$\eta_{\text{IP}} = \frac{l_{\text{eff}}}{l_{\text{IP}}} \tag{18.16}$$

18.8 Examples

Figures 18.7–18.9 show a set of experiment results obtained for a 16-node bidirectional ring topology and a 4×4 mesh-torus network, respectively. Additional experiments are for the ARPANET topology. The traffic generation algorithm for a ring, a mesh, and ARPANET are the same. The traffic is uniformly

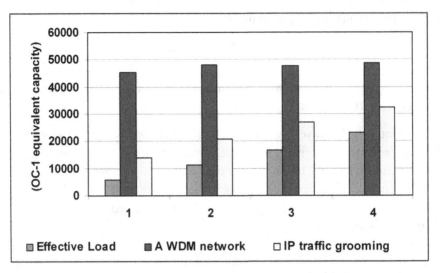

Fig. 18.7. Resource requirement in a 16-node bidirectional ring network.

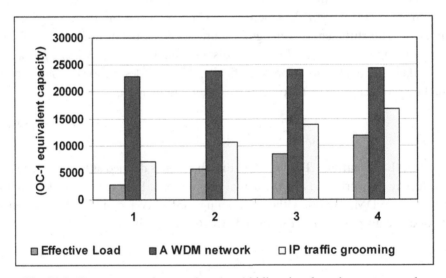

Fig. 18.8. Resource requirement in a 4 × 4 bidirectional mesh-torus network.

distributed among all source–destination pairs. For each s-d pair, an integer number between 0 and the maximum allowable traffic (max) is randomly generated. Thus the mean is $(max - 1)/2$. By increasing the value of max, one can increase the value of the mean traffic in the network. The wavelength utilization is shown in Figs. 18.10 and 18.11 for the two topologies, ring and mesh, respectively. The bars in Figs. 18.7 and 18.8 represent the number of equivalent OC-1 capacity units that are required in different network topologies with different traffic matrices. Only

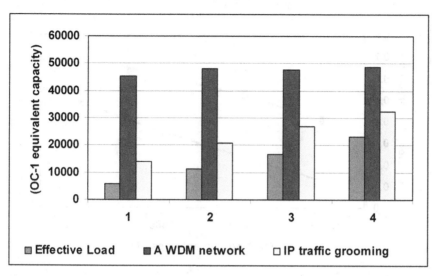

Fig. 18.9. Resource requirement in the 20-node, 31-link bidirectional ARPANET.

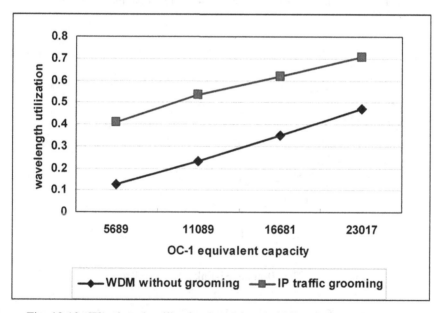

Fig. 18.10. Wavelength utilization in a 16-node bidirectional ring network.

subrate traffic is considered in the experiments. The traffic matrix is randomly generated and the effective load is increased by increasing the mean value of the subrate traffic capacity. Ten experiments are performed for each traffic pattern and the average values are presented as the final results.

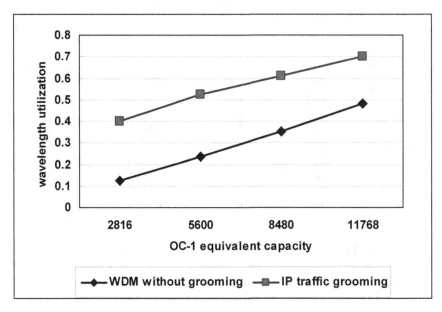

Fig. 18.11. Wavelength utilization in a 4 × 4 bidirectional mesh-torus network.

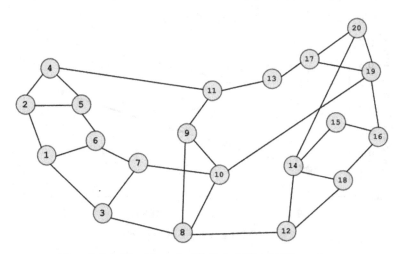

Fig. 18.12. The 20-node, 31-link ARPANET topology.

Simulations on a 20-node, 31-link ARPANET topology (shown in Fig. 18.12) are conducted and the corresponding results are shown in Figs. 18.9 and 18.13.

In the absence of traffic grooming, the capacity required in a WDM network (the middle bar) does not change much as the subrate traffic requests vary. This is because each connection is assigned an entire wavelength irrespective of whether it actually requires a full-wavelength or a fractional wavelength capacity. A significant improvement in the reserved capacity is observed when there are a greater number

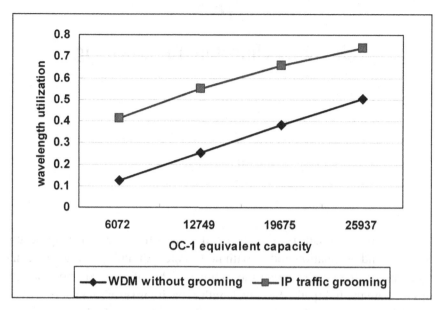

Fig. 18.13. Wavelength utilization in the 20-node, 31-link bidirectional ARPANET.

of finer granularity requests in the traffic matrix. This is because the wavelengths are severely underutilized in WDM networks without traffic grooming when most traffic requests are subrate traffic.

In comparison to the WDM networks without traffic grooming, the capacity reserved in IP traffic grooming networks goes up as the effective load increases. This reflects wavelength sharing among subrate traffic streams, which results in an improvement in wavelength utilization.

Generally, given the same traffic matrix, more wavelength links are required in a ring topology compared with a mesh-like topology. This can be observed from Figs. 18.7 and 18.8 where the same traffic matrices are tested: the effective load in a 16-node ring is almost twice that of a 4 × 4 mesh-torus network. This is partly due to the longer average path length in a ring topology compared to a mesh network with the same number of nodes. Besides, in a ring topology, each s-d pair only has two alternate paths, when establishing the same number of lightpaths. Thus more wavelengths are required in order to satisfy the wavelength continuity constraint. In these experiments, the performance of the developed algorithm for the ring topology is almost as good as it is for the mesh-torus topology.

19

Light trail architecture for grooming

The conventional lightpath is an end-to-end system that is exclusively occupied by its source and destination nodes, with no wavelength multiplexing between the multiple intermediate nodes along the lightpath. Thus if there are not enough IP streams to share the lightpath, the wavelength capacity is severely underutilized for low-rate IP bursts unless the wavelength is filled up by the efficiently aggregated IP traffic. The *light trail* is an architecture concept that has been proposed as a novel architecture designed for carrying finer granularity IP traffic. A light trail is a unidirectional *optical trail* between the start node and the end node [111, 173]. It is similar to a lightpath with one important difference in that the intermediate nodes can also access this unidirectional trail. Moreover, the light trail architecture, as detailed later on, does not involve any active switching components. However, these differences make the light trail an ideal candidate for traffic grooming. In light trails, the wavelength is shared in time by the nodes on the light trail. Medium access is arbitrated by a control protocol among the nodes that have data ready to transmit at the same time. In a simple algorithm, upstream nodes have a higher priority compared to the nodes downstream.

Current technologies that transport IP-centric traffic in optical networks are often too expensive, due to their reliance on an expensive optical and opto-electronic approach. Consumers generate diverse granularity traffic and service providers need technologies that are affordable and seamlessly upgradable. The exclusion of fast switching at the packet/burst level, combined with the flexibility in provisioning for diverse traffic granularity make light trails an attractive option for conventional circuit and burst-switched architectures.

19.1 Light trail

A four-node light trail is depicted in Fig. 19.1. The light trail starts from node 1, passes through nodes 2 and 3, and ends at node 4. Each of nodes 1–3 are allowed

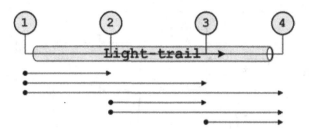

Fig. 19.1. A light trail and possible traffic streams.

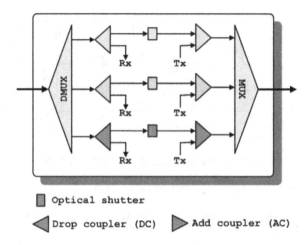

Fig. 19.2. An example node structure in a light trail framework.

to transmit data to any of their respective downstream nodes without the need for optical switch reconfiguration. Every node receives data from upstream nodes, but only a requested destination node(s) accepts the data packets while other nodes ignore them. An out-of-band control signal carrying information pertaining to the setup, tear down, and dimensioning of light trails is dropped and processed at each node in the light trail. Since a light trail is unidirectional, a light trail with N_T nodes can be utilized by up to $N_T(N_T - 1)/2$ optical connections along the trail. The six paths for the four-node light trail are shown in Fig. 19.1.

19.2 Node structure

Figure 19.2 provides a node structure that can be deployed in a light trail framework. In Fig. 19.2, the multiple wavelengths from the input link are demultiplexed and then sent to corresponding light trail switches. A portion of the signal power is directed to the local receiver and the remaining signal power passes through an optical shutter. Such a shutter can be realized using various technologies as an

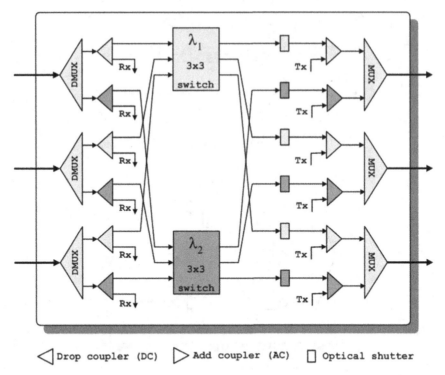

◁ Drop coupler (DC)   ▷ Add coupler (AC)   ☐ Optical shutter

Fig. 19.3. An example light trail node structure with three input fibers with two wavelengths on each fiber.

acousto-optic tunable filter (AOTF). Thus, a node receives signals from all wavelengths. If a particular wavelength is not being used by an upstream node (the incoming fiber has no signal), the local host can insert its own signal, otherwise it does not use the trail. The local signal is coupled with the incoming signal as shown in Fig. 19.2.

Figure 19.3 provides a detailed light trail node structure with three input and three output fibers and two wavelengths on each fiber. The input signal is first demultiplexed, a portion of it is dropped, and the remainder goes to the corresponding 3×3 wavelength switch, as depicted in Fig. 19.3. The outputs of the wavelength switches go through the optical shutter and along with the local added signals, are sent to the output ports of the light trail node. Notice that the optical shutter can be located either before the wavelength switch or after it at the output side.

Figure 19.4 depicts the connection of a four-node light trail in a network and the corresponding ON/OFF switch configurations. The direction of communication is from node 1 to node 4. The light trail on that wavelength is shown separately in Fig. 19.5. The optical shutter is set to an OFF state at the start and end nodes of the light trail such that the signal is blocked from traveling further. For an intermediate

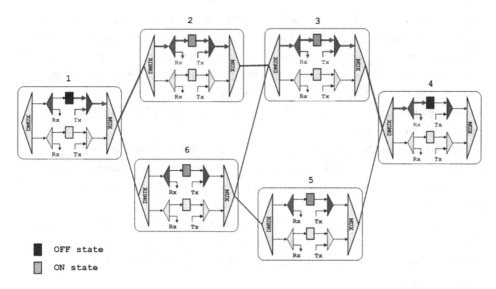

Fig. 19.4. An example of using a light trail in a general topology network.

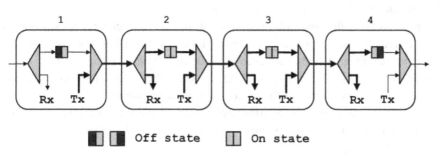

Fig. 19.5. An example node configuration in a light trail framework.

node along the light trail, the optical shutter is set to an ON state to allow the signal to pass through the node.

A unidirectional light trail is thereby obtained from the start node to the end node as shown in Fig. 19.5. No switch reconfiguration is required after the initial light trail setup. Due to the power loss within the light trail, which mainly comes from the power splitting at each node, the length of a light trail is limited and is estimated in terms of the hop length. The expected length of a light trail is four to six hops [168].

19.3 Light trail characteristics

There is no need to dynamically configure any switches when using light trails to carry IP bursts. This leads to an excellent provisioning time. Moreover, the major advantage of using light trails for burst traffic, as compared to optical burst switching (OBS), is the improved wavelength utilization. Here utilization is defined as the ratio of the capacity used over time for actual data transmission to the total reserved capacity. It has been shown that the utilization in OBS is severely degraded compared to that in light trails as the network load increases [168]. More specifically, the utilization of light trails is an order of magnitude better than that in OBS under similar conditions.

Multicasting in the optical layer is another salient feature of the light trail architecture. Nodes in a light trail are able to send the same quanta of information to a set of downstream nodes without the need for special processing or control arbitration.

In general, a light trail offers a technologically exclusive solution that enables a number of salient features and is practical. It exhibits a set of properties that distinguishes and differentiates it from other platforms. The following four characteristics are key properties.

- The light trail provides a way to groom traffic from many nodes to share a wavelength path to transmit their subwavelength capacity traffic.
- The light trail is built using mature components that are configured in such a way as to allow extremely fast provisioning of network resources. This allows for dynamic control of the fluctuating bandwidth requirements on the nodes connected to a trail.
- The light trail offers a method to group a set of nodes at the physical layer to create optical multicasting – a key feature for the success of many applications.
- The maturity of components leads to the implementation of the light trail in a cost-effective manner, resulting in economically viable solutions for mass deployment.

The light trail architecture brings up various issues in designing optical networks for transporting IP-centric traffic. These questions are as follows.

- How to identify a set of light trails at the design phase for the given traffic?
- How hard is this problem?
- What are the new constraints introduced by the light trail architecture?
- How good can wavelength utilization be in light trail networks?
- How to achieve survivability in light trail networks?

These questions are answered in the following.

19.4 Light trail design

To identify a set of light trails to carry the given traffic is one of the key issues in setting up light trails in a WDM network. The performance of a light trail in terms of wavelength utilization depends upon the locations of the light trail. The goal of the design problem is therefore to develop an effective method of grooming traffic in a light trail architecture and to come up with a set of light trails. The light trail design problem is stated as follows. *Given a graph $G(V, E)$, where $|V| = N$, and traffic matrix $D_{N \times N}$, define a minimum number of light trails to carry the given traffic.*

The design problem is expected to be a hard problem. The approach to identifying a set of light trails to be set up in a network presented here consists of two steps. The first step is called the *traffic matrix preprocessing* step. As stated earlier, due to the power losses on the lines, a long light trail may not be advisable. The length of a light trail is limited and is specified in terms of the hop length, denoted by Tl_{max}. A reasonable hop length of a light trail is set to 5. Therefore in the first step, single long-hop traffic is recursively divided into multiple hops.

The second step is to formulate the design problem and to solve it as an ILP optimization problem, for a given network topology and refined traffic matrix obtained from step one. The objective is to find the minimum number of light trails that are required for the system to carry the traffic.

19.4.1 Step I: traffic matrix preprocessing

In preprocessing of a given traffic matrix, a single long-hop traffic is divided into multiple hops to satisfy the hop-length constraint. For a given network physical topology $G(V, E)$, with N nodes and E links, one can apply Dijkstra's shortest-path algorithm to find the shortest path between all s-d pairs. This step results in a distance matrix $H_{N \times N} = \{h_{st}\}$, where h_{st} denotes the physical distance from node s to node t.

The length of a light trail is the main constraint due to the loss both at nodes and over the links. Let Tl_{max} be the maximum length of a light trail. For traffic between an s-d pair (i, j), where $h_{st} > Tl_{max}$, it is not possible to accommodate this traffic on a direct light trail. Thus, this traffic needs to go through multiple hops. Here one light trail is counted as one "hop". This necessitates the first step in this approach, namely *traffic matrix preprocessing*.

Let $D_{N \times N} = \{d_{st}\}$ denote the estimated traffic matrix. Traffic matrix preprocessing returns a modified traffic matrix that satisfies $D_{N \times N} = \{d_{st} : h_{st} \leq Tl_{max}, \forall d_{st} > 0\}$. Figure 19.6 provides pseudocode for the traffic matrix preprocessing algorithm.

In this step, traffic on an s-d pair (s, t) with $h_{st} > Tl_{max}$ is reallocated on multiple hops. The goal is to find a node n such that the path from node s to node n forms the first hop, which is less than Tl_{max} in distance. The next intermediate node n is found recursively for the new source node. Among all possible intermediate nodes, n is chosen to be as close to the destination node t as possible, as shown in step

---

Input: graph $G = (V, E)$ and a traffic matrix $D_{N \times N}$.
Output: rearranged traffic matrix $D_{N \times N}$ and the distance matrix $D_{N \times N}$.
Algorithm:
Step 0: Apply Dijkstra's shortest-path algorithm, calculate the distance matrix $D_{N \times N}$
While (find $(s, t) : d_{st} > 0, h_{st} > Tl_{\max}$)
{

    (i) Pick an intermediate node n:
           $n = \arg \min_{v \in V} \{d_{vt} \mid d_{sv} \le Tl_{\max}\}$;
    (ii) Update traffic matrix $D_{N \times N}$:
           (a) $d_{sn} \leftarrow d_{sn} + d_{st}$;
           (b) $d_{nt} \leftarrow d_{nt} + d_{st}$;
           (c) $d_{st} \leftarrow 0$.

}

Fig. 19.6. Light trail establishment step 1: traffic matrix preprocessing.

(i) in Fig. 19.6. This is done in order to reduce the number of hops that the original traffic has to take.

After preprocessing of the traffic matrix, each non-zero element in the modified traffic matrix would have a corresponding distance that is less than $Tl_{\max}$, the maximum length allowed for a light trail.

19.4.2 Step II: ILP formulation

Given the network topology $G_p(V, E)$, and the modified traffic matrix obtained from Step I, the next step is to list all possible paths within the hop-length limit for each s-d node-pair. This can be accomplished by applying a *breadth-first search* for each node. These eligible paths form a set of *all possible light trails*. Among all of these possible choices, the next step is to choose an optimal set of paths to form the light trail network, such that the total number of light trails is minimized. This problem is formulated as an ILP optimization problem. It is also assumed that each request cannot be divided into different parts and transferred separately.

For the given directed graph $G_p(V, E)$, $N = |V|$, let LT be the set of all possible light trails within a hop-length limit $Tl_{\max}$, and $\tau = 1, 2, \ldots, |LT|$ be the number assigned to each light trail in the LT.

Let C denote the full-wavelength capacity, represented as an integer which is a multiple of the smallest capacity requests. The smallest capacity request is denoted as 1. The integer entry in the traffic matrix $D_{N \times N}$, represented by d_{st}, denotes the requested capacity from node s to node t in units of the smallest capacity request.

A single fiber network with fractional wavelength capacity is considered. Hence, $d_{st} \le C$. In the absence of wavelength converters, the wavelength continuity constraints must be satisfied for light trail networks. The grooming itself helps to increase the wavelength utilization and reduces the total number of wavelengths that are required to satisfy the traffic needs. The following notation is used in the problem formulation.

19.4.2.1 Variables

- μ_{st}^{τ}: (binary variable) route indicator, takes a value of 1 if request (s, t) takes light trail τ and is zero otherwise. This also implies that nodes s and t are on trail τ and s is the upstream node for t.
- δ^{τ}: (binary variable) light trail usage indicator, takes a value of 1 if trail τ is used by any request and is zero otherwise.

19.4.2.2 ILP formulation

(i) *Objective:*

$$\min \sum_{\tau} C_{\tau} \times \delta^{\tau} \tag{19.1}$$

When $C_{\tau} = 1$, the objective is to minimize the number of light trails that are required in the network. When C_{τ} is defined as the *hop length* of light trail τ, the problem becomes to minimize the total number of wavelength links in the networks, which represent the total reserved capacity in the networks. This can be used to optimize the wavelength capacity utilization, although that might consume more light trails.

(ii) *Assignment constraint:* each request is assigned to one and only one light trail:

$$\sum_{\tau} \mu_{st}^{\tau} = 1 \quad \forall (s, t): d_{st} \in D, d_{st} > 0 \tag{19.2}$$

(iii) *Light trail capacity constraint:* the aggregated request capacity on a light trail should not exceed the full wavelength capacity:

$$\sum_{(s,t)} \mu_{st}^{\tau} d_{st} \leq C \tag{19.3}$$

(iv) *Light trail usage constraint:* if any of the s-d pairs is assigned on light trail τ, δ^{τ} is set to 1; otherwise, if none of the s-d pairs picked light trail τ, $\delta^{\tau} = 0$. Recall that δ^{τ} is a binary variable:

$$\delta^{\tau} \geq \mu_{st}^{\tau} \quad \forall (s, t): d_{st} \in D \tag{19.4}$$

$$\delta^{\tau} \in \{0, 1\} \tag{19.5}$$

19.5 Solution considerations

The light trail design is a challenging problem for the following reasons.

First, in order to use a wavelength fully, one would like to *groom* near full-wavelength capacity traffic onto the wavelength. This is similar to a normal traffic grooming problem, which is often formulated as a *knapsack problem* and is known to be an NP-complete problem. However, it might be infeasible to simply set up a light trail for any set of traffic requests that add up to C. For example, given that $d_{12} + d_{13} + d_{16} = C$, it might not be possible to establish the desired light trail due to the physical hop-length constraint. As a matter of fact, the light trail hop-length limit introduces complexity to the problem.

Secondly, the ILP formulation of the light trail design problem is similar to the *bin packing* problem [157], which is an NP-hard problem. However, if light trails are treated as the "bins" and elements in the given traffic matrix as the "items" in the bin packing problem, this problem differs from a normal bin packing problem due to a potential physical route constraint that an item cannot be put in any of the given bins, but only a subset of the bins. More specifically, an s-d pair can be assigned to the routes that satisfy: (i) nodes s and t belong to the route and (ii) node s is the upstream node of node t along the route. Hence, the approximate algorithms for solving normal bin packing problems cannot be directly applied here for solving the light trail design problem.

Light trail design: heuristic approaches. Since the study of [124] proves that the light trail design problem is NP-hard, the following heuristic algorithm for light trail design is proposed. It is well known that the first-fit and best-fit algorithms are two common and effective heuristic algorithms for solving bin packing problems. In the following, the best-fit algorithm is used to solve the light trail design problem.

19.5.1 The best-fit approach

Recall that after traffic matrix preprocessing, each request in the newly obtained traffic matrix satisfies the light trail hop-length limit, that is, the shortest hop length for each s-d pair is no greater than Tl_{max}.

The goal of the second step is to identify a set of light trails for carrying the given traffic. To do this, first pick the s-d pair that has the longest distance in the distance matrix H_{st}, since a light trail between this s-d pair is eventually required.

Once an s-d pair with the longest physical hop length is found, the *head* and *tail* of a light trail are decided. The goal now is to find the *best* eligible light trail between these two end nodes. This is analogous to fully packing a "bin" in the bin packing problem. There are two subproblems that need to be solved. First, selection of a path (within the hop-length limit) between these two nodes is required. Secondly, assignment of requests to this light trail needs to be identified.

In order to find the best light trail between the known *head* and *tail* nodes, an exhaustive search among all the possible paths between the two nodes is performed. Here the best-fit algorithm tries to pick up the path between the head and tail nodes that is the *best* among all the paths available between the *head* and *tail* nodes. This is still a *local* search, therefore the final results might not be globally optimal.

For each eligible path between the known *head* and *tail* nodes, all possible s-d pairs along this path are sorted according to their required capacities, before the routing decision is made. There are two different ways of packing them onto a path rather than doing it randomly. One is to allocate the smallest requests first, which is called *increasing packing order*, and the other way is to allocate the biggest requests first, which is called *decreasing packing order*.

- Increasing packing order tries to allocate finer requests first, so that the number of requests that can be packed onto this path is maximized. There might still be some capacity left on this light trail, but that is not sufficient for the next smallest request. This approach grooms as many requests as possible onto the light trail, thereby leaving the rest of the network with fewer requests that still need to be allocated. The expectation is that this will contribute to a saving in the total number of light trails that are needed in the network. However, for each light trail, the packing might not be the most efficient, or the spare capacity might not be minimized.
- Decreasing packing order tries to allocate bigger requests first, and leaves the light trail with minimum spare capacity. However, since the big requests are allocated first, the total number of requests that can be carried by the light trail might be smaller than that of the allocation in *increasing packing order*. Therefore, it could leave more requests unallocated in the network and more light trails might need to be set up later on in order to carry all the requests. The spare capacity on each light trail is minimized in this approach at the time of allocating the capacity.

It is not clear which approach works better and always gives the minimum number of light trails required in the network. It depends on the traffic patterns. A preferred approach is to try both and choose the one that yields a better solution for given data.

19.5.2 Algorithm design

For the given graph, all possible paths for each s-d pair can be computed. The path information may be stored in the data structure called $KSPath[N][N][NRoute_{max}]$ that contains the path information for each route in the network.

For efficient usage, paths are sorted according to their physical hop length, such that $KSPath[head][tail][1]$ contains the shortest-path information (*hop length, intermediate nodes along this path*) *head* to *tail*, and so on.

Figure 19.7 gives pseudocode for the best-fit algorithm. In this pseudocode, *seq* is used to denote a route among all valid routes from which *head* and *tail* are chosen to be the trail. Also notice that only subwavelength level requests are considered here. Therefore by default a shortest path is chosen as the light trail to carry a given request if no better path can be found; that is, initially *seq* = 1.

Input: graph $G = (V, E)$, the rearranged traffic matrix $D_{N \times N}$ and distance matrix $H_{N \times N}$.
Output: a collection of light trails.
Algorithm:
Initializations: $d = 0$, $R = \{(m, n): d_{m,n} > 0\}$.
Do {

    (i) $(m, n) = \arg \max \{h_{m,n}: (m, n) \in R\}$.
         $head = m$, $tail = n$.
    (ii) $Trail_{\mathrm{cap}} = d_{m,n}$, $newstream = Trail_{\mathrm{cap}}$,
         $best = 0$, $seq = 1$.
    (iii) **for**$(\tau = 1; \tau \leq NRoute_{\max}; \tau + +)$
         if$(KSP[head][tail][\tau].length \leq Tl_{\max})$

         (a) Copy all s-d pairs along path $KSP[head][tail][\tau]$ that need to be allocated to array $AllRequest[\,]$.
             The length of $AllRequest[\,]$ is known and denoted by NSD;
         (b) Sort $AllRequest[\,]$ according to the capacities;
         (c) **for**$(tmp = 1; \ tmp \leq NSD; tmp + +)$
             if $(newstream + AllRequest[tmp].cap \leq C)$
                 $newstream = newstream + AllRequest[tmp].cap$;
         (d) **if** $(newstream > best)$
             {

                 $best = newstream$;
                 $seq = \tau$;

             }

    (iv) Copy all s-d pairs along path $KSP[head][tail][seq]$ that need to be allocated to array $AllRequest[\,]$.
         The length of $AllRequest[\,]$ is known and denoted by NSD;
    (v) **for**$(tmp = 1; \ tmp \leq NSD; tmp + +)$
         if $(newstream + AllRequest[tmp].cap \leq C)$
         {

            $Trail_{\mathrm{cap}} = Trail_{\mathrm{cap}} + AllRequest[tmp].cap$;
            $d_{AllRequest[tmp].src, AllRequest[tmp].dst} = 0$;

         }

} *While* $(R \neq \Phi)$

Fig. 19.7. Light trail design step 2: best-first approach.

When there is a tie in route selection, the path that can accommodate most requests is chosen. It is possible to design and apply different criteria. As mentioned earlier, sorting *AllRequest[ ]* in different ways yields different algorithms, namely *best-fit decreasing packing order* and *best-fit increasing packing order*.

19.5.3 Discussions

The proposed heuristic algorithm has two steps, as shown in Figs. 19.6 and 19.7. Both the first step and the second step would need information on the paths between

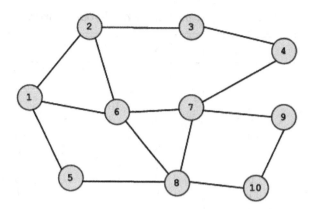

Fig. 19.8. A 10-node example network.

each s-d pair. Therefore, one can first find all possible paths for each s-d pair. The worst-case complexity of an exhaustive search for each s-d pair is $O(N^3)$. The total running time for finding all possible routes is $O(RN^3)$, where R is the number of s-d pairs (requests). In fact, instead of searching for all paths, it is preferable to search among the K-shortest paths with K being big enough. This could reduce the complexity to $O(N(E + N \log N + KN))$ for all node-pairs. This may be a promising choice for big networks.

In the best-fit packing of step 2, for each s-d pair, the best-fit route is chosen among all K paths. For path τ with n_τ nodes, there are a maximum of $t = (n_\tau - 1) + (n_\tau - 2) + \cdots + 1 = O(n_\tau^2)$ s-d pairs, where n_τ is bounded by $T l_{max}$. Hence $t = O(T l_{max}^2)$. The sorting takes $O(t \log t)$ loops, and packing takes another t loops. Thus the total complexity is $O(t \log t)$ loops for each path. There are K paths, and the same procedure is performed on the selected best-fit path. Therefore, a total of $O(K(t \log t)) = O(K(T l_{max}^2 \log T l_{max}))$ loops are needed for each s-d pair. At least one s-d pair is eliminated from matrix R in Fig. 19.7 in each step and the program stops when R is empty.

19.5.4 Algorithm performance

To evaluate the performance of the above ILP formulations and heuristic algorithms, experiments are performed on a physical topology given in Fig. 19.8. To simplify the problem, it is assumed that each physical link is bidirectional with the same length.

Table 19.1 gives a randomly generated traffic matrix for this example. The integer numbers indicate the requested capacity in units of OC-1 (51.84 Mbit/s). The entire wavelength capacity is OC-48. As aforementioned, only the fractional wavelength capacity is considered for traffic grooming in light trail networks. Intuitively, if

Table 19.1. *Traffic matrix for a 10-node network*

	1	2	3	4	5	6	7	8	9	10
1	0	5	8	11	3	8	5	7	8	10
2	3	0	8	4	0	5	1	2	3	1
3	9	3	0	7	3	10	11	8	0	6
4	6	0	8	0	2	5	5	2	1	1
5	0	6	10	4	0	2	11	10	5	2
6	11	3	4	4	3	0	2	6	8	3
7	0	2	10	2	11	5	0	1	6	0
8	0	5	6	2	3	1	11	0	5	0
9	4	5	11	8	8	2	3	1	0	5
10	0	9	9	3	7	10	1	2	1	0

Table 19.2. *ILP: Resulting light trails $Tl_{max} = 4$*

No.	Light trails	Hops	Accommodated s-d pairs	Load
1	{2, 3, 4, 7, 9}	4	(3, 7) (3, 4) (2, 7) (2, 9) (4, 9)	23
2	{3, 2, 6, 8, 10}	4	(2, 6) (2, 8) (2, 10) (3, 6) (3, 8) (3, 10)	32
3	{4, 3, 2, 1, 5}	4	(4, 1) (4, 3) (4, 5) (3, 5) (1, 5) (3, 1) (2, 1)	34
4	{4, 7, 6, 8, 10}	4	(6, 8) (6, 10) (4, 6) (4, 7) (4, 8) (4, 10)	22
5	{5, 1, 2, 3, 4}	4	(1, 2) (1, 3) (1, 4) (5, 2) (5, 3) (5, 4) (2, 4)	48
6	{5, 1, 6, 7, 9}	4	(1, 7) (1, 9) (6, 9)	21
7	{5, 1, 6, 8, 10}	4	(1, 8) (1, 10) (1, 6) (5, 6)	27
8	{5, 8, 7, 9, 10}	4	(9, 10) (8, 9) (5, 9) (5, 8) (5, 7) (7, 9) (5, 10)	44
9	{9, 7, 4, 3, 2}	4	(9, 2) (9, 3) (9, 4) (7, 3) (7, 2) (3, 2)	39
10	{9, 7, 6, 1, 5}	4	(7, 6) (6, 5) (9, 1) (9, 6) (6, 1)	25
11	{10, 8, 6, 2, 3}	4	(10, 3) (10, 2) (8, 3) (8, 2) (6, 3) (6, 2) (2, 3)	44
12	{10, 8, 6, 7, 4}	4	(10, 6) (10, 4) (7, 4) (6, 4) (6, 7) (8, 4) (8, 6) (8, 7)	35
13	{10, 9, 7, 8, 5}	4	(10, 9) (10, 8) (10, 7) (10, 5) (9, 8) (9, 7) (9, 5) (8, 5) (7, 8) (7, 5)	38

every s-d pair requires a capacity greater than half of the full-wavelength capacity, no two requests can be groomed on a light trail. Thus, it is assumed that most s-d pairs request a small fractional capacity of the full-wavelength channel. Hence, integer numbers between 0 and 11 are randomly generated as requested capacities in the experiments. The resulting traffic matrix is shown in Table 19.1.

Table 19.3. *Local best-fit: resulting light trails* $Tl_{max} = 4$

No.	Light tails	Hops	Accommodated s-d pairs	Load
1	{3, 2, 6, 8, 10}	4	(3, 10) (2, 10) (2, 8) (3, 2) (6, 10) (2, 6) (6, 8) (3, 8) (3, 6)	44
2	{10, 8, 6, 2, 3}	4	(10, 3) (8, 6) (10, 8) (6, 2) (6, 3) (8, 2) (8, 3) (2, 3) (10, 2)	47
3	{1, 6, 2, 3, 4}	4	(1, 4) (6, 4) (2, 4) (1, 2) (3, 4) (1, 3) (1, 6)	47
4	{1, 5, 8, 10, 9}	4	(1, 9) (10, 9) (5, 10) (1, 5) (8, 9) (5, 9) (1, 8) (1, 10)	41
5	{2, 6, 8, 7, 9}	4	(2, 9) (2, 7) (6, 7) (7, 9) (6, 9) (8, 7)	31
6	{3, 4, 7, 8, 5}	4	(3, 5) (7, 8) (4, 5) (4, 8) (8, 5) (4, 7) (7, 5) (3, 7)	38
7	{4, 3, 2, 6, 1}	4	(4, 1) (2, 1) (4, 6) (4, 3) (3, 1) (6, 1)	42
8	{4, 7, 9, 10}	3	(4, 10) (4, 9) (9, 10)	7
9	{5, 8, 7, 4, 3}	4	(5, 3) (8, 4) (7, 4) (5, 4) (5, 8) (7, 3)	38
10	{9, 7, 6, 2, 1}	4	(9, 1) (9, 6) (7, 2) (9, 7) (7, 6) (9, 2)	21
11	{9, 7, 4, 3}	3	(9, 3) (9, 4)	19
12	{9, 10, 8, 5}	3	(9, 5) (9, 8) (10, 5)	16
13	{10, 8, 6, 7, 4}	4	(10, 4) (10, 6) (10, 7)	14
14	{1, 5, 8, 6, 7}	4	(1, 7) (5, 6) (5, 7)	18
15	{5, 1, 2}	2	(5, 2)	6
16	{6, 1, 5}	2	(6, 5)	3

19.6 Light trail hop-length limit: $Tl_{max} = 4$

CPLEX Linear Optimizer 7.0 is used to solve the ILP formulation proposed. It is assumed that each candidate path can be used once, that is, $u = 1$. Assume that the hop-length limit $Tl_{max} = 4$, from the topology it is observed that all s-d pairs have paths within this hop-length limit. Hence, the *traffic matrix preprocessing* does not make any change in the given traffic matrix.

Table 19.2 presents the results obtained by solving the ILP formulation with a hop-length limit of $Tl_{max} = 4$. It is observed that $W = 4$ is sufficient on each link, although there is no constraint imposed on the number of wavelengths.

Table 19.2 shows that 13 light trails are needed to carry the given traffic. The traffic assignment obtained from solving the ILP formulation is also listed. For each light trail, the summation of all the traffic it carries is calculated and shown in the rightmost column in Table 19.2.

Table 19.3 depicts the results from solving the local best-fit heuristic algorithm proposed in Section 19.5.1. In this example, the local best-fit increasing packing approach requires 16 light trails.

Table 19.4. *Traffic matrix for a 10-node network: after traffic matrix preprocessing*

	1	2	3	4	5	6	7	8	9	10
1	0	5	8	11	3	8	5	7	8	10
2	3	0	17	4	0	5	1	2	3	1
3	9	3	0	7	3	10	11	14	0	0
4	6	0	8	0	2	5	5	2	1	1
5	0	6	10	4	0	2	11	10	5	2
6	11	3	4	4	3	0	2	6	8	3
7	0	2	10	2	11	5	0	1	6	0
8	0	5	6	2	3	1	11	0	5	6
9	4	5	11	8	8	2	3	1	0	5
10	0	18	0	3	7	10	1	2	1	0

Table 19.5. *ILP: resulting light trails $Tl_{max} = 3$*

No.	Light trails	Hops	Accommodated s-d pairs	Load
1	{1, 6, 7, 4}	3	(1, 4) (6, 4)	15
2	{1, 6, 7, 9}	3	(1, 6) (1, 7) (1, 9) (6, 7) (7, 9)	29
3	{1, 5, 8, 10}	3	(1, 8) (1, 10)	17
4	{2, 3, 4, 7}	3	(2, 4) (3, 4) (3, 7)	22
5	{2, 6, 7, 9}	3	(2, 6) (2, 7) (2, 9) (6, 9)	17
6	{2, 6, 8, 10}	3	(2, 8) (2, 10) (6, 8) (6, 10) (8, 10)	18
7	{3, 2, 1, 5}	3	(3, 2) (3, 1) (3, 5) (2, 1) (1, 5)	21
8	{3, 2, 6, 8}	3	(3, 6) (3, 8)	24
9	{4, 7, 6, 1}	3	(4, 6) (4, 1)	11
10	{4, 7, 8, 5}	3	(4, 7) (4, 8) (4, 5) (7, 8) (7, 5) (8, 5)	24
11	{4, 7, 9, 10}	3	(4, 9) (4, 10)	2
12	{5, 1, 2, 3}	3	(5, 2) (5, 3) (1, 2) (1, 3) (2, 3)	46
13	{5, 8, 7, 4}	3	(5, 8) (5, 7) (5, 4) (8, 7) (8, 4)	38
14	{5, 8, 6}	2	(5, 6)	2
15	{5, 8, 10, 9}	3	(5, 10) (5, 9) (8, 9)	12
16	{6, 8, 5, 1}	3	(6, 5)	3
17	{8, 6, 2, 3}	3	(8, 3) (6, 3)	10
18	{9, 7, 6, 1}	3	(9, 1) (7, 6) (6, 1)	20
19	{9, 7, 6, 2}	3	(9, 6) (9, 2) (7, 2)	9
20	{9, 7, 4, 3}	3	(9, 7) (9, 4) (9, 3) (7, 4) (7, 3) (4, 3)	42
21	{9, 10, 8, 5}	3	(9, 10) (9, 8) (9, 5) (10, 5)	21
22	{10, 8, 6, 2}	3	(10, 8) (10, 6) (10, 2) (8, 6) (8, 2) (6, 2)	39
23	{10, 9, 7, 4}	3	(10, 9) (10, 7) (10, 4)	5

Table 19.6. *Local best-fit: resulting light trails $Tl_{max} = 3$*

No.	Light trails	Hops	Accommodated s-d pairs	Load
1	{1, 6, 7, 4}	3	(1, 4) (1, 6) (1, 7) (6, 4) (7, 4) (6, 7)	32
2	{1, 6, 7, 9}	3	(1, 9) (6, 9) (7, 9)	22
3	{1, 5, 8, 10}	3	(1, 10) (5, 8) (1, 8) (8, 10) (1, 5) (5, 10)	38
4	{2, 6, 7, 9}	3	(2, 9) (2, 6) (2, 7)	9
5	{2, 6, 8, 10}	3	(2, 10) (2, 8) (6, 8) (6, 10)	12
6	{3, 2, 1, 5}	3	(3, 2) (3, 1) (3, 5) (2, 1)	18
7	{3, 4, 7, 8}	3	(3, 8) (3, 7) (3, 4) (4, 7) (4, 8) (7, 8)	40
8	{4, 7, 6, 1}	3	(4, 6) (4, 1) (6, 1) (7, 6)	27
9	{4, 7, 8, 5}	3	(4, 5) (7, 5) (8, 5)	16
10	{4, 7, 9, 10}	3	(4, 9) (4, 10) (9, 10)	7
11	{5, 1, 2, 3}	3	(5, 2) (5, 3) (1, 2) (1, 3) (2, 3)	46
12	{5, 8, 7, 4}	3	(5, 4) (5, 7) (8, 7) (8, 4)	28
13	{5, 8, 10, 9}	3	(5, 9) (8, 9) (10, 9)	11
14	{8, 6, 2, 3}	3	(8, 3) (8, 6) (8, 2) (6, 3) (6, 2)	19
15	{9, 7, 6, 1}	3	(9, 1) (9, 7) (9, 6)	9
16	{9, 7, 6, 2}	3	(9, 2) (7, 2)	7
17	{9, 7, 4, 3}	3	(9, 3) (9, 4) (7, 3) (4, 3)	37
18	{9, 10, 8, 5}	3	(9, 5) (9, 8) (10, 8) (10, 5)	18
19	{10, 8, 6, 2}	3	(10, 2) (10, 6)	28
20	{10, 8, 7, 4}	3	(10, 4) (10, 7)	4
21	{2, 3, 4}	2	(2, 4)	4
22	{3, 2, 6}	2	(3, 6)	10
23	{5, 1, 6}	2	(5, 6)	2
24	{6, 1, 5}	2	(6, 5)	3

19.6.1 Light trail hop-length limit: $Tl_{max} = 3$

When the light trail hop-length limit is set to $Tl_{max} = 3$, requests between some node-pairs in the network shown in Fig. 19.8 have to be divided and allocated to multiple light trails. More specifically, the shortest paths between nodes 3 and 10 have hop lengths of 4. Therefore, the request between these two nodes cannot be accommodated on a single light trail. The *traffic matrix preprocessing* heuristic rearranges the original traffic $d_{3,10}$ onto $d_{3,8}$ and $d_{8,10}$. Similarly, the request from node 10 to node 3 is aggregated onto the node-pair (10, 2) and (2, 3). The resulting traffic matrix is shown in Table 19.4.

Solving the ILP formulation with this modified traffic matrix gives an optimal solution consisting of 23 light trails as shown in Table 19.5. Experiments using both local best-fit increasing and decreasing packing algorithms are performed, and the better solution with a result of 24 light trails is chosen. The detailed results are shown in Table 19.6.

Table 19.7. *ILP: resulting light trails* $Tl_{max} = 5$

No.	Light trails	Hops	Accommodated s–d pairs	Load
1	{1, 2, 3, 4, 7, 6}	5	(1, 4) (1, 6) (1, 7) (3, 6) (4, 6)	39
2	{2, 1, 6, 7, 9, 10}	5	(2, 1) (2, 6) (2, 7) (2, 9) (2, 10) (1, 9) (1, 10) (9, 10) (6, 9) (6, 10)	47
3	{3, 4, 7, 8, 10, 9}	5	(3, 7) (3, 8) (3, 10) (4, 7) (4, 8) (4, 9) (4, 10) (7, 8) (7, 9) (10, 9)	42
4	{4, 3, 2, 1, 5, 8}	5	(4, 1) (4, 5) (3, 2) (3, 1) (3, 5) (1, 5) (1, 8) (2, 8)	35
5	{5, 1, 6, 2, 3, 4}	5	(5, 6) (5, 4) (1, 2) (1, 3) (6, 2) (6, 3) (2, 3) (2, 4) (3, 4)	45
6	{5, 8, 10, 9, 7, 6}	5	(5, 10) (5, 9) (8, 9) (8, 7) (8, 6) (10, 6) (9, 6) (7, 6)	41
7	{6, 1, 5, 8, 7, 4}	5	(6, 1) (6, 5) (6, 8) (6, 7) (6, 4) (5, 8) (5, 7)	47
8	{9, 7, 8, 5, 1, 2}	5	(9, 7) (9, 5) (9, 1) (9, 2) (7, 5) (7, 2) (8, 2) (5, 2)	44
9	{9, 10, 8, 7, 4, 3}	5	(9, 8) (9, 4) (9, 3) (10, 8) (10, 7) (10, 4) (8, 4) (7, 4) (7, 3) (4, 3)	48
10	{10, 8, 5, 1, 2, 3}	5	(10, 5) (10, 2) (10, 3) (8, 5) (8, 3) (5, 3)	44

19.7 Light trail hop-length limit: $Tl_{max} = 5$

When Tl_{max} increases to 5, the running time for solving the ILP formulation increases dramatically. This is because, as mentioned earlier, the number of candidate paths increases very fast as Tl_{max} increases. This increase introduces a significant number of variables and constraints in the ILP formulation. The optimal solution contains 10 light trails: the detailed results are shown in Table 19.7. The heuristic algorithms give solutions in a timescale of seconds. The better solution obtained from using both best-fit increasing packing order and best-fit decreasing packing order consists of 13 light trails as shown in Table 19.8.

19.7.1 Discussions

An observation from the optimal solutions obtained by solving ILP formations is that only the longest candidate paths are chosen as light trails. This is due to the fact that only the number of light trails is being minimized. The program stops searching further once the number of light trails does not decrease, even though it is possible to substitute some light trails with the other shorter paths.

The problem becomes unmanageable in the case of the ILP approach as the problem size increases. In such a scenario, the use of relaxation techniques would be a preferred choice. When the traffic is uniform or the variations among different requests are small enough that they can be approximately treated as uniform traffic,

Table 19.8. *Local best-fit: resulting light trails* $Tl_{\max} = 5$

No.	Light trails	Hops	Accommodated s-d pairs	Load
1	{3, 2, 6, 7, 8, 10}	5	(3, 10) (3, 6) (3, 8) (6, 8) (2, 6) (3, 2) (6, 10) (2, 8) (6, 7) (2, 7) (2, 10) (7, 8)	48
2	{10, 8, 7, 6, 2, 3}	5	(10, 3) (2, 3) (8, 3) (8, 2) (7, 6) (6, 3) (6, 2) (10, 8) (7, 2) (10, 7) (8, 6)	46
3	{1, 5, 8, 6, 7, 4}	5	(1, 4) (1, 6) (1, 8) (1, 7) (5, 4) (6, 4) (1, 5) (5, 6) (8, 4) (7, 4)	48
4	{1, 2, 3, 4, 7, 9}	5	(1, 9) (1, 3) (3, 4) (7, 9) (1, 2) (4, 7) (2, 4) (2, 9) (4, 9)	47
5	{1, 5, 8, 7, 9, 10}	5	(1, 10) (8, 7) (5, 8) (5, 9) (8, 9) (9, 10) (5, 10)	48
6	{3, 4, 7, 6, 8, 5}	5	(3, 5) (3, 7) (7, 5) (4, 6) (6, 5) (8, 5) (4, 5) (4, 8)	40
7	{4, 3, 2, 6, 1}	4	(4, 1) (6, 1) (3, 1) (4, 3) (2, 1)	37
8	{4, 7, 8, 10}	3	(4, 10)	1
9	{5, 8, 7, 6, 2, 3}	5	(5, 3) (5, 7) (7, 3) (5, 2)	37
10	{9, 10, 8, 6, 2, 1}	5	(9, 1) (10, 6) (10, 2) (9, 2) (9, 6) (9, 8)	31
11	{9, 10, 8, 7, 4, 3}	5	(9, 3) (9, 4) (9, 7) (10, 4)	25
12	{9, 10, 8, 5}	3	(9, 5) (10, 5)	15
13	{6, 8, 10, 9}	3	(6, 9) (10, 9)	9

$D_{N \times N} = \{d_{s,t} = \bar{d} | \forall(s, t)\}$. LP relaxation is a very effective means of obtaining fast solutions. This can be achieved by modifying the light trail capacity constraint in the ILP formulation in Section 19.4.2.2 as follows. The rest of the formulation remains the same:

$$\sum_{(s,t)} \mu_{s,t}^{\tau} \leq \lfloor C/\bar{d} \rfloor \tag{19.6}$$

$$0 \leq \delta^{\tau} \leq 1 \tag{19.7}$$

$$0 \leq \mu_{s,t}^{\tau} \leq 1 \tag{19.8}$$

In this formulation, the coefficient matrix of variables is totally unimodular. Hence, LP relaxation still yields integer solutions. This effect is caused by the same reason as noted earlier in LP to ILP in Section 6.1.1. Thus, this formulation can be applied to solve the light trail design problem where the traffic requests have similar capacities.

19.8 Restoration in the light trail architecture

As stated throughout the book, survivability is a critical issue in the design of light trails for an optical network to survive failures as well due to the fact that a single link failure disrupts all the light trails that use the link. Each light trail carries multiple connections. Therefore the effects would be catastrophic. For instance, if a failed link has W wavelengths, it can carry up to W light trails. Each light trail contains up to $\binom{N_T^w}{2}$ s-d pairs, where N_T^w denotes the number of nodes in the wth light trail, $w = 1, 2, \ldots, W$. Therefore, in the worst case a link failure may disrupt up to $\sum_{w=1}^{W} \binom{N_T^w}{2}$ connections. To provide 100% protection in the WDM layer of a light trail architecture implies that a backup light trail needs to be provided at the time of establishment.

Recall that a key difference between the light trail and lightpath architectures is that the intermediate nodes in a light trail can also have access to the medium. Thus the restoration model for a light trail architecture is different from that in a lightpath architecture as all node-pairs on the light trail need to be protected and an assigned alternate path may not be possible. However, as in the case of lightpaths, two possible protection schemes are possible, namely *connection-based protection* and *link-based protection*. In the following it is assumed that there is no more than one link failure at any time.

19.8.1 Connection-based protection

For each connection request d_{st}, the resources are allocated to a primary connection in a light trail LT_1 and a backup connection in another light trail LT_2. LT_1 and LT_2 are link-disjoint. The primary connection is a working connection when there is no link failure. If a link on LT_1 fails, the failure information is propagated through the control channel. The source node s of the request receives the failure information; it starts to transmit the data on LT_2 to the destination t through the backup connection.

The scheme is explained through an example, shown in Fig. 19.9. Suppose that there are two light trails: LT_1, $1 \rightarrow 2 \rightarrow 3 \rightarrow 4$ and LT_2, $2 \rightarrow 6 \rightarrow 5 \rightarrow 4$. LT_1 and LT_2 are link-disjoint. There is a connection request from node 2 to node 4. In this case, a primary connection between nodes 2 and 4 is established on light trail LT_1 with the backup connection on light trail LT_2. Suppose link $2 \rightarrow 3$ on LT_1 fails, then light trail LT_1 cannot be used. When this failure information reaches source node 2, source node 2 starts to transmit data using the backup connection LT_2.

19.8.2 Link-based protection

In this case, for each link on a light trail, a backup sub-light trail is provided. When a link on a light trail fails, a light trail is rerouted around the failed link and the backup sub-light trail is used.

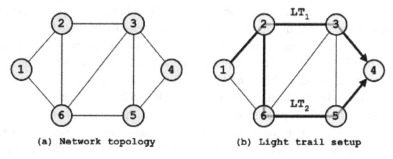

Fig. 19.9. An example for a connection-based scheme.

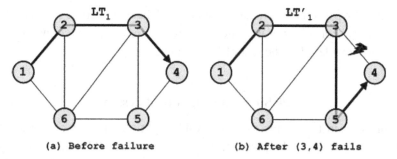

Fig. 19.10. An example for the link-based protection scheme.

Consider the same example again. Suppose for each link on LT_1 that there is a backup sub-light trail. The backup sub-light trail for link $3 \rightarrow 4$ is $3 \rightarrow 5 \rightarrow 4$ as shown in Fig. 19.10. If link $3 \rightarrow 4$ fails, information about the failure and the sub-light trail of link $3 \rightarrow 4$ is sent along the control channel (it is assumed that there exists a bidirectional control channel). When the message reaches source node 1, it will send a fault management message, which is similar to the light trail setup message, along the control channel to intermediate nodes (nodes 2, 3, and 5) and the end node 4. These nodes then configure the optical shutter and form a new light trail $1 \rightarrow 2 \rightarrow 3 \rightarrow 5 \rightarrow 4$.

19.8.3 Comparison of connection-based and link-based protection

The connection-based protection has the following advantages over link-based protection.

- *Restoration time*: in connection-based protection, as soon as the failure information message reaches the source node of a connection that is using the light trail, the source node can immediately start using the backup connection on another light trail to continue the transmission. The maximum restoration time is the transmission time of the control message.

A link-based approach requires the ability to identify a failed link at both ends, which makes restoration more difficult when node failure happens. In link-based protection, after the failure information reaches the source node of the failed light trail, the source node will have to initiate a light trail setup process, i.e. setting up a light trail that includes the remaining part of the original light trail and the nodes on the backup sub-light trail of the failed link. This takes much more time than restoration with connection-based protection.

- *The length of the light trail*: as shown in the example, the restored light trail in a link-based protection scheme is longer than the original light trail. As discussed previously, the length of the light trail is an important parameter that is related to the signal-to-noise ratio and the bit-error rate. This again limits the choice for link-based protection. Thus, link-based restoration is likely to perform poorly in this regard.

From the above discussion, it may be concluded that connection-based protection is more practical for the light trail architecture. Therefore, only connection-based protection is considered in the rest of the chapter.

19.9 Survivable light trail design

The major issue in the design of a survivable light trail network is to identify a set of light trails to carry the given traffic and to provide 100% protection against single-link failure. The survivable light trail network design problem is defined as follows. *Given a graph $G(V, E)$, where $|V| = N$, and a traffic matrix $D_{N \times N}$, identify the minimum number of light trails to carry the given traffic in such a way that for each connection request, there is a primary connection established in one light trail and resources are reserved in another light trail for backup connection. Two light trails for each s-d pair are link-disjoint.*

Recall that the maximum hop length of a light trail is denoted by $Tl_{\max}$. The *traffic matrix preprocessing* algorithm proposed in Fig. 19.6 still applies here in which a long hop is recursively divided into multiple hops in order to satisfy the light trail hop-length constraint. It is necessary that for each recovering s-d pair there is more than one path available within $Tl_{\max}$.

The next step is to develop an ILP formulation to optimize the capacity utilization in terms of the number of light trails, with the given network topology and refined traffic matrix obtained from *traffic matrix preprocessing*. The objective is to find the minimum number of light trails that are required for the system.

19.10 ILP formulation: connection-based protection

Given the network topology $G(V, E)$, and the traffic matrix obtained from *traffic matrix preprocessing*, one first lists all possible paths with the hop-length limit constraint for each s-d node-pair. This can be accomplished by using a *breadth-first search* for each node. These eligible paths form a set of all possible light trails. Among all the possible choices, an optimal set of paths is chosen to form the light trail network such that the total number of light trails are minimized and the demand constraint and the protection constraint are met. This problem is formulated as an ILP optimization problem. It is also assumed that each request cannot be split into multiple parts.

19.10.1 Notation

The network topology is represented as a directed graph $G(V, E)$ with $|V| = N$ nodes and $|E| = L$ links with W wavelengths on each link. The following notation is used.

- $n = 1, 2, \ldots, N$: number assigned to each node in the network.
- $p, p_1, p_2 = 1, 2, \ldots, P$: number assigned to a path in the network.
- $i, j, k = 1, 2, \ldots, N(N - 1)$: number assigned to a node-pair. The source and destination nodes of a connection request form a node-pair.

The following notation is used for path-related information.

- δ_p^i: path indicator. This takes a value of 1 if the primary connection for request i is established on light trail p and zero otherwise (binary variable).
- v_p^i: path indicator. This takes a value of 1 if the backup connection for connection request i is established on light trail p and zero otherwise (binary variable).
- ψ_p^i: node-pair indicator. This takes a value of 1 if node-pair i is on path p and zero otherwise (data).
- h_p: this takes a value of 1 if path p is used by some primary connection or (and) backup connection and zero otherwise (binary variable).
- d^i: demanded capacity of connection request i (data).
- I_{p_1, p_2}: this takes a value of 1 if p_1 and p_2 are link-disjoint and zero otherwise (binary data).

19.10.2 Objective

Minimize the number of light trails:

$$\min \sum_{p=1}^{P} h_p \tag{19.9}$$

19.10.2.1 Constraints

(i) *On demand constraint for each node-pair:* for each request, there is one primary connection on one light trail, and a backup connection in another light trail:

$$\sum_{p=1}^{P} \delta_p^i \psi_p^i = 1 \qquad 1 \le i \le N(N-1) \tag{19.10}$$

$$\sum_{p=1}^{P} v_p^i \psi_p^i = 1 \qquad 1 \le i \le N(N-1) \tag{19.11}$$

(ii) *On topology diversity of primary and backup connections:* the primary and backup connection for a request are established on two link-disjoint light trails:

$$(\delta_p^i \psi_p^i + v_p^i \psi_p^i)(1 - I_{p_1,p_2}) \le 1 \qquad 1 \le i \le N(N-1), 1 \le p_1, p_2 \le P \tag{19.12}$$

(iii) *On link capacity constraints:* the total demand of all connections on one light trail cannot exceed one wavelength capacity. The first term represents the capacity used by primary connections on light trail p. The second term represents the capacity used by backup connections on light trail p:

$$\sum_{i=1}^{N(N-1)} \delta_p^i \psi_p^i d_i + \sum_{i=1}^{N(N-1)} v_p^i \psi_p^i d_i \le C \qquad 1 \le p \le P \tag{19.13}$$

(iv) *On light trail identification constraints:* if one or more of the primary or backup connections uses a path, then this path is a light trail:

$$h_p \le \sum_{i=1}^{N(N-1)} \delta_p^i \psi_p^i + \sum_{i=1}^{N(N-1)} v_p^i \psi_p^i \qquad 1 \le p \le P \tag{19.14}$$

$$2N(N-1)h_p \ge \sum_{i=1}^{N(N-1)} \delta_p^i \psi_p^i + \sum_{i=1}^{N(N-1)} v_p^i \psi_p^i \qquad 1 \le p \le P \tag{19.15}$$

19.10.3 Solutions

The above formulation is solved for some example networks: a six-node network shown in Fig. 19.9(a) and a 10-node network shown in Fig. 19.11.

To simplify the problem, the links are assumed to be bidirectional with the same length. As observed earlier, the number of potential light trails, i.e. possible paths in a network, increases rapidly as the number of nodes in the network increases. For example, for the two example networks, the number of possible paths are 120 and 448, respectively. This makes it difficult to solve ILP for a large network and for a large number of connection requests. In such cases, heuristic strategies have to be adopted to solve the survivable light trail design problem.

Table 19.9. *Requests matrix for a six-node network*

	1	2	3	4	5	6
1	0	11	6	0	14	8
2	0	0	0	0	5	0
3	0	0	0	0	0	0
4	0	0	0	0	0	0
5	0	0	0	0	0	0
6	0	0	0	0	19	0

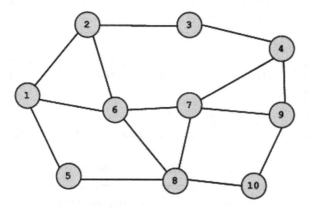

Fig. 19.11. A 10-node, 15-link network.

19.10.4 A simple example

A simple example is presented to understand the solution obtained from the ILP formulation, which optimally identifies the light trails covering all the working connections and corresponding backup connections. For the network shown in Fig. 19.9, a traffic matrix shown in Table 19.9 is to be routed. The integer numbers indicate the request capacity in units of OC-1 (51.84 Mbit/s), while the entire wavelength capacity is OC-48. The hop-length limit is set to be 3, i.e. $Tl_{max} = 3$. Table 19.10 gives the resulting light trails that cover all the connection requests and their corresponding backup connections. The notation (s, d) and $(s, d)_b$ in column 4 denote primary and backup connections, respectively.

In order to provide 100% protection for a single link failure, each request is allocated resources for a working connection in one light trail and a reserved resource for its backup connection in another light trail. The two light trails are disjoint. In Table 19.10, the working connection for request $(1, 2)$ uses the light trail $1 \rightarrow 2 \rightarrow 3$, while the backup connection for this request is accommodated

Table 19.10. *Resulting light trails for example request matrix I*

No.	Light trails	Hops	Accommodated s-d pairs	Load
1	$\{1, 2, 3\}$	2	$(1, 2), (1, 3)_b$	17
2	$\{1, 2, 6, 5\}$	3	$(2, 5), (6, 5), (1, 5)_b, (1, 6)_b$	46
3	$\{1, 6, 2\}$	2	$(1, 6), (1, 2)_b$	19
4	$\{1, 6, 3, 5\}$	3	$(1, 3), (1, 5), (6, 5)_b$	39
5	$\{2, 3, 5\}$	2	$(2, 5)_b$	5

Table 19.11. *Requests matrix for a six-node network*

	1	2	3	4	5	6
1	0	7	6	9	19	17
2	27	0	3	6	28	2
3	14	3	0	19	31	9
4	26	5	29	0	5	23
5	27	20	20	17	0	14
6	9	30	1	1	1	0

Table 19.12. *Resulting light trails*

No.	Light trails	Hops	Accommodated s-d pairs	Load
1	$\{1, 2, 3, 4\}$	3	$(1, 2), (1, 4), (1, 3)_b, (2, 4)_b, (3, 4)_b$	47
2	$\{1, 6, 3, 2\}$	3	$(1, 3), (1, 2)_b, (6, 2)_b$	43
3	$\{1, 6, 5, 4\}$	3	$(1, 4)_b, (1, 5), (1, 6), (6, 4), (6, 5)$	47
4	$\{2, 3, 6, 1\}$	3	$(2, 1), (2, 6), (3, 1), (2, 3)_b$	46
5	$\{1, 2, 6, 3\}$	3	$(2, 3), (1, 6)_b$	20
6	$\{2, 6, 5, 4\}$	3	$(2, 4), (2, 5)_b, (2, 6)_b$	36
7	$\{1, 2, 3, 5\}$	3	$(2, 5), (1, 5)_b$	47
8	$\{3, 2, 6, 5\}$	3	$(3, 2), (3, 5), (3, 6)$	43
9	$\{4, 3, 2, 1\}$	3	$(3, 1)_b, (4, 1), (4, 2)_b$	45
10	$\{4, 5, 6, 2\}$	3	$(4, 2), (4, 6), (5, 2)_b$	48
11	$\{4, 3, 5, 6\}$	3	$(4, 3), (4, 5), (5, 6)_b$	48
12	$\{5, 6, 2, 1\}$	3	$(5, 1), (6, 1)$	36
13	$\{5, 3, 2, 1\}$	3	$(5, 2), (2, 1)_b$	47
14	$\{5, 6, 3, 4\}$	3	$(5, 3), (6, 3), (5, 4)_b, (6, 4)_b$	39
15	$\{3, 5, 4\}$	2	$(3, 4), (5, 4)$	36
16	$\{4, 5, 3, 6\}$	3	$(5, 6), (4, 5)_b, (4, 3)_b$	48
17	$\{4, 3, 6, 2\}$	3	$(6, 2), (3, 2)_b$	33
18	$\{6, 2, 3, 5\}$	3	$(3, 5)_b, (6, 3)_b, (6, 5)_b$	33
19	$\{4, 3, 6, 1\}$	3	$(3, 6)_b, (6, 1)_b, (4, 6)_b$	41
20	$\{4, 5, 6, 1\}$	3	$(4, 1)_b$	26
21	$\{5, 3, 6, 1\}$	3	$(5, 3)_b, (5, 1)_b$	47

Table 19.13. *Traffic matrix for a 10-node network*

	1	2	3	4	5	6	7	8	9	10
1	0	10	0	20	7	0	8	0	0	20
2	0	0	10	9	0	10	3	22	0	0
3	6	0	0	5	0	4	0	7	21	0
4	8	0	11	0	0	12	11	0	0	8
5	11	0	0	12	0	11	0	4	0	18
6	15	0	14	0	7	0	13	16	0	0
7	0	14	0	18	2	0	0	9	13	0
8	12	0	10	0	4	8	0	0	0	9
9	4	19	0	5	12	0	0	17	0	0
10	11	0	21	0	0	10	0	22	8	0

on light trail $1 \to 6 \to 2$. Similarly, for the request $(1, 3)$, two link-disjoint light trails $1 \to 6 \to 3 \to 5$ and $1 \to 2 \to 3$ are used. Five light trails and a total of 12 wavelength links are used for this request matrix.

Table 19.11 provides a randomly generated dense traffic matrix for the six-node network. It is assumed that the hop-length limit $Tl_{max} = 3$. From the topology it is observed that all s-d pairs have paths within this hop-length limit; hence, *traffic matrix preprocessing* is not needed. Since the experiments are performed on small fractional wavelength requests, the number of wavelengths on each link is not a critical constraint. For this example $W = 4$ is sufficient, although there is no constraint being put on the number of wavelengths. Table 19.12 presents the results from solving the ILP formulation with a hop-length limit of $Tl_{max} = 3$.

Table 19.12 shows 21 light trails that are needed to carry the primary and backup connections. The traffic assignment obtained from solving the ILP formulation is also listed. For each light trail, the amount of total traffic carried is shown in the rightmost column in Table 19.12. It can be noted that most of the light trails are fully or almost fully occupied. Hence, the resource utilization is quite high.

19.10.5 Example 2

Table 19.13 is a randomly generated traffic matrix for the 10-node network. The request capacity for each node-pair is uniformly distributed between 0 and 22. It is assumed that the hop-length limit $Tl_{max} = 4$: Table 19.14 shows the resulting light trails.

Compared with Table 19.12, some of the light trails in Table 19.14 are not very efficiently occupied. This is due to the fact that the average number of requests

Table 19.14. *Resulting light trails*

No.	Light trails	Hops	Accommodated s-d pairs	Load
1	$\{1, 2, 6, 8, 5\}$	4	$(1, 5)_b, (6, 5)_b, (6, 8)_b, (8, 5)_b$	34
2	$\{1, 5, 8, 7, 4\}$	4	$(5, 4), (1, 7)_b$	20
3	$\{1, 6, 7, 4, 3\}$	4	$(6, 3)_b, (1, 4)_b$	34
4	$\{1, 6, 7, 9, 10\}$	4	$(1, 7), (1, 10)$	28
5	$\{2, 3, 4, 7, 6\}$	4	$(2, 4), (2, 3)_b, (2, 7)_b, (3, 4)_b, (2, 6)_b$	37
6	$\{2, 3, 4, 7, 8\}$	4	$(3, 8)_b, (2, 8)_b$	29
7	$\{3, 4, 9, 7\}$	3	$(4, 7)_b, (3, 9)_b$	32
8	$\{2, 6, 7, 4, 3\}$	4	$(2, 3), (2, 6)$	20
9	$\{3, 2, 6, 7, 4\}$	4	$(3, 4), (2, 4)_b, (6, 7)_b, (7, 4)_b$	45
10	$\{3, 2, 6, 7, 9\}$	4	$(3, 6)_b, (3, 9), (2, 7)$	28
11	$\{3, 4, 7, 6, 1\}$	4	$(3, 6), (4, 1), (4, 6), (3, 1)_b$	30
12	$\{4, 3, 2, 6, 8\}$	4	$(2, 8), (3, 8), (4, 6)_b$	41
13	$\{4, 7, 6, 2, 3\}$	4	$(6, 3), (4, 3)_b, (7, 2)_b$	39
14	$\{4, 7, 6, 8, 10\}$	4	$(4, 10)_b, (8, 10), (7, 8), (4, 7)$	37
15	$\{4, 9, 10 \}$	2	$(4, 10)$	8
16	$\{5, 1, 2, 3, 4\}$	4	$(5, 4)_b, (1, 4)$	32
17	$\{5, 1, 6, 2\}$	3	$(5, 1), (5, 6), (1, 2)_b$	32
18	$\{5, 1, 6, 8, 10\}$	4	$(5, 8)_b, (5, 10)_b$	22
19	$\{5, 8, 7, 6, 1\}$	4	$(5, 6)_b, (5, 1)_b, (8, 6), (8, 1)$	42
20	$\{5, 8, 7, 9, 10\}$	4	$(5, 8), (5, 10), (7, 9), (8, 10)_b$	44
21	$\{6, 1, 5, 8, 10\}$	4	$(1, 10)_b, (6, 8)$	36
22	$\{6, 8, 7, 9, 4\}$	4	$(7, 4), (6, 7)$	31
23	$\{7, 8, 5, 1, 2\}$	4	$(7, 8)_b, (7, 2)$	23
24	$\{7, 9, 10, 8\}$	3	$(10, 8)_b$	22
25	$\{8, 6, 2, 1, 5\}$	4	$(8, 1)_b, (8, 6)_b, (8, 5)$	24
26	$\{7, 4, 9\}$	2	$(7, 9)_b$	13
27	$\{8, 10, 9, 4, 3\}$	4	$(10, 3)_b, (9, 4)_b, (8, 3), (4, 3)$	47
28	$\{9, 4, 3, 2, 1\}$	4	$(9, 1)_b, (9, 2)_b, (4, 1)_b, (3, 1)$	37
29	$\{9, 4, 7, 8, 5\}$	4	$(9, 8), (9, 5)_b, (7, 5)_b$	31
30	$\{9, 7, 4 \}$	2	$(9, 4)$	5
31	$\{9, 7, 6, 1, 2\}$	4	$(9, 1), (9, 2), (1, 2)$	33
32	$\{9, 7, 6, 1, 5\}$	4	$(1, 5), (9, 5), (7, 5), (6, 5)$	28
33	$\{10, 8, 6, 2, 3\}$	4	$(8, 3)_b, (10, 3), (10, 6)_b$	41
34	$\{10, 8, 6, 2, 1\}$	4	$(10, 1), (6, 1)_b$	26
35	$\{10, 8, 7, 4, 9\}$	4	$(10, 9)_b$	8
36	$\{10, 9, 7, 6, 1\}$	4	$(10, 6), (10, 1)_b, (6, 1)$	36
37	$\{10, 9, 7, 6, 8\}$	4	$(9, 8)_b, (10, 8), (10, 9)$	47

per node-pair $(50/(10 \times 9) \approx 0.55)$ in this traffic matrix is smaller than that in the traffic matrix for the six-node network $(30/(6 \times 5) = 1)$. Since the resulting light trail networks shown in Table 19.14 still have spare capacity, even if the number of requests increases, some of the new requests could still be accommodated using the existing light trails.

Appendix 1

Optical network components

Optical components are devices that transmit, shape, amplify, switch, transport, or detect light signals. The improvements in optical component technologies over the past few decades have been the key enabler in the evolution and commercialization of optical networks. In this appendix, the basic principles behind the functioning of the various components are briefly reviewed. In general, there are three groups of optical components.

(i) *Active components*: devices that are electrically powered, such as lasers, wavelength shifters, and modulators.
(ii) *Passive components*: devices that are not electrically powered and that do not generate light of their own, such as fibers, multiplexers, demultiplexers, couplers, isolators, attenuators, and circulators.
(iii) *Optical modules*: devices that are a collection of active and/or passive optical elements used to perform specific tasks. This group includes transceivers, erbium-doped amplifiers, optical switches, and optical add/drop multiplexers.

A1.1 Fiber optic cables

The backbone that connects all of the nodes and systems together is the optical fiber. The fiber allows signals of enormous frequency range (25 THz) to be transmitted over long distances without significant distortion in the information content. While there are losses in the fiber due to reflection, refraction, scattering, dispersion, and absorption, the bandwidth available in this medium is orders of magnitude more than that provided by other conventional mediums such as copper cables. As will be explained below, the bandwidth available in the fiber is limited only by the attenuation characteristics of the medium at low frequencies and its dispersion characteristics at high frequencies.

357

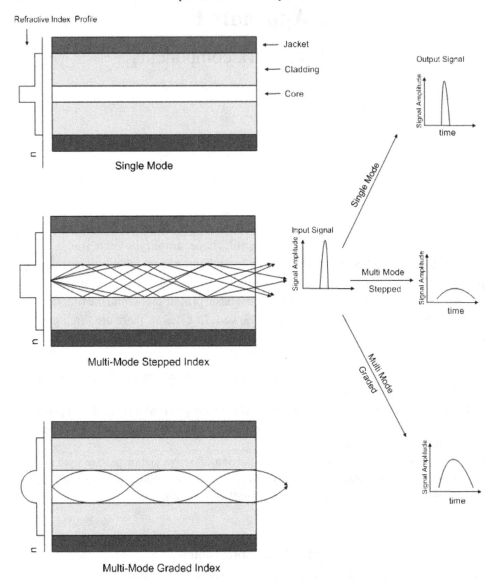

Fig. A1.1. Propagation of light in different types of fibers: single-mode, multimode step index, and multimode graded index.

A1.1.1 Light propagation in optical fibers

In general, there are two major types of fibers: multimode and single mode. The multimode fiber comes in two kinds: step index and graded index. Figure A1.1 depicts the main idea behind these types.

The multimode fiber has a large core with a typical diameter of 30–100 μm with a thin cladding of refractive index that is lower than that of the core. Light propagation in multimode fibers can be explained in ray optics by the principle of total internal reflection (TIR). According to TIR, if light traveling in a denser medium (core) is incident on a lower refractive index medium (cladding), beyond the critical angle, it is completely reflected back into the core instead of being refracted into the cladding. The critical cone, also known as the acceptance cone, is defined as a cone of angles within which all rays that are launched into the fiber are guided by the fiber. For a fiber with a core refractive index of 1.5 and a cladding index of 1.485, the acceptance angle is about 12°.

A1.1.1.1 Modal dispersion

Multimode fibers are limited by what is called the *modal dispersion*. Modes are determined from differential equations used to analyze propagation in fibers in a fashion similar to that of propagation in cylindrical waveguides. Modes may be thought of as specific path directions. Since rays incident at different directions traverse different path lengths to reach a particular destination, modes have different velocities. Consequently, as the signals travel down the fiber, the velocity variation among the modes causes an undesirable spreading of the signal which leads to intersymbol interference at high data rates. For example, the bit-rate–distance product of an optical communication system is constrained to 10 Mbit/s km for a fiber with a core refractive index of 1.5 and a cladding refractive index of 1.485.

In digital transmission, acceptable dispersion is $\tau \leq T/k$ and the bit-rate limit is expressed by $B \leq 1/k\tau$, where τ is the difference in the width between the initial pulse and the broadened pulse, B is the bit rate, T is the bit period, and k is the dispersion factor (typically set to 4). In Fig. A1.2, $\tau = t_2 - t_1$.

There are two remedies for modal dispersion. The first is achieved by creating the core material with a graded refractive index to form a multimode graded index fiber. It should be recalled that modal dispersion is caused because light rays traveling in the periphery of the fiber cover longer distances and take longer, while rays traveling along the axes of the fiber cover a shorter distance and take a shorter time to reach the same destination point. This situation can be countered by making the refractive index high in the core and reduce it gradually towards the periphery. The core thus acts like a lens focusing light toward the center of the core. The signal spreading is improved compared to a regular multimode fiber with a step index. With an optimum graded-index profile, with a nearly quadratic decrease of refractive index from core to the cladding, the bandwidth–distance product is 8 Gbit/s km for a multimode fiber with a core index of 1.5 and a cladding index of 1.485.

Alternatively, modal dispersion can be avoided by making the core of the fiber smaller, of the order of 5–10 μm, and only allowing a single mode to propagate

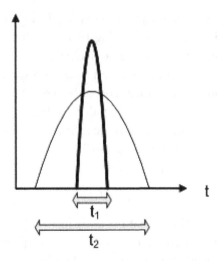

Fig. A1.2. Modal dispersion causes the pulse to spread from an initial width of t_1 to a final width of t_2.

in the fiber. This is referred to as a single-mode fiber. Many of the long-distance transatlantic and transpacific telephone lines are carried on single-mode fibers. It should be noted that even in a single-mode fiber there is some signal spreading, but it is limited.

A1.1.2 Fiber dispersion

The propagation characteristics of a wavelength are affected by the refractive index of the medium and non-linearity of the propagation constant. This dependence, which leads to an undesired broadening of data signals, is called dispersion. An important type of dispersion called chromatic dispersion sets fundamental limits on the performance of optical communication systems. Chromatic dispersion includes contributions from material dispersion and wavelength dispersion. The dependence of the dielectric constant on frequency is known as material dispersion and the non-linear dependence of the propagation constant on frequency is known as waveguide dispersion.

The speed of light in a medium depends on the refractive index of the medium. The refractive index is directly related to the dielectric constant of the medium which in turn depends on the wavelength. So, each wavelength travels at a different speed and hence there is signal spreading resulting in material dispersion.

The origin of waveguide dispersion can be described as follows. As light propagates in the fiber, part of the power is in the core and part is in the cladding. The effective index seen by the wavelength is neither the index of the core nor the index of the cladding but some value in between the two, depending on the power distribution in these regions. Since this power distribution depends on the wavelength, different wavelengths see different effective index values, thereby changing the propagation characteristics.

Another dispersion effect that severely degrades transmission at 10 Gbit/s and higher is polarization mode dispersion (PMD). PMD arises when light polarized along one axis moves faster than light polarized along another axis. This effect can spread the pulse enough to make it overlap with other pulses or change its own shape such that the original signal is completely distorted. The difference in the propagation characteristics occurs because of the asymmetry in the fiber core formed due to manufacturing defects. For a typical single-mode fiber, the accumulated time spread due to PMD is 50 ps after traveling a distance of 100 km, which can be a serious problem for systems operating at rates over 10 Gbit/s.

A1.1.3 Fiber non-linearities

In addition to dispersion effects, several second-order non-linear effects such as stimulated Raman scattering (SRS), stimulated Brillouin scattering (SBS), four-wave mixing (FWM), and self-phase modulation (SPM) are also observed which impede transmission at high speeds (over 10 Gbit/s) and high transmitter powers. SRS and SBS arise due to the interaction of light with photons in the fiber medium. Due to scattering, energy is transferred from a light wave of shorter wavelength (pump wave) to one of longer wavelength (Stokes wave). In the case of SRS, the pump wave is the signal wave while in the case of SBS, the Stokes wave is the signal wave. As the wave progresses in the fiber, the pump wave loses power while the Stokes wave gains power.

SPM and FWM arise due to the dependence of the refractive index on the intensity of the applied field which is directly related to the square of the field amplitude. SPM leads to pulse chirping and consequent broadening while FWM leads to signals at new frequencies that may interfere with existing signals in the system. Another effect called cross-phase modulation (CPM) is also observed where chirp is induced in one channel depending on the variation of refractive index with intensity in another channel.

In general, all of these non-linear effects are weak and depend on long interaction lengths to build up to a substantial level. The effect of non-linearities can be significantly reduced by increasing the wavelength spacing between the channels. For instance, in the 1550 nm band with 100 GHz spacing, the non-linear effects are negligible. However, by increasing channel spacing, the total number of channels that can be packed into the low-loss window is reduced.

A1.1.4 Fiber loss

The power loss or attenuation is the reduction of the signal power as it travels along the fiber. The output power P_o at the end of a fiber of length L is related to the input power P_{in} by

$$P_o = P_{in}e^{-\alpha L}$$

where α is attenuation of the fiber expressed in dB/km. Currently, fiber manufacturers have been able to achieve losses as low as 0.1–0.2 dB/km in optical fiber.

The two main loss mechanisms are material absorption and Rayleigh scattering. Absorption is due to silica as well as the impurities in the medium while scattering arises because of fluctuations in the density of the medium at the microscopic level. The loss due to absorption has been reduced so much so that scattering is mainly responsible for fiber attenuation.

A1.1.5 New optical fiber types

The dependence of the refractive index of silica fiber is non-linear and at some wavelength, the material dispersion becomes zero. The wavelength corresponding to this is called the zero-dispersion wavelength. A conventional silica fiber has its zero-dispersion wavelength at 1.3 μm. Fibers based on non-linear optical effects with the zero-dispersion wavelength shifted to 1.55 μm have been engineered since this corresponds to the minimum attenuation window. However, dispersion-shifted fibers suffer from heavy penalties due to non-linearities. These penalties are reduced if a little chromatic dispersion is present in the fiber that minimizes the interaction between the interfering waves. This led to the development of non-zero-dispersion-shifted fibers. Other types of fibers called dispersion-compensated fibers have also been designed. Signals could be carried in a conventional fiber for long distances and finally compensated by using a small section of fiber having material with a dispersion of opposite sign.

A1.2 Filters, multiplexors and demultiplexors

Optical filters are realized based on the principles of absorption, diffraction, and interference. The filters are either fixed or tunable. Good filters have a low insertion loss which is defined as the input-to-output loss ratio. The losses should be independent of the state of polarization and should not vary with temperature. Isolation, which is defined as the relative power passed through from the adjacent channels is an important parameter. Tunable filters are characterized by the tuning range and the tuning time. The tuning range refers to the gamut of frequencies over which the filter can be operated, while the tuning time refers to the time it takes for the filter to switch from one frequency to another.

An important application of optical filters is in the design of demultiplexors and multiplexors. A demultiplexor is a device that receives multiple signals from a fiber and splits it into its constituent wavelengths which are then directed into individual fibers or photodetectors as required. A multiplexor combines signals from multiple fibers and outputs it on one fiber for transmission. Multiplexors and demultiplexors are reciprocal devices in that if one of them is used with its ports reversed, the other component functionality is obtained. The optical multiplexors/demultiplexors can be classified as passive or active components. Passive components are based on prisms, diffraction gratings, and frequency filters. Active components are based on a combination of passive components and tunable detectors with each detector tuned to a specific frequency.

A1.2.1 Fabry–Perot etalon filter

The Fabry–Perot filter is based on the principle of partial beam interference, where a beam is split and made to interfere with itself to produces crests and troughs. The Fabry–Perot etalon consists of a cavity enclosed by two parallel mirrors. When light enters the cavity, it is reflected a number of times between the mirrors. The length of the cavity determines the frequency that is selected by the filter. By adjusting the length of the cavity, either by moving the walls physically or by varying the refractive index, a selected wavelength can be made to pass through while the others can be made to interfere destructively. While the Fabry–Perot filter can be made to access the entire low-attenuation window of the fiber, the tuning times are of the order of milliseconds and hence are not suitable for high-speed packet-switching applications.

A1.2.2 Dielectric thin-film filter

Dielectric thin-film filters consist of alternate layers of high and low refractive index material, each layer being $\lambda/4$ thick, as seen in Fig. A1.3. Light reflected by regions of low index does not shift phase while light reflected by high-index regions shifts in phase by 180°. Reflections from successive layers lead to constructive interference of certain wavelengths while the output power for other wavelengths is very low.

A1.2.3 Mach–Zehnder filter

In a Mach–Zehnder interferometer, a signal is split into two equal halves at the input, sent through waveguides of unequal length, and later recombined at the output.

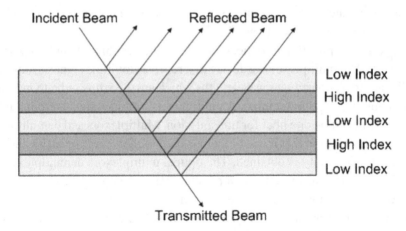

Fig. A1.3. A dielectric thin-film filter made of alternate layers of high and low refractive index. It can be used as a high-pass filter, a low-pass filter, or a high-reflectance layer.

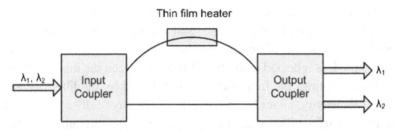

Fig. A1.4. A Mach–Zehnder filter made by splitting the wavelength into two parts and making them go through variable lengths of fiber. The thin film heater element is used to control the wavelength that appears on the output ports.

The two halves arriving at the output have different phases due to the path length difference. Signals that differ in phase by 180° are filtered out. By constructing a chain of these interferometers, a specific wavelength can be selected. The filter is made tunable by using a piezoelectric crystal to alter the physical length of the waveguide or by altering the refractive index thermally as shown in Fig. A1.4.

A1.2.4 Acousto-optic filter

The acousto-optic filter is based on the interaction of sound and light. When radio waves are passed through a transducer, the refractive index of the crystal changes, enabling the material to act as a grating. Light incident on the grating will diffract differently depending on the wavelength and the angle of incidence. By changing the radio-frequency (RF) wave, the selected wavelength can be changed. Typical acousto-optic filters have a tuning range of 250 nm and a tuning time of about 10 μs. By having multiple RF waves, multiple wavelengths can be filtered simultaneously. However, the drawback of acousto-optic filters is that the transfer functions are broad and hence susceptible to cross-talk. This imposes a constraint on the separation between adjacent channels and hence limits the number of supported channels.

A1.2.5 Electro-optic filter

The tuning time of the acousto-optic filter is limited by the time it takes for the surface acoustic wave to traverse the crystal. Crystals for which the refractive index can be changed by application of electrical signals can operate faster. Such filters where electrical current can be applied to select different wavelengths are called electro-optic filters. Since the tuning time in electro-optic filters is only limited by the speed of the electronics, tuning times of the order of nanoseconds are possible, though the tuning range has been observed to be limited.

A1.2.6 Liquid crystal Fabry–Perot filter

The design and operational principles of liquid crystal filters are similar to those of Fabry–Perot filters. The cavity consists of a liquid crystal and the refractive index of the liquid crystal can be modulated by an electric current. Thus, the selected wavelength can be changed as in the electro-optic filters. These filters have fast tuning times (1 μs), low power consumption, and are inexpensive to fabricate. Liquid crystal technology offers good promise for use in high-speed applications.

A1.2.7 Absorption filter

Absorption filters consist of a thin film made of material that exhibits an absorption peak at a selected frequency. Since this absorption peak is a material property, tuning the filter operation frequency or sharpening absorption edges becomes difficult.

A1.2.8 Fiber Bragg gratings

A fiber Bragg grating consists of a segment of fiber where regions of high and low refractive index alternate, as observed in Fig. A1.5. A periodic variation in the index can be introduced by exposing the core of the fiber to an intense ultraviolet optical interference pattern of periodicity equal to that of the periodicity of the grating to be formed. The incident wave is reflected from each period of the grating. The reflections add in phase only when the Bragg condition is satisfied, $\lambda = 2\Lambda/n$, where Λ is the period of the grating, λ is the selected wavelength and n is the order

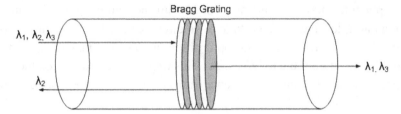

Fig. A1.5. A fiber Bragg grating made by inscribing alternate layers of low and high refractive index. These can be used as band stop filters, dispersion compensators, and as aids for network monitors.

of the grating. The wavelengths that do not satisfy Bragg's condition interfere destructively and hence are not selected. The fiber Bragg grating can be tuned by changing the grating period of the fiber, which can be achieved either mechanically or thermally. They can be fabricated with losses as low as 0.1 dB and channel cross-talk supression as high as 40 dB.

There are many applications of fiber Bragg gratings. If placed at the output of a circulator, it reflects back only the wavelength that it is designed for, thereby acting as a band stop filter. When placed at the output of a laser, it reflects a portion of the power back onto the source to be detected and monitored for possible laser failure. Fiber Bragg gratings with a linearly variable pitch may be used for dispersion compensation by inducing travel time variations.

A1.2.9 Diffraction gratings

When light is incident on a diffraction grating, each wavelength is diffracted to a different region in space. A fiber is placed at each of these focal points. Controlled focusing of individual wavelengths can be done with a lens system.

A1.2.10 Arrayed waveguide grating

An arrayed waveguide grating (AWG) is the generalization of a Mach–Zehnder interferometer grating. It consists of an input coupler and an output coupler connected by waveguides of varying length. An incoming signal is split and each part is made to traverse waveguides of varying lengths before they are finally combined at the output coupler. The lengths of the waveguides are designed such that an incoming wavelength on a specific port interferes constructively on exactly one of the output ports and interferes destructively on all the other ports, as illustrated in Fig. A1.6.

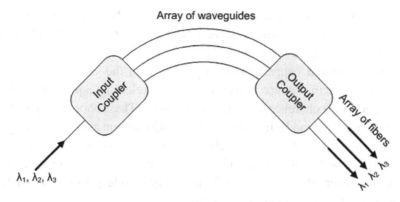

Fig. A1.6. A three-channel arrayed waveguide based on the principles of interferometry.

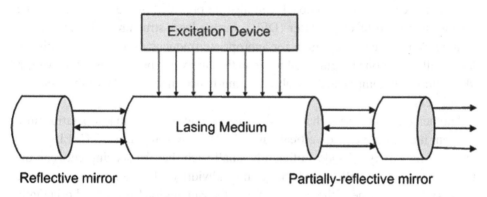

Fig. A1.7. The general structure of a laser that includes the excitation device, the lasing medium, and the reflective mirrors.

A1.3 Light sources

In optical communications, light sources should be compact, monochromatic, and consume little power. The generated optical power, however, must be adequately high so that it is detectable with a specified accuracy at the receiver end after undergoing attenuation and dispersion in the medium. It must be possible to modulate the source signal at the required rate, wavelength drift must be absent, and the dependence of the output intensity on time and temperature should be negligible. There are two kinds of popular light sources: light-emitting diodes and lasers.

A1.3.1 Laser

Laser stands for light amplification by stimulated emission of radiation. The Fabry–Perot laser consists of two mirrors, a cavity between the mirrors, a lasing medium, and an excitation device (Fig. A1.7). The excitation device typically applies an electrical signal to the lasing medium, which in turn leads to population inversion. The population is said to be inverted when there are a greater number of electrons in the excited state than in the ground state.

Stimulated emission happens when a photon is in the vicinity of an excited electron and the electron jumps to the ground level releasing a photon that has the same frequency and coherence as the incident photon. These photons will reflect off the walls of the cavity and be incident on the lasing medium again, triggering more stimulated emissions. Photons for which the frequency is an integral fraction of the cavity length are preferentially selected by the cavity to interfere constructively. It should be noted here that, for a given laser, many wavelengths could satisfy this condition and are called longitudinal modes.

In a Fabry–Perot laser, the feedback of light is localized and it occurs at the walls of the cavity. Light feedback can also be provided in a distributed manner as in a distributed feedback laser (DFB). In this laser structure, a Bragg grating is embedded in the active region for appropriate frequency selection. By suitable design, all except one longitudinal mode in the cavity can be suppressed. However, DFBs are more complicated to fabricate and hence more expensive than Fabry–Perot lasers.

Another class of lasers that achieve single-longitudinal mode operation in a different manner are called vertical cavity surface emitting lasers (VCSELs). The length of the cavity is made sufficiently small such that the spacing between the modes increases and falls outside the gain bandwidth of the laser. Such lasers have outputs taken from one of the surfaces. Since the gain region has a very short length, the walls need to be highly reflective and such reflectivities are not possible with metallic walls. So, a stack of low and high refractive index dielectrics are deposited to serve as a wavelength-selective mirror.

Though laser diodes have the advantages of high output power (10 mW), high speeds (Gbit/s), and narrow spectral width, they are very sensitive to temperature variations and wavelength drifts over time. As the injected current is altered, the charge carrier concentration changes and consequently the refractive index of the medium varies, leading to a shift in the selected wavelength. This is called wave-length chirping.

A1.3.2 Light-emitting diode

The light-emitting diode (LED) is a p–n semiconductor device that emits light when voltage is applied across its terminals. The emitted light is incoherent due to spontaneous emission. LEDs are fabricated in two basic structures: edge emitting (ELED) and surface emitting (SLED). SLEDs emit light over a wide angle and they are almost exclusively used only in multimode systems. They have a broad spectral width of about 100 nm. They can be modulated at bit rates of up to 100 Mbit/s, but higher speeds can be achieved at lower power levels. They are inexpensive, reliable, robust, produce more power and are simple to design. ELEDs have a structure similar to that of laser diodes with the reflectors removed. The spectral characteristics are narrow but are more complex to design.

A1.4 Optical amplifiers

Optical amplifiers are systems that amplify signals in the optical domain as opposed to repeaters which amplify after conversion to the electrical domain. This type of amplification, called 1R (regeneration), does not perform reshaping or reclocking and provides total data transparency. A single amplifier can simultaneously amplify all wavelengths and consequently avoids the overhead of one amplifier per channel. Optical amplification uses the principle of stimulated emission as used in a laser. The three basic types of amplifiers are erbium-doped fiber amplifiers (EDFAs), semiconductor optical amplifiers (SOAs), and Raman amplifiers.

A1.4.1 Semiconductor optical amplifier

A semiconductor optical amplifier is a laser structure that is modified to function as an amplifier. A weak optical signal is incident on the active region which upon stimulation releases an amplified signal. SOAs can achieve gains of 25 dB with a polarization sensitivity of 1 dB and a bandwidth range of 40 nm. SOAs based on quantum wells have been extensively researched. These devices provide high-bandwidth, high-gain saturation, and fast switching times at the expense of being highly polarization sensitive.

A1.4.2 Erbium-doped fiber amplifier

An erbium-doped fiber amplifer is a fiber segment, a few meters long, that is heavily doped with the rare-earth element erbium, as seen in Fig. A1.8. The most convenient excitation wavelengths are 980 and 1480 nm. When this wavelength is pumped into the doped fiber, erbium ions are excited and stimulated emission takes place, releasing photonic energy corresponding to the low attenuation window. EDFAs are polarization independent and have a large dynamic range. However, they are vulnerable to cross-talk due to spontaneous light emission. Typical gains achieved by EDFAs are of the order of 25 dB, although gains as high as 51 dB have been demonstrated.

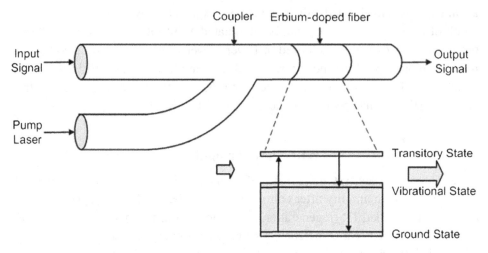

Fig. A1.8. EDFA consisting of an erbium-doped silica fiber, an optical pump, and a coupler.

A1.4.3 Raman amplifiers

While EDFAs provide a large gain over a short distance, the Raman amplifier provides a small gain over a large distance. The Raman amplifier is based on the non-linear optical effect. The most important feature of these amplifiers is that they have a wide bandwidth range. However, they require long fibers with high-power pump lasers.

A1.4.4 Modulators

Modulators, seen in Fig. A1.9, encode the data in the signal by turning a source on and off. This can be done internally within the laser structure or externally.

A1.4.4.1 Internal modulators

The modulation on the optical signal is achieved by controlling the current injected into the laser. Such lasers can be used up to a few Gbit/s. The disadvantage of direct modulation is that the resulting pulses are chirped. The frequency of the internally modulated signals changes with time, leading to adverse effects because of pulse broadening.

A1.4.4.2 External modulators

An alternate mode of operation is to operate the light source continuously and to place an external modulator in front of the source. This is called external modulation

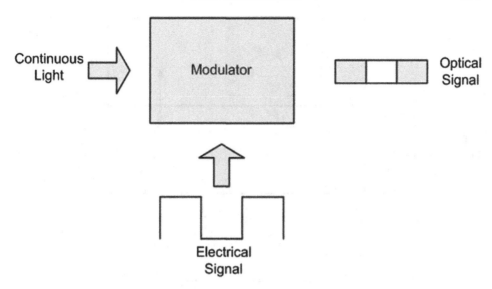

Fig. A1.9. A modulator using an on–off keying scheme modulates the intensity of the light based on the injected electrical signals.

and has been able to achieve 10 Gbit/s and beyond. There are several semiconductor-type modulators and among them the integrated Mach–Zehnder modulator is the most popular. Here, the signal is split into two halves, passed through electrically actuated phase controllers, and then recombined. By controlling the voltage levels on the phase controllers appropriately, the output signal can be switched on or off.

A1.5 Photodetectors

A photodetector converts an optical signal into an electrical signal. Photodetectors are made from semiconductor materials. The most popular photodetectors include the avalanche photodiode (APD) and the positive intrinsic negative photodiode (PIN).

A1.5.1 PIN

The PIN photodiode is a semiconductor device that consists of an intrinsic region sandwiched between the p-type and the n-type region. When light is incident on the reverse-biased junction, the electrons in the conduction band absorb the light, get excited to the valence band, causing an electrical current to flow in the circuit, as illustrated in Fig. A1.10. The output current generated is proportional to the input optical power.

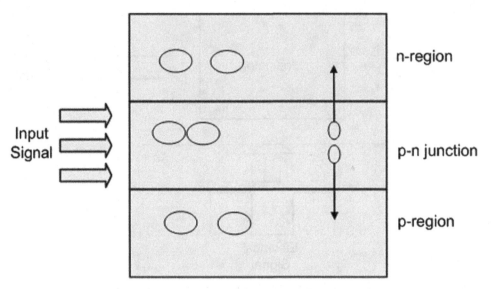

Fig. A1.10. Light incident on a reverse-biased p–n junction creates free electrons and holes leading to current in the external circuit.

A1.5.2 APD

The APD is a semiconductor device that creates strong fields in the junction region. Due to the avalanche process, a substantial amount of current is produced even with a few initial photons. The APD has the advantage of providing gain before the first electronic amplification stage in the receiver and hence reduces circuit noise.

The parameters that characterize the photodetectors are responsivity, which is the ratio of the output current to the input optical power, and quantum efficiency, which is the ratio of the number of output electrons to the number of input photons.

A1.6 Couplers

Couplers are devices that split or combine signals in the network. A 2×2 coupler consists of two input ports and two output ports. A typical coupler takes a fraction α of the power on input 1 and places it on output 1 while it takes a fraction $1 - \alpha$ of the power from input 1 and places it on output 2. α is the coupling ratio and may or may not be designed to depend on the incident wavelength. The other parameter that is of interest is the excess loss, which is the loss introduced by the coupler over and above the loss introduced by the coupling ratio α. The simplest application of a coupler is to split the incoming signal into two equal halves and such a coupler is called a 3 dB coupler. In addition to the loss of 3 dB, this coupler may introduce an excess loss of 0.2 dB. Couplers can also be used to tap off a small portion of power from a source for monitoring and supervision purposes.

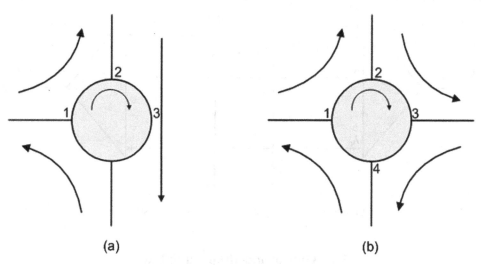

Fig. A1.11. Representation of (a) three-port and (b) four-port circulators. They are used to construct optical add–drop elements.

A1.7 Isolators

The main function of isolators is to transmit optical power in one direction more than in the reverse direction. Isolators are characterized by their insertion loss, which is the loss of optical power through it, and by isolation, which is defined to be the ratio of the transmitted power in one direction to that in the reverse direction. Isolators typically utilize the Faraday rotation principle to rotate the direction of polarization and prevent back reflections. The insertion loss is around 1 dB and the isolation is around 40–50 dB.

A1.8 Circulators

Circulators, shown in Fig. A1.11, are similar to isolators, except that they have multiple ports. Circulators are used for routing traffic and to add or drop certain signals. A circulator can be designed using a combination of couplers and isolators. Circulators operate on the same principles as isolators.

A1.9 Repeaters

A repeater, also called a transceiver, is a combination of a transmitter and a receiver. A repeater module first receives the weak signal from the incoming cable. The receiver converts the information in the electronic domain. Then the signal is shaped, reclocked, and amplified. Finally, the clean version of the signal modulates the transmitter to generate an optical signal.

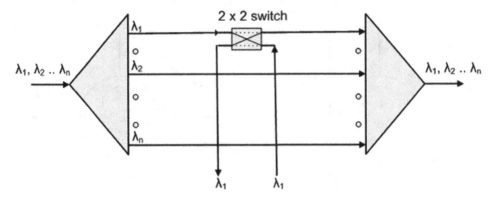

Fig. A1.12. A simple optical add–drop system.

A1.10 Optical add–drop multiplexors

OADMs selectively remove a wavelength from a set of wavelengths in a fiber and then add the same wavelength with a different data content in the same direction. A simple add–drop module can be fabricated using a multiplexor (mux), a 2×2 switch, and a demultiplexor (demux). Figure A1.12 shows the typical block diagram operation of such a device. OADMs are classified as fixed-wavelength and dynamically wavelength selectable OADMs. In dynamically selectable OADMs, the wavelengths between the optical demux/mux may be dynamically directed from the outputs of the demux to any of the inputs of the mux.

A1.11 Optical switches

Cross-connecting channels is a key function in networks and is done using optical switches. Optical switches come in two forms, photonic switches and hybrid switches. Photonic switches cross-connect channels purely in the optical domain while hybrid switches convert the signals into the electrical domain before performing the switching function. These devices provide reconfiguration at the wavelength level or at the fiber level. A simple 2×2 switch based on the Mach–Zehnder interferometer principle is shown in Fig. A1.13.

The newest kind of photonic switches are those that are based on microelectromechanical systems also known as MEMs. Nanotechnology has been employed to micro-machine tiny mirrors on a substrate that switches by reflecting optical beams. MEM switches, as shown in Fig. A1.14, have demonstrated a low-loss, compact design, and a high on–off contrast ratio. However, their speeds are of the order of milliseconds since they involve physical movement of the mirrors.

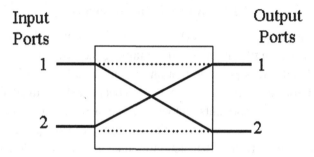

Fig. A1.13. A simple 2 × 2 switch constructed based on Mach–Zehnder interferometry. The switch can be in one of two states: the bar state or the cross state, depending on the control electronics.

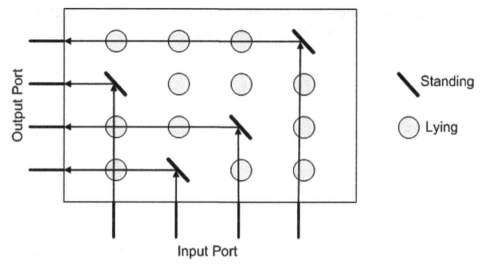

Fig. A1.14. In a crossbar arrangement, a mirror lies flat to allow a light beam to pass through and tilts up to a vertical position to steer light to a specific port.

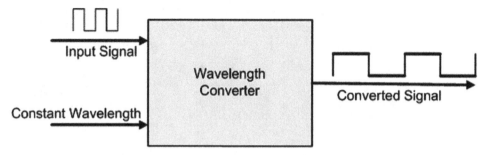

Fig. A1.15. A wavelength converter based on a non-linear property of the material.

A1.12 Wavelength converters

Wavelength converters are devices (Fig. A1.15) that convert the incoming signal of a particular wavelength into a signal containing the same information but on a different wavelength. It is possible that an incoming call cannot be accepted in a portion of a network on a particular wavelength because that wavelength is already busy or because other components that work in this wavelength range are not available. In such situations, the data using the incoming wavelength can be switched onto an idle, available wavelength to accommodate the call. Wavelength conversion thus enables efficient spatial reuse of wavelength resources in the network, adding to the flexibility of multi-wavelength systems.

Conversion may be achieved by employing the non-linear properties of semiconductors. SOAs are used as wavelength converter devices. Various methods based on cross-gain modulation, four-wave mixing, and other interferometric techniques have been explored to achieve wavelength conversion.

Appendix 2

Network design

An interconnection network connects various sources of information using a set of point-to-point links. A link is a connection using a copper wire or an optical fiber, or may be wireless. The nodes are autonomous data sources and can request to transfer any amount of information to any other node. Figure A2.1 shows an example network consisting of four nodes. Node A has a link connected to nodes B and C. Node B is connected to nodes A and D. Nodes C and D are connected to nodes A and B, respectively. If node C desires to send some information to node B, it sends it to node A which in turn routes it to node B. Node A thus acts as an intermediate node. The *capacity* of a node is the amount of information it can transmit (also called its *source capacity*) or receive (also called its *sink capacity*). The capacity of a link is the amount of information that can be transferred over the link in one unit of time.

The network design deals with the interconnection of various nodes and how to transmit information from one node to another. Network architecture and design both have multiple meanings. The most commonly used interpretation relates to the decisions one needs to make to design a network. The four most important aspects of network architecture and design are described here.

A2.1 Network topology

A topology defines how nodes are interconnected. For example, the topology of the NSF network is shown in Fig. A2.2. Most network topologies are hierarchical in nature. The design involves developing the structure of the hierarchy, structures of nodes at each level, and detailed designs of the nodes. It also involves assigning link and node capacities to transport the desired traffic. A hierarchical topology is depicted in Fig. A2.3. We will be studying the decision making process and related algorithms and examples in detail in this appendix.

377

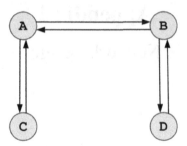

Fig. A2.1. A four-node network.

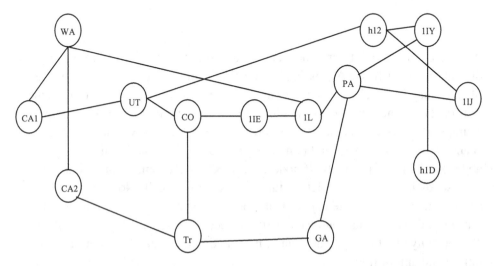

Fig. A2.2. Topology of the NSF network.

A network node is placed in a hierarchical fashion in such a way that it is "close" enough to several data sources. The closeness is described in terms of suitable performance metrics such as the physical distance, the cost of connection, and so on. A network node serves as a service point for all the data sources connected to it. Such a node is called a "gateway" as it connects data sources to network nodes. Nodes A, B, C, etc. are such nodes. Gateways connect to data sources, such as nodes 1, 2, ..., 12, at the next lower level in the hierarchy and to routers or switches such as nodes X, Y, etc. at the next higher level. Switches and routers route information to other switches and routers on the way to other gateways from where the data are delivered to destinations. There may be more levels in the hierarchy.

Node and link placement, and their capacities, in a network topology depends on the desired or required traffic flow that is defined by the traffic characteristics. This is a well-studied problem and more information can be found in [21, 32, 45, 53, 182, 265]. In principle, ideal locations for both network nodes and links may

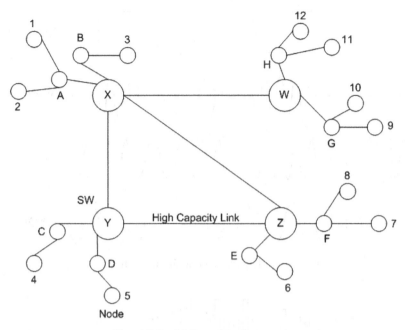

Fig. A2.3. A hierarchical network.

be specified using (possibly complex) algorithms that would optimize the network design using the performance metrics of interest. In practice, these placements are also governed by factors such as the existing network, the ease of operation, the convenience of management, and so on, which are not always easy to accommodate in the design process.

A2.2 Transmission technology

Physical layer transmission technologies describe the characteristics of the physical medium. These involve signal processing techniques, modulation and demodulation techniques, coding and decoding of information, multiplexing, and demultiplexing techniques employed to enhance the utilization of each link, and issues related to these techniques. The physical medium can be a wire, such as a copper link, a coaxial link, an optical fiber, or a wireless link using microwave or radio frequencies. The signals being transmitted over the physical medium can be modulated and demodulated using amplitude modulation, frequency modulation, or phase modulation, where the information being transmitted modifies the shape of the waveform being transmitted. Figure A2.4 demonstrates examples of modulation techniques. Multiplexing techniques such as *time-division, frequency-division*, and *code-division* multiplexing techniques are used to mix and transmit information from various sources on a single link. In time-division multiplexing, each source

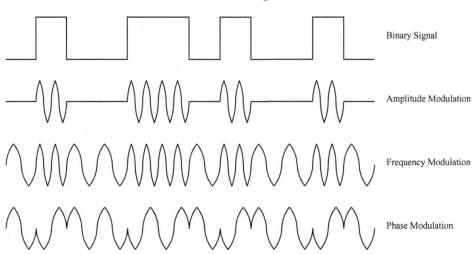

Binary Signal

Amplitude Modulation

Frequency Modulation

Phase Modulation

Fig. A2.4. Different modulation techniques.

1 2 3 1 2 3 1 2 3 1 2 3 1 2 3 1 2 3   Time

Fig. A2.5. Time multiplexing.

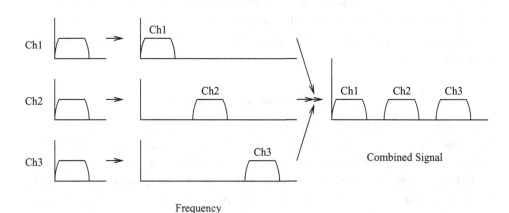

Frequency

Fig. A2.6. Frequency multiplexing.

is given a fraction of time in a given interval, called a *frame*. In frequency- and code-division multiplexing, frequencies and the bandwidth available on a channel are effectively partitioned so that all sources can use parts of the capacity of the channel simultaneously. Figures A2.5 and A2.6 show different multiplexing techniques.

A2.3 Traffic control and network management techniques

This aspect of the network architecture involves the control of the switching technology, flow control algorithms for smooth flow of loads offered to the network,
control message flow to set up paths, connection requests and response protocols,
collection of data on actual usage, fault detection and management algorithms, and
effective resource utilization algorithms.

Connection requests between nodes in a network are realized by employing a
routing algorithm. Routing algorithms are designed to utilize the existing network
capacity and the switching methodology. Either *circuit switching* or *packet switching* can be employed for data transfer in a communication network.

In a circuit-switched network, a complete, dedicated path from a source to a destination is established through the network before communication begins. Dedicated
physical resources are reserved for the communication to take place. A signal is
sent from the source to the destination node, through intermediate nodes, requesting
a connection. An acknowledgment signal is sent back from the destination to the
source accepting or rejecting the request. A request is accepted only if all nodes on
the path have the required resources available and they can reserve these resources
statically. If no free path exists from end to end, the traffic is blocked and has to wait
for transmission. Path establishment may take a substantial period of time. Once
the path is established, information is transmitted freely from the source node to
the destination node using the path. The sender and receiver may use any format for
data transfer and bit rate subject to the constraints of the physical channel. When
the transfer is completed, the path is removed.

In a packet-switched transmission, no physical path is established in advance.
Instead, when a source has information to transmit, it assembles it in a packet. A
packet consists of data to be sent and a header. The header contains the source
and destination addresses, possibly intermediate node addresses through which the
packet must be routed, and some error-correcting codes to check the correctness of
the information. The packets are forwarded from node to node, one hop at a time.
The packets are queued in buffers at the intermediate nodes along a route between a
source and a destination, traveling from node to node, releasing links and switching
elements immediately after using them. A packet is received by a node, checked
for correctness and retransmitted, if required. During the transmission, store and
forward operations at each node increase the overhead and time delay of data
packets.

Most networks use packet switching for smaller messages. In a packet-switched
network, it is possible that the path may change from packet to packet between the
same source–destination pair. Circuit switching is used when a source/destination
has a substantial amount of information to transmit. The main advantage of circuit

switching is that the links on a path are always available for communication. The only delay is the propagation delay. However, if no information transfer takes place for some time during the existence of a connection, the capacity on the path is wasted. To reduce this waste, it is possible to statistically multiplex the path. A path is established in advance between a source and a destination that is used by every packet from that source to that destination. However, actual transmission occurs as in a packet-switched network. Such a path is called a *virtual path*, as opposed to a physical path that is established in a circuit-switched network. Such a service, in which a path is established in advance, is also called a *connection oriented service*. A pure packet-switched service is called a *connectionless* service. To set up a path, the network control sends messages to nodes on the path to request a connection. If all nodes on the path from a source node to a destination node agree, then the path is established. Flow control algorithms are employed to control the actual flow of information so that each node on the path is not overwhelmed by the information. If a fault occurs in a node on the path, then the path is established again. In the case of a connectionless service, data packets are just sent to the next node along the path. If a faulty node is encountered on the path, the routing is changed on the fly. The internet datagram service uses a pure packet-switching protocol whereas the telephone system, for the most part, uses circuit switching. Networks employing asynchronous transfer mode (ATM) (that are currently being developed) use the concept of a virtual path for ATM cell transmission. An ATM cell is a small, 53 byte long packet that includes 48 bytes of actual data and 5 bytes of control information. In this case, a path is established that is used by all the cells but actual transmission is cell by cell from point to point similar to a packet-switched network.

A2.4 Cost

The cost of the network is viewed differently by different people. The cost includes parameters such as installation of links and nodes, including the cost of the facility to house the nodes and laying the links (copper or fiber). Laying out links is very expensive and includes buying/leasing land, digging land, laying out conduits, the cost of cables, wires, or optical fibers, end-interfaces, buffering, processing hardware at each end of a link, and management of links. There are additional operational and maintenance costs in keeping the hardware up and running, replacing faulty components and cables, and managing the resources.

For a network service provider, the cost consists of laying out and operating the network. On the other hand, the users or consumers of network resources do not concern themselves with these costs. The costs they account for are the cost quoted to them by the network providers in terms of tariffs for different *qualities of service (QoS)* at different times. These tariffs are usage sensitive and depend on the volume of data being transported, time of day (morning, evening, or night), priority of transmission, tolerable delay and loss of data, and several such factors. These factors together are called QoS parameters. To provide and guarantee a specific quality of service, the network service provider has to dedicate some network resources such as bandwidth on individual links, buffer spaces at various nodes on the path, time slots for transmission of specific data, and alternate resources in case of a failure for that service. The costs of these resources form the basis for tariffs. The development of a cost model for a link is a difficult problem. Often good approximations and simplification of cost structures are used by the network service providers to keep the complexity of the network design and service tariffs under control. In our examples, we consider both models (the actual physical network cost model and the consumer network cost model) of cost in our designs.

A2.5 Approaches to network design

If a new network is being designed from scratch with no existing capacity, well-defined traffic requirements (traffic intensities), and full freedom in selecting network components, then the designers can make the best possible decisions by balancing the cost and QoS requirements, such as the throughput, the delay, and other performance measures. However, more often than not, most real designs are incremental, that is, the resources are added or upgraded over the existing capabilities as required by the new demands. The network really evolves with the needs and, in general, is in response to the new requirements. This restricts the optimality in design as the existing design governs the final output.

Inputs for network designs are based on the best estimates of the anticipated traffic between various sources and destinations. Such data are available in the form of a traffic matrix. A large number of networks are designed using current and additional anticipated needs and certain rules of thumb in an incremental fashion. The decisions are based on the experience of the designer. It is possible to make serious mistakes as part of a new design. For example, when the information transmission is from point to point as in packet switching, intermediate nodes store and forward the incoming information. By not providing enough buffer space or control for incoming traffic streams, losses may be excessive and/or delays may exceed the acceptable limits. A loss may or may not be tolerable. For example, in voice communication a small loss may almost go unnoticed, but loss of even a byte may not be tolerable in a computer file transfer application. For a voice or real-time video communication, any significant delay may mean that the information is no longer relevant at the destination.

The design process could be manual or automated using exact or heuristic-based algorithms. An automatic design process can avoid such serious design flaws in the network. Unfortunately, most of the known properties and optimization techniques relate to networks that are designed from new and not incrementally. Heuristic algorithms are used as part of an automated design process to incorporate design principles used in manual algorithms. One of the most commonly used heuristics is the *greedy* algorithm. Sometimes a greedy algorithm may find an optimal solution. It selects a feature that appears to be of immediate benefit. Consider a situation in which there are several nodes that communicate with each other. Providing a direct link between two nodes that have the maximum amount of traffic flowing between them is a greedy approach. This may have other effects later on in the algorithm. Similarly, incorporating and using the cheapest link in a network is also a good design practice. However, this may have serious cost implications at a later stage in the design and a greedy algorithm may fail to account for them. A greedy algorithm may not always yield the best result, but nonetheless is the most commonly used heuristic algorithm.

To fully understand the network design process and the algorithms necessary in network design, we first develop a graph model of the network. Graph models capture the exact behavior of a network and simplify the task of analysis.

A2.6 Quality of service requirements

Unlike conventional packet- or circuit-switched networks, some applications such as broadband integrated services digital network (B-ISDN) require the network to provide not only connectionless traffic transportation but also connection-oriented operation for real-time data transfer between end users with multiple bit rates. Broadband packet switching based on asynchronous transfer mode (ATM) which has a fixed packet length, has been proposed for multimedia and multi-bit-rate communications for end users by using the network resources efficiently. The most important aspect of these networks is to satisfy the quality of service (QoS) requirements. These features require a different approach to network design in comparison with the conventional packet-switched network design. For example, the cell loss probability has to be considered in an ATM network design. In circuit-switched networks, the call blocking probability is an important metric for determining the design of the circuit-switched networks.

Connection-oriented services have certain maximum delay requirements in exchanging information between the end users as given by the QoS requirements. The delay in a packet-switched network includes switching, queuing, transmission, and propagation delays. Due to the high data rate of fiber optic links, the propagation delay and the node queuing delay are the dominant delay factors. In the conventional packet-switched network design problem, the average network delay and throughput have usually been used as the metrics to optimize the network cost and performance. In multimedia networks, services may have critical delay requirements; instead of the average network delay requirement, the end-to-end delay must be considered while determining the network topology.

In addition to these new requirements, the high data rates require special attention to fault management or fault tolerance. Compared with low-speed data networks, it is possible to lose many data packets if a data link fails even for a short time. Fault management requires that the network has a control mechanism which ensures that the existing traffic is affected as little as possible due to a failed link, and the traffic on the failed link is rerouted through the spare capacity on other links. This rerouting of traffic from a failed link to the other links [37] can be performed by a special facility such as a digital cross-connect system (DCS) [295].

Fault tolerance in high-speed networks is greatly needed even for short-time link failures due to the large cell loss possibility. An alternative route may be longer than the original path. If a service, such as a data file transfer, is not sensitive to propagation delay, the reconfiguration can be done using arbitrary available spare capacity on the other links. Since voice and video service are sensitive to end-to-end delay, the reconfiguration path must be selected such that the end-to-end delay requirements are met. This performance requirement restricts the logical reconfiguration that can be embedded into the physical network. Therefore, while designing the network topology, possible failures of network links must be considered in advance.

To maintain the QoS requirements in services [136–139], we also have to consider the cell loss probability during a burst transfer. Burst cell loss can occur in several stages of the network: switch buffer overflow, cells discarded for congestion control, and physical link errors. The optical fiber link has negligible physical link errors. However, the switch buffers for each link may be of a fixed size and the packet contention for the same link may cause the output buffer to overflow in each link. Thus, we have to find the optimal link capacity assignment to meet the cell loss restrictions.

A2.7 Probability distributions

We will first describe three important probability distribution functions that are used in the analysis of network systems. More details can be found in [7].

Normal distribution. A random variable, x, is normally distributed if its probability density function is of the form

$$f(x) = \frac{1}{\sqrt{2\pi\sigma^2}} e^{-(x-\mu)^2/2\sigma^2}$$

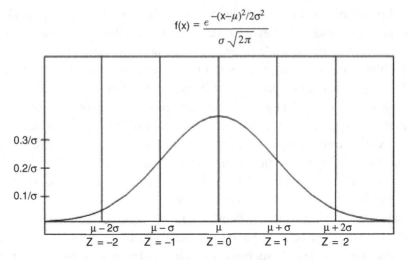

Fig. A2.7. A bell-shaped curve.

This is a bell-shaped curve density function, as shown in Fig. A2.7. The peak of the bell occurs at $x = \mu$ and the width of the bell depends on the variable σ. The random variable x is completely characterized by two variables: the mean μ and the variance σ^2. The variable σ is the standard deviation. Three standard deviations from the mean cover about 99% of the area under the curve. That is why most of the time we are interested in $\mu + 3\sigma$ variations in the value of a random variable x.

Binomial distribution. The number of ways in which k out of n objects can be selected is given by

$$C(n, k) = \frac{n!}{k!(n-k)!}$$

If the probability of selecting a particular type of object is p (and the probability of selecting the other object(s) is $(1 - p)$), then the probability of selecting k such objects out of a total of n objects is given by

$$P(n, k, p) = C(n, k)\, p^k\, (1 - p)^{n-k}$$

The mean value of this statistic is $E(n, p) = np$ and the variance is $V(n, p) = np(1 - p)$. The standard deviation, σ, is given by $\sqrt{V(n, p)} = \sqrt{np(1 - p)}$.

Exponential distribution. A random variable x is exponentially distributed with parameter λ if the probability of $x \leq t$ is given by

$$P(x \leq t) = 1 - e^{-\lambda t} \qquad t \geq 0$$

The mean and variance of x are $1/\lambda$ and $1/\lambda^2$, respectively.

An arrival of a request for service is usually modeled using a random process. Two requests are often assumed to be independent of each other. A process in which interarrival times between two consecutive requests are independent and distributed according to an exponential distribution with parameter λ is called a Poisson process (with parameter λ).

Design. How we use these distributions can be demonstrated using the following example. Suppose a network has ten nodes that want to communicate among themselves. Suppose the probability that a node originates a data request is $p = 0.1$. The switching network can connect a call if all required links are free or not in use. How many links should we provide so that a communication request can be satisfied with high probability? In this case, the average number of requests is $E(n, p) = 10 \times 0.1 = 1.0$ and the deviation is $\sigma = \sqrt{10 \times 0.1 \times 0.9} = \sqrt{0.9} = 0.95$. To satisfy most of the requests with high probability, we may like to provide $\mu + 3 \times \sigma = 1 + 3 \times 0.95 \approx 4$ links.

A2.8 Delays in networks

A communication link can be viewed as a bit pipe over which a given number of bits are transmitted over a unit of time. This number is called the transmission capacity of the link and depends on the physical channel and the interface at the two ends of the link. The bit pipe (link) is used to serve all traffic streams that need to use the link. The traffic on all streams may be merged into a single queue and transmitted on a first-come-first-served basis. This is called statistical multiplexing. It is also possible to maintain several queues for a link, one for each traffic stream or one for each priority if the incoming traffic streams have multiple priority levels assigned to them. If a packet length is L and the link capacity is C bits/s then it takes L/C seconds to transmit a packet.

In the case where all incoming communication requests for a link are assigned to a queue and serviced as the resources become available, there are four different kinds of delays a packet suffers on a link. If the packet has to travel through multiple links, then the total delays will be the sum of delays on all the links.

(i) *Queuing delay.* The queueing delay is the delay between when a packet is assigned to a queue and when it is ready to be processed for transmission. During this time that packet simply waits in a queue. This time depends on the number of packets waiting ahead of this packet in the queue.

(ii) *Processing time delay.* The processing time is the time between events when the packet is ready to be processed and the time it is assigned to the link for the transmission. The processing delay depends on the speed of the link processor and the actions the processor needs to take to schedule the transmission.

(iii) *Transmission delay.* The time difference between the transmission of the first and last bit of the packet is referred to as the transmission delay. This delay depends on the bit transmission rate of the link.

(iv) *Propagation delay.* The propagation delay refers to the time difference between the instances when the last bit is transmitted by the head of the link (source) and it is received by the tail of the link (destination). This delay depends on the physical distance of the link and the speed of propagation, and can be substantial for a high-speed link.

A2.9 Queuing models

To compute the queueing delay for a packet, we have to understand the nature of the packet arrival process to a link, the kind of service time it needs (the amount of transmission time) and the number of links we have from the source to the destination. In most queueing systems [154, 155], we assume that the arrival process is a Poisson one. We also assume that the holding time (the amount of time a request requires to service) follows an exponential distribution with parameter μ. The mean service time is then given by $1/\mu$. If two nodes i and j are connected by m links, then m packets can be transmitted from node i to node j at the same time. Generally $m = 1$ and therefore packets are transmitted one at a time. In the case of circuit switching, it can be seen as one request being established at a time.

M/M/m queue. A queueing system with m servers, a Poisson arrival process, and an exponentially distributed service time, is denoted by an M/M/m queueing system. The first letter M stands for memoryless. It can also be G for the general distribution of interarrival times or D for deterministic interarrival times. The second letter stands for the type of probability distribution of the service times and can again be M, G, or D. The last number indicates the number of servers.

In an M/M/1 queueing system, the average number of requests in the system in the steady state is given by $\lambda/(\mu - \lambda)$ and the average delay per request (waiting time plus service time) is given by $1/(\mu - \lambda)$. Utilization of the system is denoted by $\rho = \lambda/\mu$ and the average time for a request in the system is given by the average service time$/(1 - \rho)$. The average waiting time, T_w, is given by the difference of the average time in the system and the average service time. This time is equal to $1/(\mu - \lambda) - 1/\mu$. The average number of requests in the queue is given by $\lambda \times T_w$. Also, the probability that there are exactly k requests waiting is given by $P_k = (1 - \rho)\rho^k$.

Performance metrics. When a request for service arrives, the server (link) may be busy or free. If the server is free, the request is serviced. If the server is busy, then there are two possibilities. (i) The request is queued and serviced when the server becomes available. In this case we are interested in finding out how long, on average, a request may have to wait before it is serviced. In other words, we need to find out how many requests are pending in a queue or the average length of the queue. This has implications in designing queues to store requests. (ii) The incoming request is denied service. This is called blocking. We would like the blocking probability to be as small as possible.

Appendix 3

Graph model for network

A network is represented by a *graph* $G = (V, E)$, where V is a finite set of elements called *nodes* or *vertices*, and E is a set of unordered pairs of nodes called *edges* or *arcs* [85]. This is an *undirected* graph. A *directed* graph is also defined similarly except that the arcs or edges are ordered pairs. For both directed and undirected graphs, an arc or an edge from a node i to a node j is represented using the notation (i, j). Examples of five-node directed and undirected graphs are shown in Fig. A3.1. In an undirected graph, an edge (i, j) can carry data traffic in both directions (i.e. from node i to node j and from node j to node i), whereas in a directed graph, the traffic is only carried from node i to node j.

Graph representations. A graph is stored either as an adjacency matrix or an incidence matrix, as shown in Fig. A3.2. For a graph with N nodes, an $N \times N$ 0–1 matrix stores the link information in the adjacency matrix. The element (i, j) is a 1 if node i has a link to node j. An incidence matrix, on the other hand, is an $N \times M$ matrix where M is the number of links numbered from 0 to $M - 1$. The element (i, j) stores the information on whether link j is incident on node i or not. Thus, the incidence matrix carries information about exactly what links are incident on a node. If a graph has more than one link from a node to another node, the incidence matrix will be able to carry this information exactly, whereas the adjacency matrix will require additional information to store the number of links.

The following terms associated with a graph are used throughout this appendix.

(i) The *degree* of a node is the number of links incident on a node. In the case of a directed graph, we count both the number of incoming links, or the *in-degree*, and the number of outgoing links, or the *out-degree*, of a node. For example in Fig. A3.1(a), node 1 has a degree of three, whereas in Fig. A3.1(b), node 1 has an in-degree of one and an out-degree of two.

(ii) A *walk* in a graph $G = (V, E)$ is a sequence of nodes $w = [v_1, v_2, \ldots, v_k], k > 1$, such that $(j, j + 1) \in E, j = 1, 2, \ldots, k - 1$. A walk is *closed* if $k > 1$ and $v_1 = v_k$.

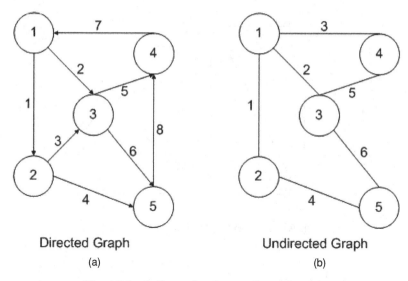

Directed Graph   Undirected Graph

(a)   (b)

Fig. A3.1. A directed and an undirected graph.

i \ j	1	2	3	4	5
1	1	1	1	1	0
2	1	1	0	0	1
3	1	0	1	1	1
4	1	0	1	1	0
5	0	1	1	0	1

N \ L	1	2	3	4	5	6
1	1	1	1	0	0	0
2	1	0	0	1	0	0
3	0	1	0	0	1	1
4	0	0	1	0	1	0
5	0	0	0	1	0	1

Adjacency Matrix   Incidence Matrix

Fig. A3.2. Graph representations for the graph shown in Fig. A3.1(a).

(iii) A walk without any repeated nodes in it is called a *path.*

(iv) A closed walk without any repeated intermediate nodes is called a *cycle.* An acyclic network does not contain any cycles, as shown in Fig. A3.3.

(v) A node s is said to be *connected* to node t if node s has a path to node t in the graph. This path is called an (s, t) path.

(vi) The length of a path is the number of links on the path.

(vii) An (s, t) path is called the *shortest path* if there is no other path of length shorter than the length of the given path.

(viii) $\delta(i, j)$ denotes the length of a shortest path between nodes i and j. In a network, it is a measure of the maximum communication delay.

(ix) The *diameter* (the longest shortest path between any pair of nodes) of a graph is given by $\text{Max}\{ \delta(i, j) \, \forall \, i, j \in V\}$.

(x) A graph is said to be connected if a path exists between any pair of nodes, s and t.

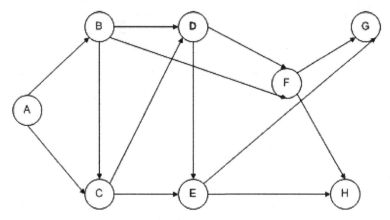

Fig. A3.3. An example of an acyclic graph.

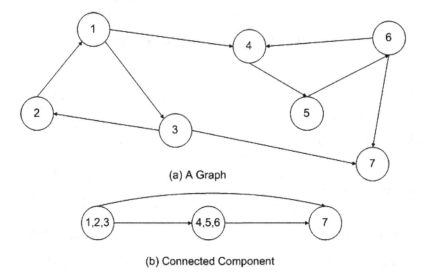

(a) A Graph

(b) Connected Component

Fig. A3.4. A graph and its connected components.

(xi) A graph is said to be *strongly connected* if $\forall i, j, \in V$ there exists a path from node i to node j.

(xii) A *connected component* of a graph (V, E) is a subgraph $G' = (V', E')$, $V' \subseteq V$, and $E' \subseteq E$, with every $(i, j) \in E'$, $i, j \in V'$ such that G' is strongly connected.

    An example of connected components of a graph is shown in Fig. A3.4 where nodes 1, 2, and 3 form one connected component and nodes 4, 5, and 6 form another. Node 7 is a component by itself.

(xiii) The *node connectivity* of a graph is the minimum number of nodes which should be removed from the graph in order to partition it into two disjoint subgraphs, that is, the number of node-disjoint paths. The node connectivity is a measure of the reliability of communication.

Appendix 4

Graph algorithms

Once we select the graph model of a network, various algorithms can be used to efficiently design and analyze a network architecture. Some of the most fundamental algorithms among them are finding trees in a graph with minimum cost (where cost is defined appropriately) or finding a minimum spanning tree, visiting nodes of a tree in a specific order, finding connected components of a graph, finding the shortest paths from a node to another node, from a node to all nodes, and from all nodes to all nodes in a distributed or centralized fashion, and assigning flows on various links for a given traffic matrix.

In the following we describe some useful graph algorithms that are important in network design. Recall that N represents the number of nodes and M represents the number of links in the graph.

A4.1 Shortest-path routing

Shortest-path routing, as the name suggests, finds a path of the shortest length in the network from a source to a destination [19, 73, 224, 241, 249]. This path may be computed statically for the given graph regardless of the resources being used (or assuming that all resources are available to set up that path). In that case, if at a given moment all resources on that path are in use then the request to set a path between the given pair is blocked. On the other hand, the path may be computed for the graph of available resources. This will be a reduced graph that is obtained after removing all the links and the nodes that may be busy at the time of computing from the original graph. In either case, computation of the shortest path is based on the following concepts.

Suppose for a given graph G, each arc (i, j) is also assigned a weight (or length) denoted by a_{ij}. We are interested in finding a path of the shortest length from a given source node to a given destination node. The path will not have any node repeated. This type of problem is fundamental in graphs, and in particular networks, as we may be interested in searching a path from a source to a destination that is least expensive in terms of traversal. The weight or length of an arc may represent the actual cost of traveling on that edge. The cost may be in terms of delay, or dollars, or any other metric of importance.

One of the key ideas in computing the shortest path is that of dynamic programming. It is also based on the principle of optimality. For the shortest-path computation, it has been shown that if all edge weights are positive, then an undirected graph can be treated as a directed graph by replacing each undirected edge (i, j) by two directed edges (i, j) and (j, i). With negative edge weights, this transformation introduces cycles of negative weights and the shortest path may go through the cycle as often as necessary to bring the total path lengths to zero or negative. Thus, it is not desirable. In the following, we will assume that all edge weights are positive. With that assumption the shortest path can be computed using the following formulation.

A4.1.1 Bellman's equations

To compute the shortest path from source s to destination t, it turns out that we end up computing the shortest path from the source node s to all destinations [224]. Let a_{ij} be the weight of edge (i, j) if the edge exists. Otherwise, it is ∞. Let u_j be the weight of the shortest path from origin s to node j. For simplicity we assume that the nodes are numbered from 1 to n and the source node is node 1. We can always renumber the nodes. It is clear that $u_1 = 0$. Let node k be the last node on the shortest path from node 1 to node j. Then we can say that $u_j = u_k + a_{kj}$. This also implies that the path from node 1 to node k with path length u_k must also be the shortest path from node 1 to node k. Otherwise, the path we selected is not the shortest path. This is from the "principle of optimality." Now, we only have a finite number of choices for k. Bellman's equations use this principle to search for shorter paths to other nodes by using the known shortest path to node k and edge weights of direct links from node k to other nodes for all such k values. The equations state that

$$u_1 = 0$$

and

$$u_j = \min_{k \neq j}\{u_k + a_{kj}\} \qquad j = 2, 3, \ldots, N$$

Using these equations, we can find a shortest path to a node as follows. First, find a node k with edge (k, j) such that $u_j = u_k + a_{kj}$. Then find an arc (l, k) such that $u_k = u_l + a_{lk}$, and continue in this fashion. Eventually, we would reach node 1. Unfortunately, Bellman's equations do not lead to a solution directly.

A4.2 Shortest path in an acyclic network

In an acyclic network, as shown in Fig. A4.3, it is easy to use Bellman's equations to find a shortest path. The nodes in such a network can be renumbered in such a fashion that an edge (i, j) exists if and only if $i < j$. In this case we can rewrite Bellman's equations as

$$u_1 = 0$$

and

$$u_j = \min_{k<j}\{u_k + a_{kj}\} \qquad j = 2, 3, \ldots, N$$

These equations can then be solved as u_1 is known, u_2 only depends on u_1, u_3 only depends on u_1 and u_2, and so on. The complexity of this problem is $O(N^2)$.

A4.2.1 Dijkstra method

For non-cyclic graphs, we need another method given by Dijkstra [73]. This method is applicable to a graph for which edge weights are positive. This algorithm starts with labeling nodes in stages. At each stage of computation, some labels are designated permanent and others remain tentative. A permanent label on a node represents the true length of the shortest path from that node. After including the new labeled nodes, distances to all other nodes are computed again.

Let d_{ij} denote the distance from node i to node j. Let i be the source node. Then d_{ii} is set to zero and d_{ij}, $i \neq j$ is set to a large value if j is not a neighbor of i. Otherwise, it is set equal to the weight of the direct link, a_{ij}. Next, the algorithm finds a node j with minimum d_{ij} and labels it *permanent*. It then uses it to improve distances to other nodes by computing

$$d_{ik} \leftarrow \min(d_{ik}, d_{ij} + a_{jk})$$

At each stage in the process, the value of d_{ik} represents the best known shortest distance from i to k. Using these labels of the nodes, the algorithm then marks another unlabeled node with a minimum value of d_{ik} as permanently labeled. The same computation is carried out again. Since all edge weights are positive, in the next iteration, none of the marked nodes can have any smaller value.

An example of the execution of the algorithm is shown in Fig. A4.1. Node A is the source node. A dark node is a permanently labeled node. At each step, one node is marked labeled and the value associated with a node is its shortest distance from the source thus far with L being a large value. The algorithm terminates in $N - 1$ steps.

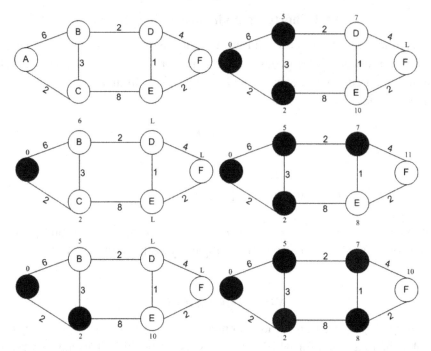

Fig. A4.1. Example execution of Dijkstra's shortest-path algorithm.

A4.3 Shortest paths between all pairs of nodes

Now, suppose we want to compute the shortest path between all pairs of nodes. This may be necessary as communication may occur between any pair of nodes. It is desirable to use the shortest path as this reduces the requirements for network resources. Sometimes this may cause congestion, as has been shown by many researchers. For example, suppose that the network graph is such that it can be partitioned into two parts, A and B, and the two parts are only connected by two links, one with a low-weight link and the other with a high-weight link, as shown in Fig. A4.2. All communication between the two halves will use the low-weight link and the other link remains unused. The second link should not have been included in the design, but if it exists then its use will reduce the congestion on the low-weight link. The shortest-path routing algorithm does not utilize the second link at all.

Coming back to the all-to-all communication problem, we can compute paths from every node to every other node. Thus, we need to solve the problem N times. Alternately, we may use an integrated procedure developed separately, that may be more advantageous. We investigate the latter approach next. Let u_{ij} denote the length of the shortest path from node i to node j and let u_{ij}^m be the shortest path such that the path contains no more than m edges. It is clear that u_{ij}^N will be u_{ij}, the

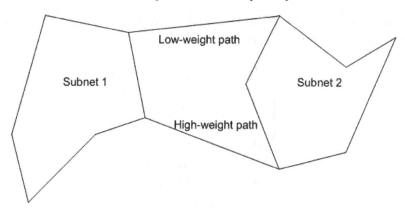

Fig. A4.2. Two parts of a network connected by only two links.

length of the shortest path from node i to node j. Also,

$$u_{ii} = 0$$
$$u_{ij}^0 = \infty \quad i \neq j$$

and

$$u_{ij}^{m+1} = \min_k \left(u_{ik}^m + a_{kj} \right)$$

The last equation computes the shortest path lengths for the paths that contain up to $m + 1$ edges given that we know the shortest path lengths for paths that contain up to m edges. This can be seen to be equivalent to the matrix multiplication $C = A \times B$, where element c_{ij} is computed using

$$c_{ij} = \sum_k a_{ik} b_{kj}$$

We modify the computation of c_{ij} as

$$c_{ij} = \min_k (a_{ik} + b_{kj})$$

by replacing multiplication by addition and summation by the minimum function. We know that $A = (a_{ij})$ is the matrix of arc lengths and let U^0 be the identity matrix, then $U^0 \times A = A$. Let $U^{m+1} = U^m \times A$. Then $U^{N-1} = A^{N-1}$ gives us the desired shortest path length matrix. It turns out that this type of matrix multiplication is also associative. Thus, we can compute $A^{2k} = A^k \times A^k$ and once $2k > n - 1$, we have U^{N-1}. A single matrix multiplication has $O(N^3)$ complexity, and we need to perform $\log N$ matrix multiplications. Therefore, the overall complexity is $O(N^3 \log N)$. This is more complex than Dijkstra's algorithm but in practice may run faster. An example computation for the five-node graph in Fig. A4.1(a) for a given edge weight matrix A is given next:

$$A = \begin{pmatrix} 0 & 100 & 40 & 30 & \infty \\ 100 & 0 & \infty & \infty & 20 \\ 40 & \infty & 0 & 20 & 30 \\ 30 & \infty & 20 & 0 & \infty \\ \infty & 20 & 30 & \infty & 0 \end{pmatrix}$$

$$A^2 = \begin{pmatrix} 0 & 100 & 40 & 30 & 70 \\ 100 & 0 & 50 & 130 & 20 \\ 40 & 50 & 0 & 20 & 30 \\ 30 & 130 & 20 & 0 & 50 \\ 70 & 20 & 30 & 50 & 0 \end{pmatrix}$$

$$A^4 = \begin{pmatrix} 0 & 90 & 40 & 30 & 70 \\ 90 & 0 & 50 & 70 & 20 \\ 40 & 50 & 0 & 20 & 30 \\ 30 & 70 & 20 & 0 & 50 \\ 70 & 20 & 30 & 50 & 0 \end{pmatrix}$$

A4.3.1 Floyd–Warshall method

Another method to compute the shortest paths between all node-pairs is due to Floyd and Warshall and has a computational complexity of $O(N^3)$. In this method, u_{ij}^m defines the length of the shortest path from node i to j such that it does not pass through nodes numbered greater than $m - 1$ except nodes i and j. Then $u_{ij}^1 = a_{ij}$ and $u_{ij}^{m+1} = \min\{u_{ij}^m, u_{im}^m + u_{mj}^m\}$. u_{ij}^{N+1} is the shortest path length matrix. Also, $u_{ii}^m = 0$ for all i and for all m.

This procedure has $N(N - 1)(N - 2)$ equations, each of which can be solved by using $N(N - 1)(N - 2)$ additions and $N(N - 1)(N - 2)$ comparisons. This is the same order of complexity as that for Bellman's method (also known as the Bellman–Ford method as it was independently discovered by two researchers) which yields the shortest path only from a single origin. Dijkstra's method can also be applied N times, once from each source node, to compute the same shortest path length matrix. This takes only $N(N - 1)/2$ additions for each pass, for a total of $N^2(N - 1)/2$ additions, but again housekeeping functions in Dijkstra's method make it non-competitive.

Computation in the Floyd–Warshall method proceeds with $u^1 = A$ and U^{m+1} is obtained from U^m by using row m and column m in U^m to revise the remaining elements. That is, u_{ij} is compared with $u_{im} + u_{mj}$ and is replaced if the latter is smaller. Thus, the computation can be performed in place and is demonstrated in the following for the graph in Fig. A4.1(a).

$$A^0 = \begin{pmatrix} 0 & 100 & 40 & 30 & \infty \\ 100 & 0 & \infty & \infty & 20 \\ 40 & \infty & 0 & 20 & 30 \\ 30 & \infty & 20 & 0 & \infty \\ \infty & 20 & 30 & \infty & 0 \end{pmatrix} \qquad A^1 = \begin{pmatrix} 0 & 100 & 40 & 30 & \infty \\ 100 & 0 & 140 & 130 & 20 \\ 40 & 140 & 0 & 20 & 30 \\ 30 & 130 & 20 & 0 & \infty \\ \infty & 20 & 30 & \infty & 0 \end{pmatrix}$$

$$A^2 = \begin{pmatrix} 0 & 100 & 40 & 30 & 120 \\ 100 & 0 & 140 & 130 & 20 \\ 40 & 140 & 0 & 20 & 30 \\ 30 & 130 & 20 & 0 & \infty \\ 120 & 20 & 30 & \infty & 0 \end{pmatrix} \qquad A^3 = \begin{pmatrix} 0 & 100 & 40 & 30 & 70 \\ 100 & 0 & 140 & 130 & 20 \\ 40 & 140 & 0 & 20 & 30 \\ 30 & 130 & 20 & 0 & 50 \\ 70 & 20 & 30 & 50 & 0 \end{pmatrix}$$

$$A^4 = \begin{pmatrix} 0 & 100 & 40 & 30 & 70 \\ 100 & 0 & 140 & 130 & 20 \\ 40 & 140 & 0 & 20 & 30 \\ 30 & 130 & 20 & 0 & 50 \\ 70 & 20 & 30 & 50 & 0 \end{pmatrix} \qquad A^5 = \begin{pmatrix} 0 & 90 & 40 & 30 & 70 \\ 90 & 0 & 50 & 70 & 20 \\ 40 & 50 & 0 & 20 & 30 \\ 30 & 70 & 20 & 0 & 50 \\ 70 & 20 & 30 & 50 & 0 \end{pmatrix}$$

A4.4 Multiple shortest paths

Many times it is useful to be able to compute additional shortest paths between a node-pair which may be longer than the first shortest path but are still short in case the first shortest path is not available. The first path may be congested or may have a failed link or a node. The problem can be constrained by specific requirements such as allowing or not allowing repeated nodes and links, or specific nodes and/or links. Specific methods exist to compute alternate shortest paths for all cases (see, for example, [255]). One specific case with respect to fault tolerance is non-availability of a node or a link. Such a path can be computed by removing the specific node or link in the original graph (removal of a node also removes all the associated links) and then using the same shortest-path algorithm. In another scenario, we may want another path that is mutually exclusive of the first path. In that case, all the nodes and links have to be removed from the original graph before computing another shortest path. The algorithm to be used in these cases is the same as that already stated.

A4.5 Minimum spanning tree (MST)

The minimum spanning tree is the "best" tree one can identify in a given graph with edge weights. Recall that edge weights represent some "cost" of communicating on that edge. The cost may be delay, or expense in terms of real dollars to use the link.

The MST problem is to find a set of edges with a total minimum cost so that the nodes in the graph remain connected. A greedy algorithm can be used to find this set of edges, called MSTE. The algorithm starts with one edge with minimum weight. Then it finds an edge "e," the best candidate that has not yet been considered and adds it if it is feasible. An edge can only be added to this set if it does not create a cycle in the graph with the same set of nodes as the original graph and set of edges MSTE. MSTE is complete when it contains $N - 1$ edges in an N-node graph. It is known that a greedy algorithm indeed finds an MSTE.

There are several algorithms to find an MST. We will consider two algorithms here based on the greedy approach, but their complexities may differ slightly.

A4.5.1 Kruskal algorithm

This algorithm essentially requires one to sort all edges, shortest first. Then the edges are included in the set MSTE, one at a time, in an order such that the edges do not form a cycle. The test for forming a cycle can be efficiently made by maintaining a proper data structure of edges included thus far. The complexity of sorting is $O(M \log M)$ and the test is of complexity $O(M + N)$ as suggested by Tarjan [249]. Since the process terminates once the set MSTE includes $N - 1$ edges, one may not have to sort all edges (the first few may be sufficient). This can be achieved by putting all the edge weights in a heap that can be created in $O(M)$ time. An edge with the smallest weight can be removed from the heap in $O(\log M)$ time. If k edges have to be considered to select $N - 1$ edges for inclusion in MSTE, then the complexity of the selection process is $O(M + k \log M)$. Therefore, the total complexity is $O(M + N + k \log M)$. An example of the execution of Kruskal's algorithm is shown in Fig. A4.3. Each edge is labeled with its weight and its number (shown in parentheses). In each pass, the selected edge and the included nodes are shown in the table.

A4.5.2 Prim's algorithm

For a dense network, when M is of $O(N^2)$, an alternate method to find an MST is from Prim [241]. This algorithm maintains a tree and adds additional nodes to the tree using minimum cost edges. For this purpose, the minimum distance of each node that is out of the tree is maintained from the tree nodes. Each time a new node is added, the distances of nodes that are not yet in the tree from the tree changes. Therefore, these distances need to be revised. In fact, distances of nodes outside the partial tree from the newly inserted node only need to be

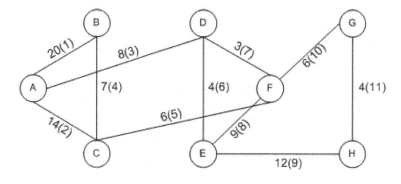

Edge Length	03	04	04	06	06	07	08	09	12	14	20
Edge Number	07	06	11	05	10	04	03	08	09	02	01

Pass	Edges	Nodes
1	7	B,C
2	7,6	B,C,F
3	7,6,11	B,C,F,G,H
4	7,6,11,5	C,D,E,F,G,H
5	7,6,11,5,10	C,D,E,F,G,H
6	Rejected Edge 4	C,D,E,F,G,H
7	...	...

Fig. A4.3. Kruskal's MST algorithm.

considered as that is the only change in the tree. The algorithm has a complexity of $O(N^2)$. We need N passes, one each to select N nodes to be included in the tree. Each time we need to find a node with minimum distance (this is an $O(N)$ procedure) and update distances of all other nodes after considering the new node (another $O(N)$ procedure if the distances are maintained in the adjacency list). Both $O(N)$ procedures can be performed in $O(d)$ if the maximum degree of each node is only d since we only need to consider d neighbors of the new node introduced in the tree. Thus, the overall complexity of the procedure is $O(dN)$.

A4.5.3 Constrained MST

MST computation may be constrained using some optimality criteria or requirements. In the case of constrained MST computation, selection of edges is constrained using appropriate selection criteria consistent with the specified

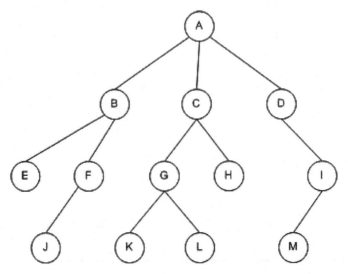

Fig. A4.4. A 13-node tree example.

constraints. For example, in the previous algorithm, it is assumed that the weight of an edge is the only criterion. But the new constraint may be that no node can have more than a certain number of edges connected to it. In that case, the algorithm may have to decide on a selectable candidate differently. If a node already has a given number of edges originating from it, then no more edges connected to that node may become part of the solution.

A4.6 Tree traversal

For a given tree graph, one may like to visit all the nodes of the tree. Recall that a tree graph has no loops and the number of edges is exactly equal to $N - 1$. A node is visited after another node along a link. We will assume that there is no more than one link between any pair of nodes. Nodes can be visited in two different ways. In the first case, once we are at a node, we visit all of its neighbors before visiting any other (non-neighbor) nodes. This is called *breadth-first order*. For example, for a given tree graph as shown in Fig. A4.4, we first visit the root node A. Following that, we visit all its children, which are B, C, and D. Then we traverse children of B, C, and D, that are E and F, G and H, and I, respectively. Finally, we visit the children of these nodes and include nodes J, K, L, and M in the list of visited nodes.

In the second case, we visit nodes in *depth-first order*. In this case, when we visit a node, we immediately visit its children first before visiting any of its siblings. In the example tree of Fig. A4.4, the nodes will be visited in the order A, B, E, F, J, C, G, K, L, H, D, I, and M.

Depending on the application, one or the other method is used. For example, if the tree nodes represent solutions of a problem and we are interested in one solution, a depth-first search is likely to yield the solution faster. On the other hand, if we are interested in all possible solutions, then a breadth-first search is more appropriate.

A4.7 Network (Max) flow

In a given network one may like to compute the available capacity on all paths from a source to a destination. In that case we need to determine the maximum information flow possible from the source to the destination. This is accomplished by using a network flow analysis algorithm [152]. The network graph is treated as a directed graph and the maximum possible flow from a source node s to a destination node t is computed. For a given directed graph, each edge (i, j) is assigned a capacity using a non-negative value C_{ij} that represents the available capacity to carry information on edge (i, j) from node i to node j. In addition, nodes themselves may have additional constraints in terms of the amount of information they can support in terms of buffer space and other factors from all incoming edges or links. This is the node capacity constraint. Let X_{ij} be the amount of actual flow through edge (i, j). At each node, information must be conserved as part of the total flow from s to t. This means that the amount of information entering a node must be the same as the amount of information leaving that node. This information must not exceed the capacity of the node, or the following constraints must be satisfied:

$$0 \le X_{ij} \le C_{ij} \quad \text{and} \quad \sum_j X_{ij} = \sum_j X_{ji}$$

Also $\sum_j X_{sj}$ is the amount of information that leaves the source node s and it is equal to $\sum_j X_{jt}$, i.e. the amount of information that arrives at the destination node t.

Any such set of flows, $\{X_{ij}\}$, that satisfies the above constraints is called a feasible flow set. Maximizing feasible flow by increasing flow on different links while satisfying all constraints yields the maximum-flow value. For a given graph this is achieved as follows.

First, we find a feasible flow from node s to node t (0 flow is trivial). Now, let P be an undirected path in the directed network from s to t. An edge on this path is called a *forward edge* if it is directed towards node t. Otherwise, it is a *backward edge*. A flow on this path can be augmented or increased if $X_{ij} < C_{ij}$ on all forward

Graph algorithms

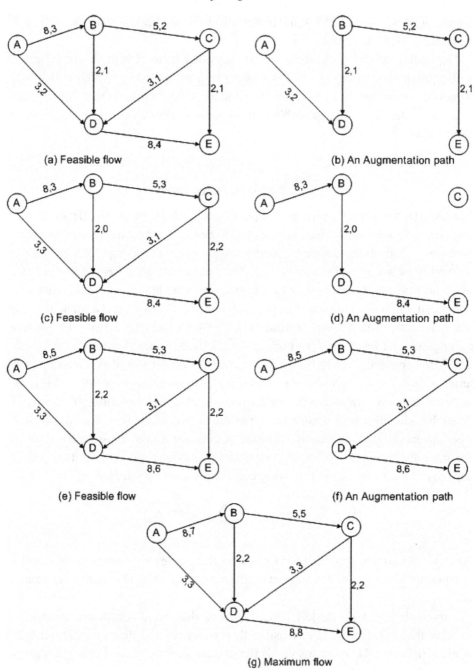

Fig. A4.5. An example demonstrating feasible flow and augmentation paths.

edges and $X_{ij} > 0$ on all backward edges. The amount of increase is given by

$$\min \left\{ \min_{\text{forward}} \{C_{ij} - X_{ij}\}, \min_{\text{backward}} \{X_{ij}\} \right\}.$$

If this value is greater than zero, then such a path is called an *augmentation path*. The process is repeated on all possible undirected paths. A flow is maximum if no augmentation path is available.

Figure A4.5 demonstrates the computation of maximum flow. Figure A4.5(a) depicts a feasible flow. Each edge is marked with its capacity and the current flow value. Figure A4.5(b) shows an augmentation path with three forward edges and one backward edge. Using the relationship described above, the amount of flow that can be increased is one. Figure A4.5(c) shows the graph again with a new feasible flow. Figure A4.5(d) shows another augmentation path with three forward edges. The flow can be increased by two on this path and the new feasible flow is shown in Fig. A4.5(e). Figure A4.5(f) shows another augmentation path with four forward edges and the flow is again increased by two to obtain a maximum flow of ten as shown in Fig. A4.5(g).

The maximum value of an *s–t* flow is equal to the minimum capacity of an *s–t cut*. A cut is defined by a set of edges that partitions the network into two parts with *s* and *t* in separate partitions. A *minimum cut set* is a cut set for which the total capacity of the edges is minimum.

Appendix 5

Routing algorithm

A routing algorithm establishes an appropriate path from any given source to a destination. The objective of network routing is to maximize network throughput with minimal cost in terms of path length. To maximize throughput, a routing algorithm has to provide as many communication paths as possible. To minimize the cost of paths, the shortest paths have to be provided. However, there is always a trade-off between these two objectives. Most routing algorithms are based on assigning a cost measure to each link in a network. The cost could be a fixed quantity related to parameters such as the link length, the bandwidth of a link, or the estimated propagation delay. Each link has a cost associated with it and in most cases it is assumed that the links have equal cost.

An interconnection network is *strictly non-blocking* if there exists a routing algorithm to add a new connection without disturbing existing connections [292]. A network is *rearrangeable* if its permitted states realize every permutation or allowable set of requests; here it is possible to rearrange existing connections if necessary [292]. Otherwise it is *blocking*.

The *store-and-forward* operation in packet switching incurs a time delay and causes significant performance degradation. If the algorithm is used in a packet-switching network, the total time delay of a data packet is obtained by summing up the time delay at each intermediate node. Since non-availability of any link along a route causes the route not to be available, the network sees a high probability of blocking under heavy traffic, which rejects the incoming request and eventually causes data loss or delay.

The routing algorithm can be centralized or decentralized. A centralized algorithm may use a global backtracking depth-first search or any other algorithm described in Appendix 5.

A5.1 Embedding arbitrary connection requests

The interconnection network should be able to embed arbitrary requests until resources are available in the network. If a set of requests is such that each node needs to communicate with a unique node, then such a set of requests is called a permutation. It is desirable to be able to satisfy this set of requests simultaneously. If each node requires communication with up to k other nodes, it may not be possible to satisfy these requests in one round and the communication requests may have to wait. Depending on the application environment, either the requests are partitioned in k disjoint permutations (some may be partial permutations) or the communication needs are satisfied in k rounds without any contention. Alternately, a network is designed to satisfy all the requests up to k requests at the same time. A better solution would probably lie in between these two extreme cases. Depending on the number of transmitters and receivers, a node should be able to source and sink those many connections. The links in the network should be able to support the traffic corresponding to the requests being serviced simultaneously. The permutation routing capability of a network is extremely useful in improving the overall performance of a system.

In a permutation routing, the messages transferred from a source to a destination can be regarded as a commodity flow. For each commodity, the required flow of commodity is 1 for a single source and a single destination. In a general network, the problem of solving multi-commodity integral flows is known to be NP-complete.

Appendix 6

Network topology design

A network can be designed using various topologies. Many interconnection networks have been proposed by the research community; some have been prototyped but few have progressed to become commercial products. A network may be static or dynamic [29, 174, 175, 191, 284]. The topologies can be divided into two categories: (i) regular and (ii) irregular. The regular topologies follow a well-defined function to interconnect nodes. The regularity, symmetry, and most often the strong connectivity of the regular network topologies make them suitable for general purpose interconnection structures where the characteristics of the traffic originating from all nodes are identical and destinations are uniformly distributed over the set of nodes. Thus, the link traffic is also uniformly distributed. The irregular topologies are optimized based on the traffic demands. If there is a high traffic flow between two nodes, then they may be connected using a direct link. If a direct link is not feasible, then an alternative is to provide a short path between the two nodes. Such designs are much more involved and need special attention.

We will first discuss regular topologies and then get into the design of irregular topologies. We will also discuss some specific regular topologies, such as a binary cube and its variations, in greater detail.

A6.1 Regular topologies

There are several regular topologies that have been proposed by various researchers in the literature. The most important among these are complete connected graphs, star, tree, ring, multi-ring, mesh, and hypercube topologies. One of the desirable properties of a structure is to be able to accommodate or embed an arbitrary permutation. We discuss various regular topologies in the following paragraphs.

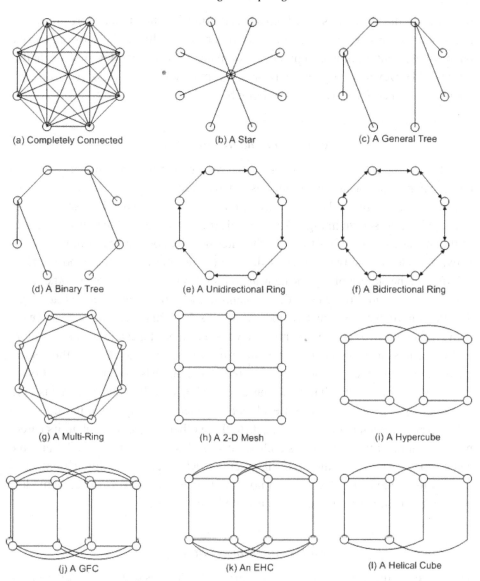

Fig. A6.1. A hierarchical network.

A6.1.1 Completely connected topologies

In a completely connected topology, every node is connected to every other node as shown in Fig. A6.1(a), that is, for every $\forall i, j \in N$, $(i, j) \in E$. Thus, there are $N(N - 1)$ links. The routing is straightforward as a node i sends messages for node j directly on the corresponding link (i, j). Each node has $N - 1$ transmitters and $N - 1$ receivers, one for each link. The diameter of the graph is one and the reliability of the network is very high as in addition to a direct link, there are $N - 2$ paths of two hops from a node to every other node. This is the most expensive but most efficient topology. In practice, not many networks are designed using this topology. However, in a given network, one may set *virtual topologies* that are equivalent to a completely connected graph.

A6.1.2 Star and tree topologies

The star and the tree are two topologies that require a minimum number of links to connect N nodes. The number of links is exactly equal to $N - 1$. A star topology has a central node to which all other nodes are connected, as shown in Fig. A6.1(b). The tree topology, as shown in Figs. A6.1(c) and A6.1(d), is hierarchical where the root of the tree at each level has to act as the intermediate node in any communication between nodes in the two halves of the tree (called the left subtree and the right subtree). In the star topology, the central node communicates with every other node using the direct link. If we consider the central node only as an intermediate node, then the routing between any two nodes is always through the central node and each path is of length two. The central node may become a bottleneck in communication. Failure of this node also causes the entire network to fail. Moreover, the central node is the most expensive node with degree $N - 1$ and has to support $N - 1$ other connections. On the other hand, the degree of each node is bounded and that is a big advantage. For example, in a star each node connects to only one other node and in a binary tree each node only connects to three other nodes, one link to its parent node and at most two links to its children nodes. The longest path in a binary tree can be up to $2 \log N$. In a hierarchical structure such as a tree, a different number of parallel links can be used to connect nodes at two adjacent levels to accommodate more traffic near the root node. This is called a fat tree [45].

A6.1.3 Rings and multi-rings

The ring and multi-ring topologies are even simpler designs with fixed node degrees. For a simple ring, each node is connected to two other nodes. If the connections are unidirectional then the simplest ring has one incoming link and one outgoing link. The diameter of the graph is $N - 1$. In a bidirectional ring, each node has two incoming links and two outgoing links. A node i has a link to node $i + 1$ and node $i - 1$ (module N). The diameter of the graph is $N/2$. The multi-ring architecture has multiple links from each node to other nodes. Each set of corresponding links from each node forms one ring. Some examples of ring topologies are shown in Figs. A6.1(e)–A6.1g.

A6.1.4 Meshes

A node in an n-dimensional mesh structure has $2n$ neighbors, two in each dimension. A two-dimensional structure is shown in Fig. A6.1(h). Each grid point is numbered using an n-*dimensional* tuple. Two- and three-dimensional meshes are most commonly used in designing interconnection structures for multiprocessor systems. A mesh can be extended or shrunk in any dimension allowing the easy reconfiguration and scalability required in many subsystem designs.

A6.1.5 The hypercube and its variations

A hypercube is an n-dimensional structure, as shown in Fig. A6.1(i). The hypercube and its variants are popular interconnection structures due to their unique properties such as symmetry, regularity, low diameter, and good fault tolerance characteristics [104]. A Boolean n-cube $Q_n = (V, E)$ has $|V| = N = 2^n$ nodes. Each node is numbered using an n-bit binary string. The *Hamming distance* between two binary strings is the number of bit positions in which they differ. A pair of nodes in a Boolean cube are connected by an edge providing a bidirectional communication path between them if the Hamming distance between their binary addresses is one. An important property of an n-cube is that it can be constructed or decomposed recursively from/to two lower-dimensional subcubes as is clear from its recursive definition as given next.

Definition: a Boolean n-cube $Q_n = (V, E)$ is defined recursively as follows.

(i) The 0-cube Q_0 is defined as a single node with no edge.
(ii) $Q_n = Q_{n-1} \uplus Q_{n-1}^d$, where the $\uplus$ operation is a *twofold operation* of the graph $G = (V, E)$ denoted by $G_t = G \uplus G^d$ that yields a graph $G_t = (V_t, E_t)$, where $V_t = V \cup V^d$ and $E_t = E \cup E^d \cup \{ (v, |V| + v) \mid \forall v \in V \}$.

The degree of each node, the diameter of the graph, and the node connectivity of the hypercube graph is n each. The length of the shortest path between any two nodes i and j in an n-cube is equal to the Hamming distance between their binary representations. There are $H(i, j)!$ shortest paths between two nodes i and j, and among them $H(i, j)$ paths are independent (node-disjoint or parallel). In a Boolean n-cube, there are no cycles of odd length. The other regular topologies discussed above can be embedded in a binary n-cube or its variations discussed below.

A6.1.6 Dynamic topologies

A dynamic topology is created by modifying an existing topology as the need arises. This is achieved by adding links between nodes to either create more paths or

point-to-point direct links to reduce delays and congestion and to improve performance. The resulting networks usually look like random graphs with possibly no
symmetry and very little fault tolerance.

A6.2 Reconfigurable topologies

There are two important issues in the design of a reconfigurable network: the
ease of embedding a given permutation and the cost of implementing the network.
An $N \times N$ crossbar can realize all permutations easily but has a cost which is
proportional to $O(N^2)$. To reduce the cost, a rearrangeable network [159] may
be acceptable. The generalized folding cube (GFC) and the enhanced hypercube
(EHC) are two such topologies derived from the binary cube architecture.

A6.2.1 Generalized folding cube

A generalized folding cube is obtained by folding a hypercube along any dimension
as follows. For a given Boolean n-cube $Q_n = (V, E)$, the *folding operation* of the
cube Q_n denoted by $f(Q_n)$ yields a graph $Q_{n-1}^1 = (V^1, E^1)$ and consists of the
following two steps.

(i) Split the cube into two subcubes by removing $(n - 1)$-dimensional links from Q_n
($Q'_{n-1} = (V', E')$ and $Q''_{n-1} = (V'', E'')$).

(ii) Overlap the two subcubes Q'_{n-1} and Q''_{n-1} in such a way that $v' \in V'$ and $v'' \in V''$
become one and the same node $v^1 \in V^1$ if v' and v'' differ by 2^{n-1}. v^1 is numbered as
$\min(v', v'')$. Either of the two links in each dimension, corresponding to v' and v'', can
be used by either nodes, v' and v'', for communication.

The kth folding operation $f^k(Q_n) = f(f^{k-1}(Q_n))$ yields a graph $Q_{n-k}^k =
(V^k, E^k)$. The kth unfolding operation $f^{-k}(Q_n^p) = f^{-1}(f^{-(k-1)}(Q_n^p)) = Q_{n+k}^{p-k} =
(V^{p-k}, E^{p-k})$, where $k \in \{1, 2, \ldots, p\}$. The GFC denoted by $Q_n^p = (V^p, E^p)$ for
$p \geq 0$, is defined from the folding operation of a hypercube: $f^p(Q_{n+p}) = Q_n^p$. The
GFC consists of 2^p pairs of links in each dimension and each node of the GFC
consists of 2^p individual nodes of the original cube and a $(n + 1)2^p \times (n + 1)2^p$
switch. The original hypercube $Q_n = (V, E)$ can be considered as a special case
of the GFC and also is denoted by $Q_n^0 = (V^0, E^0)$. Figure A6.1(j) shows a three-
dimensional GFC with 2^p pairs of links in each dimension. The rearrangeability of
the GFC is shown in [253].

A6.2.2 Enhanced hypercube

If we wish to keep only one node at each vertex position and still want to design
a rearrangeable network, then by duplicating links in any one dimension of the

original hypercube, i.e. two pairs of links are provided instead on one, we obtain a structure that can provide conflict-free routes for every permutation [252]. The EHC is shown in Fig. A6.1(k).

A reconfigurable architecture, such as an EHC or a GFC, is able to embed other structures efficiently. The EHC and the GFC concepts can be combined to design a more cost-effective network. This design methodology has been used to design and implement the Proteus multi-computer system [31].

A6.2.3 Helical cube

A binary cube grows only as an integer power of two. To remove this deficiency, a number of alternatives [103] have been suggested. An attractive option is a helical cube that removes $N - K$ nodes from a hypercube to obtain a K-node structure while preserving all the advantageous properties of the binary cube such as regularity, simplicity of routing, and fault tolerance. The degree of each node remains $n = \log K$. Only neighbors of removed nodes are affected and reconnected in such a fashion that the high graph connectivity is maintained. The links connected to nodes that are being removed are connected pairwise using a helical connection strategy, hence the name helical cube. An example of a helical cube is shown in Fig. A6.1(l). The details of the actual connection scheme are given in [28]. It has been shown that this structure can have any number of nodes while maintaining a high connectivity and the same level of fault tolerance as the original cube.

A6.3 Arbitrary topology design

If the graph structure is not constrained to be a regular topology, then the design problem can be formulated as a linear or non-linear programming problem. Suppose we are to use certain kinds of links and are also given a traffic matrix. Here we assume that only one type of link is available and will consider a more complex problem in the last section. We wish to design a network that is connected. The cost of connecting different links is different. Let X_{ij} denote if a link between nodes i and j exists, $X_{ij} \in \{0, 1\}$ and suppose the cost to lay the link is denoted by C_{ij}. Let the original nodes be numbered from 1 to N.

One of the goals of the design is to minimize the cost, which is given by

$$\text{cost} = \min \sum_{ij} C_{ij} X_{ij}$$

Then the existence of links has to be subjected to conditions that the network should satisfy. For example, each node should be connected by at least one link.

This can be specified as

$$\sum_i X_{ij} + X_{ji} \geq 1$$

Then we may have constraints to specify that there is a path from each node to another node, or the graph should be connected. It is hard to formalize this as an equation, but it can be easily checked for a given $\{X_{ij}\}$ configuration. After all the constraints are specified, one solves the problem to find a solution that is a vector of $\{X_{ij}\}$.

It may appear to be a simple problem but is generally very hard even for a moderate number of nodes. Therefore, it is usually solved using some heuristics.

A6.4 Linear programming problems

In network design, we are mostly concerned with minimizing cost or delay in the network while maximizing the performance. Such problems can be expressed as optimization problems. The statements of such problems have an objective function that is required to be minimized or maximized subject to certain constraints. In most cases, these constraints are also linear in relation. A general linear programming problem (LPP) [84, 81] is to find values of n real variables, denoted by $x_1, x_2, x_3, \ldots, x_n$ which will minimize or maximize an objective function given by

$$z = \sum_{j=1}^{n} c_j x_j$$

where c_j is a cost or reward value associated with variable x_j. The set of constraints that govern a feasible solution may vary in numbers and are also linear combinations of variables x and have a general form

$$\sum_{j=1}^{n} a_{ij} x_j \geq b_i \qquad i = 1, 2, \ldots, m$$

The values of variables may also be bounded by some lower and upper bounds as parts of constraints. For example, it may be desirable that all variables are positive or do not exceed a certain value. There are various methods to solve LPPs. The most commonly used method is the simplex method which has no more than $\binom{n}{m} = \frac{n!}{m!(n-m)!}$ solutions for $m \leq n$ possible solutions. The simplex method systematically searches for an optimal solution over this space.

A variation of this problem is when all variables are restricted to be integers only. This is called an integer programming problem (IPP) and it makes the problem more complicated. There are standard packages available to solve the two types of problems. The goal of a network designer is to formulate the problem as an LPP or IPP and then to solve it using a standard package or a heuristic algorithm. If the problem size (the number of variables and constraints and therefore the number of possible solutions to search from) becomes too large, then we use heuristic methods only to solve the problem.

References

[1] Bell Labs Lucent Technologies, http://www.bell labs.com.

[2] http://www.cplex.com, ILOG CPLEX 7.0 Reference Manual.

[3] Calient Networks, http://www.calient.net.

[4] Ciena http://www.ciena.com.

[5] Xros Nortel Networks, http://www.nortelnetworks.com.

[6] A. Agrawal, L. B. Sofman, and T. S. El-Bawab, Enhancement of bandwidth efficiency by traffic grooming in optical-cross-connect based networks. *Proceedings of the SPIE*, **5247**:1 (2003), 196–202.

[7] A. O. Allen, *Probability, Statistics, and Queueing Theory* (New York: Academic Press, 1978).

[8] A. Antoniades, S. J. B. Yoo, K. Bala *et al.* An architecture for a wavelength-interchanging cross-connect utilizing parametric wavelength converters. *Journal of Lightwave Technology*, **17**:7 (1999), 1113–25.

[9] A. Birman, Computing approximate blocking probabilities for a class of all-optical networks. *IEEE Journal on Selected Areas in Communications*, **14** (1996), 852–7.

[10] A. Birman, Computing approximate blocking probabilities for a class of all-optical networks. In *Proceedings of IEEE INFOCOM '95* (New York: IEEE Press, 1995), pp. 651–8.

[11] A. Birman and A. Kershenbaum, Routing and wavelength assignment methods in single-hop all-optical networks with blocking. In *Proceedings of IEEE INFOCOM '95* (New York: IEEE Press, 1995), pp. 431–8.

[12] A. R. B. Billah, B. Wang, and A. A. S. Wang, Efficient traffic grooming in synchronous optical network/wavelength-division multiplexing bidirectional line-switched ring networks. *Optical Engineering.* **43**:5 (2004), 1101–14.

[13] A. R. B. Billah, B. Wang, and A. A. S. Awwal, Effective traffic grooming algorithms in SONET/WDM ring networks. *Photonic Network Communications*, **6**:2 (2003), 119–38.

[14] A. L. Chiu and E. H. Modiano, Traffic grooming algorithms for reducing electronic multiplexing costs in WDM ring networks. *Journal of Lightwave Technology*, **18**:1 (2000), 2–12.

[15] A. L. Chiu and E. H. Modiano, Reducing electronic multiplexing costs in unidirectional SONET/WDM ring networks via efficient traffic grooming. In *Proceedings of IEEE Globecom '98* (New York: IEEE Press, 1998).

[16] A. Fumagalli, M. Tacca, I. Cerutti *et al.* Effects of design constraints on the total wavelength mileage in optical mesh networks with shared line protection. In *OptiComm: Optical Networking and Communications* (2000), pp. 42–53.

[17] A. Fumagalli, M. Tacca, I. Cerutti *et al.* Survivable networks based on optimal routing and WDM self-healing rings. In *Proceedings of IEEE INFOCOM '99* (New York: IEEE Press, 1999), pp. 726–33.

[18] A. Ge, F. Callegati, and L. Tamil, On optical burst switching and self-similar traffic. *IEEE Communication Letters*, **4**:3 (2000), 98–100.

[19] A. Girard, *Routing and Dimensioning in Circuit-Switched Networks* (Reading, MA: Addison-Wesley Publishing Co., 1990).

[20] A. M. Hill, S. Carter, J. Armitage *et al.* A scalable and switchless optical network structure employing 32×32 free-space grating multiplexer. *IEEE Photonic Letters*, **8**:4 (1996), 569–71.

[21] A. Kershenbaum, *Telecommunications Network Design Algorithms* (New York: McGraw-Hill, 1993).

[22] A. Kodian, W. D. Grover, J. Slevinsky, and D. Moore, Ring-mining to p-cycles as a target architecture: riding demand growth into network efficiency. In *Proceedings of the 19th Annual National Fiber Optics Engineers Conference (NFOEC 2003)*, Orlando, FL (2003).

[23] A. Kodian, A. Sack, and W. D. Grover, p-cycle network design with hop limits and circumference limits, In *Proceedings of First International Conference on Broadband Networks (BROADNETS 2004)*, San José, CA (New York: IEEE Press, 2004), pp. 244–53.

[24] A. Mokhtar and M. Azizoğlu, Adaptive wavelength routing in all-optical networks. *IEEE/ACM Transactions on Networking*, **6**:2 (1998), 197–206.

[25] A. Mokhtar and M. Azizoğlu, Performance of packet-switched WDM broadcast networks with multicast traffic. In *Proceedings of SPIE, All-Optical Communication Systems: Architecture, Control, and Network Issues III*, vol. 3230 (Bellingham, WA: SPIE, 1997), pp. 220–31.

[26] A. K. Somani and M. Mina, On trading wavelengths with fibers: a cost-performance based study. In *Thirty-Eighth Annual Allerton Conference on Communication, Control and Computing* (2000), pp. 1274–83.

[27] A. K. Somani and M. Azizoğlu, All-optical LAN interconnection with a wavelength selective router. In *Proceedings of INFOCOM '97* (New York: IEEE Press, 1997).

[28] A. K. Somani and S. Thatte, The helical cube network. *NETWORKS, an International Journal*, **26** (1995), 87–100.

[29] A. K. Somani, Design of an efficient network. *Proceedings of 7th International Parallel Processing Symposium*, Newport Beach, CA (1993), pp. 413–18.

[30] A. Sridharan and K. N. Sivarajan, Blocking in all-optical networks. *Proceedings of IEEE INFOCOM '00* (New York: IEEE Press, 2000).

[31] A. K. Somani, C. M. Wittenbrink, R. M. Haralick *et al.*, Proteus system architecture and organization. *Proceedings of 5th International Parallel Processing Symposium* (1991), pp. 287–94.

[32] A. S. Tanenbaum, *Computer Networks* (Englewood Cliffs, NJ: Prentice-Hall, 1996).

[33] A. Viswanathan, N. Feldman, B. Boivie, and R. Woundy, *ARIS: Aggregate Route-Based IP Switching*. IETF Internet Draft (1997).

[34] A. N. Washington and H. G. Perros, Call blocking probabilities in a traffic groomed tandem optical network. Special issue dedicated to the memory of Professor Olga Casals, ed. Blondia and Stavrakakis. *Journal of Computer Networks*, **45** (2004).

[35] B. V. Caenegem, W. V. Parys, F. De Turck, and P. M. Demeester, Dimensioning of survivable WDM networks. *IEEE Journal of Selected Areas in Communications*, **16**:7 (1998), 1146–57.

[36] B. T. Doshi, S. Dravida, P. Harshavardhana *et al.* Optical network design and restoration. *Bell Labs Technical Journal*, January–March (1999), 58–83.

[37] B. Gavish and I. Neuman, Routing in a network with unreliable components. *IEEE Transactions on Communications*, **40**:7 (1992), 1248–58.

[38] B. Mukherjee, *Optical Communication Networks* (New York: McGraw-Hill, 1997).

[39] B. Xiang, S. Wang, and L. Li, A traffic grooming algorithm based on shared protection in WDM mesh networks. In *Proceedings of the Fourth International Conference on Parallel and Distributed Computing, Applications and Technologies* (2003), pp. 254–8.

[40] C. Assi, A. Shami, and M. A. Ali, Integrated routing algorithms for provisioning sub-wavelength connections in IP-over-WDM networks. *Photonic Network Communications*, **4**:34 (2002), 377–90.

[41] C. De Matos, M. Pugnet, and A. Le Corre, Ultrafast coherent all-optical switching in quantum-well semiconductor microcavity. *Electronics Letters*, **36**:1 (2000), 93–4.

[42] C. A. Floudas, *Nonlinear and Mixed-Integer Optimization: Fundamentals and Applications* (Oxford: Oxford University Press, 1995).

[43] C. Guillemot, M. Renaud, P. Gambini *et al.*, Transparent optical packet switching: the European ACTS KEOPS project approach. *Journal of Lightwave Technology*, **16**:12 (1998), 2117–34.

[44] C. Kim, D. A. May-Arrioja, P. Newman, and J. Pamulapati, Ultrafast all-optical multiple quantum well integrated optic switch. *Electronics Letters*, **36**:23 (2000), 1929–30.

[45] C. Leiserson, Fat-trees: universal network for hardware-efficient supercomputing. *IEEE Transactions on Computers*, **34** (1985), 892–901.

[46] C. Lee and E. K. Park, A genetic algorithm for traffic grooming in all-optical mesh networks. In *2002 IEEE International Conference on Systems, Man and Cybernetics* (New York: IEEE Press, 2002).

[47] C. Liu and L. Ruan, Finding good candidate cycles for efficient p-cycle network design. In *Proceedings of 13th International Conference on Computer Communications and Networks* (2004), pp. 321–6.

[48] C. Ou, K. Zhu, B. Mukherjee *et al.* Survivable traffic grooming in WDM mesh networks. In *Optical Fiber Communications Conference* (Washington, DC: Optical Society of America, 2003), pp. 624–5.

[49] C. Qiao and M. Yoo, Optical burst switching (OBS) – a new paradigm for an optical internet. *Journal of High Speed Networks*, **8**:1 (1999), 69–84.

[50] C. Xin and C. Qiao, Performance analysis of multi-hop traffic grooming in mesh WDM optical networks. In *The 12th International Conference on Computer Communications and Networks* (2003).

[51] D. Banerjee and B. Mukherjee, Wavelength-routed optical networks: linear formulation, resource budgeting tradeoffs, and a reconfiguration study. *IEEE/ACM Transactions on Networking*, **8**:5 (2000), 598–607.

[52] D. Banerjee and B. Mukherjee, A practical approach for routing and wavelength assignment in large wavelength-routed optical networks. *IEEE Journal of Selected Areas in Communications*, **14**:5 (1996), 903–8.

[53] D. Bertsekas and R. Gallager, *Data Networks* (Englewood Cliffs, NJ: Prentice-Hall, 1987).

[54] D. J. Blumenthal, Photonic packet switching and optical label swapping. *Optical Networks Magazine*, **2**:6 (2001), 54–65.

[55] D. Greenfield, The well-groomed switch. *Network Magazine*, **17**:4 (2002), 50–53.

[56] D. K. Hunter and D. G. Smith, New architectures for optical TDM switching. *IEEE/OSA Journal of Lightwave Technology*, **11**:3 (1993), 495–511.

[57] D. B. Johnson, Finding all the elementary circuits of a directed graph. *SIAM Journal on Computing*, **4** (1975), 77–84.

[58] D. Li, Z. Sun, X. Jia, and K. Makki, Traffic grooming on general topology WDM networks. *IEEE Proceedings Communications*, **150**:3 (2003), 197–201.

[59] D. Li, Z. Sun, X. Jia, and S. Makki, Traffic grooming for minimizing wavelength usage in WDM networks. In *Proceedings Eleventh International Conference on Computer Communications and Networks* (2002), pp. 460–5.

[60] D. Papadimitriou, F. Poppe, S. Dharanikota *et al.*, *Inference of Shared Risk Link Groups*. Internet Draft, draft-many-inference-srlg00.txt (2001).

[61] D. A. A. Mello, J. U. Pelegrini, M. S. Savasini *et al.* Inter-arrival planning for sub-graph routing protection in WDM networks. In *11th International Conference on Telecommunications, ICT* (2004).

[62] D. A. Schupke, A. Autenrieth, and T. Fischer, Survivability of multiple fiber duct failures. In *Proceedings of Design of Reliable Communication Networks (DRCN) Workshop* (2001).

[63] D. A. Schupke, W. D. Grover, and M. Clouqueur, Strategies for enhanced dual failure restorability with static or reconfigurable p-cycle networks, *Proceedings of IEEE International Conference on Communications (ICC 2004)* (New York: IEEE Press, 2004), pp. 1628–33.

[64] D. A. Schupke, M. C. Scheffel, and W. D. Grover, Configuration of p-cycles in WDM networks with partial wavelength conversion. *Photonic Network Communications*, **6**:3 (2003), 239–52.

[65] D. A. Schupke, M. C. Scheffel, and W. D. Grover, An efficient strategy for wavelength conversion in WDM p-cycle networks. *Proceedings of Fourth International Workshop on the Design of Reliable Communication Networks (DRCN 2003)*, Banff, Alberta (2003), pp. 221–7.

[66] D. A. Schupke, C. G. Gruber, and A. Autenrieth, Optimal configuration of p-cycles in WDM network. In *Proceedings of the IEEE International Conference on Communications (ICC 2002)* (New York: IEEE Press, 2002), pp. 2761–5.

[67] D. Stamatelakis and W. D. Grover, Theoretical underpinnings for the efficiency of restorable networks using preconfigured cycles (p-cycles). *IEEE Transactions on Communications*, **48**:8 (2000), 1262–5.

[68] D. Xu, Y. Xiong, and C. Qiao, Protection with multi-segments (PROMISE) in networks with shared risk link groups (SRLGs). *IEEE/ACM Transactions on Networking*, **11**:2 (2003), 248–58.

[69] E. Bampis and G. N. Rouskas, The scheduling and wavelength assignment problem in optical WDM networks. *IEEE/OSA Journal of Lightwave Technology*, **20**:5 (2002), 782–9.

[70] E. Bouillet, J. Labourdette, G. Ellinas *et al.*, Stochastic approaches to compute shared mesh restored lightpaths in optical network architectures. In *Proceedings of IEEE INFOCOM '02* (New York: IEEE Press, 2002), pp. 801–7.

[71] E. W. Dijkstra, A note on two problems in connexion with graphs. *Numerische Mathematik*, **1** (1959), 269–71.

[72] E. Karasan and E. Ayanoglu, Effects of wavelength routing and selection algorithms on wavelength conversion gain in WDM optical networks. *IEEE/ACM Transactions on Networking*, **6**:2 (1998), 186–96.

[73] E. Karasan and E. Ayanoglu, Performance of WDM transport networks. *IEEE Journal of Selected Areas in Communications*, **16** (2003), 1081–96.

[74] E. Modiano and A. Narula-Tam, Survivable lightpath routing: a new approach to the design of WDM-based networks. *IEEE Journal of Selected Areas in Communication*, **20**:4 (2002), 800–9.

[75] E. Modiano and P. J. Lin, Traffic grooming in WDM networks. *IEEE Communications Magazine*, **39**:7 (2001), 124–9.

[76] E. Vararigos and V. Sharma, The ready-to-go virtual-circuit protocol: a loss-free protocol for multigigabit networks using FIFO buffers. *IEEE/ACM Transactions on Networking*, **5** (1999), 705–18.

[77] G. P. Austin, B. T. Doshi, C. J. Hunt *et al.*, Fast, scalable, and distributed restoration in general mesh optical networks. *Bell Labs Technical Journal*, **6**:1 (2001), 67–81.

[78] F. J. Blouin, A. Sack, W. D. Grover, and H. Nasrallah, Benefits of p-cycles in a mixed protection and restoration approach. In *Proceedings of the Fourth International Workshop on the Design of Reliable Communication Networks (DRCN 2003)*, Banff, Alberta (2003).

[79] G. B. Dantzig, *Linear Programming and Extensions* (Princeton, NJ: Princeton University Press, 1963).

[80] G. Ellinas, E. Bouillet, R. Ramamurthy *et al.*, Routing and restoration architectures in mesh optical networks. *Optical Networks Magazine*, **4**:1 (2003), 91–106.

[81] G. Ellinas, G. Halemariam, and T. Stern, Protection cycles in mesh WDM networks. *IEEE Journal of Selected Areas in Communication*, **18**:10 (2000), 1924–37.

[82] G. Hadley, *Linear Programming* (Reading, MA: Addison-Wesley, 1962).

[83] F. Harary, *Graph Theory* (Reading, MA: Addison-Wesley, 1969).

[84] G. Huiban, S. Perennes, and M. Syska, Traffic grooming in WDM networks with multi-layer switches. In *2002 IEEE International Conference on Communications* (New York: IEEE Press, 2002), pp. 2896–901.

[85] G. Jeong and E. Ayanoglu, Comparison of wavelength-interchanging and wavelength-selective cross-connects in multiwavelength all-optical networks. In *Proceedings of INFOCOM '96* (New York: IEEE Press, 1996), pp. 156–63.

[86] F. P. Kelly, Blocking probabilities in large circuit switched networks. *Advances in Applied Probability*, **18** (1986), 473–505.

[87] G .Z. Li, B. Doverspike, and C. Kalmanek, Fiber span failure protection in mesh optical networks. In *SPIE Optical Networking and Communications Conference (Opticomm)*, vol. 4599 (Bellingham, WA: SPIE, 2001), pp. 130–42.

[88] G. Li, D. Wei, C. Kalmanek, and R. Doverspike, Efficient distributed restoration path selection for shared mesh restoration. *IEEE/ACM Transactions on Networking*, **11**:5 (2003), 761–71.

[89] G. Mohan and A. K. Somani, Routing dependable connections with specified failure restoration guarantees in WDM networks. In *Proceedings of IEEE INFOCOM '00* (New York: IEEE Press, 2000), pp. 1761–70.

[90] G. Mohan, C. S. R. Murthy, and A. K. Somani, Efficient algorithms for routing dependable connections in WDM optical networks. *IEEE/ACM Transactions on Networking*, **9**:5 (2001), 553–66.

[91] G. Shen and W. D. Grover, Exploiting forcer structure to serve uncertain demands and minimize redundancy of p-cycle networks. In *Proceedings of Optical*

Networking and Communications Conference (OptiComm 2002), Dallas, TX (2003).

[92] G. Shen and W. D. Grover, Performance of protected working capacity envelopes based on p-cycles: fast, simple, and scalable dynamic service provisioning of survivable services. In *Proceedings of Asia–Pacific Optical and Wireless Communications Conference (APOC)*, vol. 5626 (2004).

[93] G. Shen and W. D. Grover, Extending the p-cycle concept to path segment protection for span and node failure recovery. *IEEE Journal on Selected Areas in Communications*, **21**:8 (2003), 1306–19.

[94] G. M. Woodruff and R. Kositpaiboon, Multimedia traffic principles for guaranteed ATM network performance. *IEEE Journal on Selected Areas in Communications*, **8** (1990), 437–46.

[95] G. Zhu, H. Ghafouri-Shiraz, and Y. Fei, Effective wavelength assignment algorithms for optimizing design costs in SONET/WDM rings. *Journal of Lightwave Technology*, **19**:10 (2001), 1427–39.

[96] H. J. S. Dorren, M. T. Hill, Y. Liu *et al.*, Optical packet switching and buffering by using all-optical signal processing methods. *Journal of Lightwave Technology*, **21**:1 (2003), 2–12.

[97] H. Frank, I. T. Frisch, and W. Chou, Topological considerations in the design of the ARPA computer network. In *Conference Record, 1970 Spring Joint Computer Conference, AFIPA Proceedings*, vol. 36 (Montvale, NJ: AFIPS Press, 1970).

[98] H. Harai, M. Murata, and H. Miyahara, Performance of alternate routing methods in all-optical switching networks. In *Proceedings of INFOCOM '97* (New York: IEEE Press, 1997).

[99] H. Huang and J. A. Copeland, A series of Hamiltonian cycle-based solutions to provide simple and scalable mesh optical network resilience. *IEEE Communications Magazine*, **40**:11 (2002), 46–51.

[100] H. F. Jordan, D. Lee, K. Y. Lee, and S. V. Ramanan, Serial array time slot interchangers and optical implementations. *IEEE Transactions on Computers*, **43**:11 (1994), 1309–18.

[101] H. P. Katseff, Incomplete hypercube. *IEEE Transactions on Computers*, **37** (1988), 604–7.

[102] H. Sullivan and T. R. Bashkow, A large scale, homogeneous, fully distributed parallel machine, I. In *Proceedings of 4th Annual Symposium on Computer Architecture* (1977), pp. 105–17.

[103] H. Zang and B. Mukherjee, Connection management for survivable wavelength-routed WDM mesh networks. *Optical Networks Magazine*, **2**:4 (2001), 17–28.

[104] H. Zang, L. Shasrabuddhe, J. P. Jue *et al.*, Connection management for wavelength-routed WDM networks. In *Global Telecommunications Conference, GLOBECOM'99*, vol. 2 (1999), pp. 1428–32.

[105] H. Zhang and B. Mukherjee, Path-protection routing and wavelength assignment (RWA) in WDM mesh networks under duct-layer constraints. *IEEE/ACM Transactions on Networking*, **11**:2 (2003), 248–58.

[106] H. Zhang, J. P. Jue, and B. Mukherjee, A review of routing and wavelength assignment approaches for wavelength-routed optical WDM networks. *Optical Networks Magazine*, **1**:1 (2000), 47–60.

[107] H. Zhang and O. Yang, Finding protection cycles in DWDM networks. In *Proceedings of IEEE International Conference on Communications*, vol. 5, New York (New York: IEEE Press, 2002).

[108] H. Zhu, H. Zang, K. Zhu, and B. Mukherjee, A novel generic graph model for traffic grooming in heterogeneous WDM mesh networks. *IEEE Transactions on Networking*, **11**:2 (2003), 285–99.

[109] I. Baldine and G. N. Rouskas, Traffic adaptive WDM networks: a study of reconfiguration issues. *IEEE/OSA Journal of Lightwave Technology*, **19**:4 (2001), 433–55.

[110] I. Chlamtac and A. Gumaste, Light-trails: a solution to IP centric communication in the optical domain. In *Second International Workshop on Quality of Service in Multiservice IP Networks (QoS-IP 2003)* (Heidelberg: Springer-Verlag, 2003), pp. 634–44.

[111] I. Chlamtac, A. Ganz, and G. Karmi, Lightpath communications: an approach to high bandwidth optical WANs. *IEEE Transactions on Communications*, **40**:7 (1992), 1171–82.

[112] I. Chlamtac, V. Elek, A. Fumagalli, and C. Szabo, Scalable WDM network architecture based on photonic slot routing and switched delay lines. In *Proceedings of INFOCOM '97* (New York: IEEE Press, 1997), pp. 769–76.

[113] I. P. Kaminow, C. R. Doerr, C. Dragone *et al.*, A wideband all-optical WDM network. *IEEE Journal on Selected Areas in Communications*, **14**:5 (1996), 780–99.

[114] I. Ramdani, J. Prat, and J. Comellas, Grooming in SDH/WDM mesh networks for different traffic granularities. In *Proceedings of 5th International Conference on Transparent Optical Networks* (2003).

[115] I. Widjaja, Performance analysis of burst admission control protocols. *IEEE Proceedings on Communications*, **142** (1995), 7–14.

[116] I. Widjaja, I. Saniee, L. Qian *et al.*, A new approach for automatic grooming of SONET circuits to optical express links. In *2003 IEEE International Conference on Communications* (New York: IEEE Press, 2003), pp. 1407–11.

[117] J. Anderson, J. S. Manchester, A. Rodrigues-Moral, and M. Veeraraghavan, Protocols and architectures for IP optical networking. *Bell Labs Technical Journal*, January–March (1999), 105–24.

[118] J.-C. Bermond and S. Ceroi, Minimizing SONET ADMs in unidirectional WDM rings with grooming ratio 3. *Networks*, **41**:2 (2003), 83–6.

[119] J. Doucette and W. D. Grover, Capacity design studies of span-restorable mesh transport networks with shared-risk link group (SRLG) effects. In *SPIE Optical Networking and Communications Conference (Opticomm)*, Boston, MA (Bellingham, WA: SPIE, 2002).

[120] J. Doucette and W. D. Grover, Influence of modularity and economy-of-scale effects on design of mesh-restorable DWDM networks. *IEEE Journal of Selected Areas in Communications*, **18**:10 (2000), 1912–23.

[121] J. Doucette, D. He, W. D. Grover, and O. Yang, Algorithmic approaches for efficient enumeration of candidate p-cycles and capacitated p-cycle network design. In *Proceedings of the 4th International Workshop on Design of Reliable Communication Networks*, Banff, Alberta (2003), pp. 212–20.

[122] J. Fang and A. K. Somani, Enabling sub-wavelength level traffic grooming in survivable WDM optical network design. In *Proceedings of IEEE Globecom* (New York: IEEE Press, 2003).

[123] J. Fang, W. He, and A. K. Somani, IP traffic grooming in light trail optical networks. *IEEE Journal of Selected Areas in Communications* (2004), submitted.

[124] J. H. Franz and V. K. Jain, *Optical Communications Components and Systems* (New Delhi: Narosa Publishing House, 2000).

[125] J. P. Jue, D. Datta, and B. Mukherjee, A new node architecture for scalable WDM optical networks. In *IEEE International Conference on Communications*, vol. 3 (New York: IEEE Press, 1999), pp. 1714–8.

[126] J. Jue and G. Xiao, An adaptive routing algorithm for wavelength-routed optical networks with a distributed control scheme. In *Proceedings of the Ninth International Conference on Computer Communications and Networks* (2000), pp. 192–7.

[127] J. Huang, Q. Zeng, J. Liu *et al.*, Dynamic distributed traffic grooming in optical mesh networks. *Proceedings of the SPIE*, **5247**:1 (2003), 528–36.

[128] J. Q. Hu, Diverse routing in mesh optical networks. *IEEE Transactions of Networking*, **51**:3 (2003), 489–94.

[129] J. Q. Hu, Optimal traffic grooming for wavelength-division-multiplexing rings with all-to-all uniform traffic. *Journal of Optical Networking*, **1**:1 (2002), 32–42.

[130] J.-W. Kang, J.-S. Kim, C.-M. Lee *et al.*, 1×2 all-optical switch using photochromic-doped waveguides. *Electronics Letters*, **36**:19 (2000), 1641–3.

[131] J. Kroculick and C. Hood, Provisioning multilayer resilience in multiservice optical networks. In *OptiComm: Optical Networking and Communications* (2000), pp. 30–41.

[132] J. Luciani, B. Rajagopalan, D. Awduche *et al.*, IP over optical networks a framework. *Internet Draft* (2001).

[133] J. Simmons, E. Goldstein, and A. Saleh, Quantifying the benefit of wavelength add–drop in WDM rings with distance-independent and dependent traffic. *Journal of Lightwave Technology*, Jan (1999), 48–57.

[134] J. Sonosky, Service applications for SONET DCS distributed restoration. *IEEE Journal of Selected Areas in Communications*, **12** (1994), 59–68.

[135] K. Song and A. K. Somani, Modeling and design of dependable high speed information networks. *The International Association of Science and Technology for Development Journal*, **18**:3 (1998), 214–23.

[136] K. S. Song and A. K. Somani, Fault tolerant ATM backbone network design considering cell loss rates and end-to-end delay constraints. In *IEEE Phoenix Conference on Computers and Communications* (New York: IEEE Press, 1994), pp. 119–25.

[137] K.-S. Song and A. K. Somani, Adaptive resource management for LAN interconnection in wide area ATM networks. In *Proceedings of International Conference on Computer Communications and Networks (ICCCN '94)*, San Francisco, CA (1994), pp. 148–52.

[138] K.-S. Song and A. K. Somani, Interworking connectionless service with ATM network for multimedia communication. In *5th IEEE COMSOC Workshop, MULTIMEDIA '94*, Kyoto (1994). .

[139] J. Weston-Dawkes and S. Baroni, Mesh network grooming and restoration optimized for optical bypass. In *National Fiber Optic Engineers Conference* (2002), pp. 1438–49.

[140] J. Y. Wei and R. I. McFarland, Just-in-time signaling for WDM optical burst switching networks. *Journal of Lightwave Technology*, **18**:2 (2000).

[141] J. Xu and Q. Zeng, Quantifying the benefits of traffic grooming in interconnected WDM rings using a two stage multiplexing scheme. *Proceedings of the SPIE*, **4604** (2001), 80–5.

[142] J. Yates, J. Lacey, and D. Everitt, Blocking in multiwavelength TDM networks. In *4th International Conference on Telecommunication Systems, Modeling, and Analysis* (1996), pp. 535–41.

[143] J. Yates, J. Lacey, D. Everitt, and M. Summerfield, Limited-range wavelength translation in all-optical networks. In *Proceedings of INFOCOM '96* (New York: IEEE Press, 1996), pp. 954–61.

[144] J. Turner, Terabit burst switching. *Journal of High Speed Networks*, **8**:1 (1999).

[145] J. Yates, Performance analysis of dynamically-reconfigurable wavelength-division multiplexed networks. The University of Melbourne (1997).

[146] K. Bala, T. E. Stern, D. S. Levi, and K. Bala, Routing in a linear lightwave network. *IEEE/ACM Transactions on Networking*, **3**:4 (1995).

[147] K. Chan and T. P. Yun, Analysis of least congested path routing in WDM lightwave networks. In *Proceedings of IEEE INFOCOM '94* (New York: IEEE Press, 1994), pp. 962–9.

[148] K. C. Lee and O. K. Li, A wavelength-convertible optical network. *IEEE Journal on Lightwave Technology*, **11** (1993), 962–70.

[149] K. Struyve and P. Demeester, Dynamic routing of protected optical paths in wavelength routed and wavelength translated networks. In *European Conference on Communications* (1997), pp. 151–4.

[150] K. Zhu and B. Mukherjee, Traffic grooming in an optical WDM mesh network. *IEEE Journal of Selected Areas in Communications*, **20**:1 (2002), 122–33.

[151] L. R. Ford and D. R. Fulkerson, *Flows in Networks* (Princeton, NJ: Princeton University Press, 1962).

[152] L. Kazovsky, S. Benedetto, and A. Willner, *Optical Fiber Communication Systems* (Norwood, MA: Artech House, 1996).

[153] L. Kleinrock, *Queueing Systems, Volume 2: Computer Applications* (New York: Wiley-Interscience, 1980).

[154] L. Kleinrock, *Queueing Systems, Volume 1: Theory* (New York: Wiley-Interscience, 1975).

[155] L. Li and A. K. Somani, A new analytical model for multifiber WDM networks. In *Proceedings of IEEE Globecom'99*, Rio de Janeiro (New York: IEEE Press, 1999).

[156] L. T. Kou and G. Markowsky, Multidimensional bin packing algorithms. *IBM Journal of Research and Development*, **21**:5 (1997), 443–8.

[157] L. Sahasrabuddhe, S. Ramamurthy, and B. Mukherjee, Fault management in IP-over-WDM networks: WDM protection versus IP restoration. *IEEE Journal on Selected Areas in Communications*, **20**:1 (2002), 21–33.

[158] L. G. Valiant, A scheme for fast parallel communication. *SIAM Journal on Computing*, **11** (1982), 350–61.

[159] L. Xu, H. G. Perros, and G. N. Rouskas, A simulation study of optical burst switching access protocols for WDM ring networks. *Computer Networks*, **41**:2 (2003), 143–60.

[160] L. Xu, H. G. Perros, and G. N. Rouskas, Access protocols for optical burst-switched ring networks. *Information Sciences*, **149**:1–3 (2003), 75–81.

[161] L. Xu, H. Perros, and G. Rouskas, Techniques for optical packet switching and optical burst switching. *IEEE Communications Magazine*, Jan (2001), 136–42.

[162] M. Alanyali and E. Ayanoglu, Provisioning algorithms for WDM optical networks. In *Proceedings of IEEE INFOCOM '98* (New York: IEEE Press, 1998), pp. 910–18.

[163] M. Azizoğlu, S. Subramaniam, and A. K. Somani, Converter placement on wavelength-routed network paths. In *All-Optical Communication Systems: Architecture, Control, and Network Issues III, Proceedings of SPIE*, vol. 3230, ed. J. M. Senior, R. A. Cryan, and C. Qiao (Bellingham, WA: SPIE, 1997), pp. 265–76.

[164] M. S. Borella and B. Mukherjee, Efficient scheduling of nonuniform packet traffic in a WDM/TDM local lightwave network with arbitrary tuning latencies. In *Proceedings of INFOCOM '95* (New York: IEEE Press, 1995), pp. 129–37.

[165] M. S. Borella, J. P. Jue, D. Banerjee *et al.*, Optical components for WDM lightwave networks. *Proceedings of the IEEE*, **85**:8 (1997), 1274–307.

[166] M. Brunato and R. Battiti, A multistart randomized greedy algorithm for traffic grooming on mesh logical topologies. In *Next Generation Optical Network Design and Modelling, IFIP TC6/WG6* (Dordrecht: Kluwer Academic Publishers, 2003), pp. 417–30.

[167] M. Clouqueur and W. D. Grover, Quantitative comparison of end-to-end availability of service paths in ring and mesh-restorable networks. In *Proceedings of the 19th Annual National Fiber Optics Engineers Conference (NFOEC 2003)*, Orlando, FL (2003).

[168] M. Clouqueur and W. D. Grover, Mesh-restorable networks with complete dual failure restorability and with selectively enhanced dual-failure restorability properties. In *SPIE Optical Networking and Communications Conference (Opticomm)*, Boston, MA (Bellingham, WA: SPIE, 2002).

[169] M. Csoppenzsky and A. K. Somani, Distributed routing algorithms and their performances for enhanced hypercube architecture. In *IEEE Phoenix Conference on Computers and Communications* (New York: IEEE Press, 1992), pp. 15–20.

[170] M. T. Frederick and A. K. Somani, A single-fault recovery strategy for optical networks using sub-graph routing. In *7th IFIP Working Conference on Optical Networks Design and Modeling (ONDM 2003)*, Budapest (2003).

[171] M. T. Frederick, P. Datta, and A. K. Somani, Evaluating dual-failure restorability in mesh-restorable WDM optical networks. In *Proceedings of International Conference on Computer Communications and Networks (ICCCN'94)* (2004), pp. 309–14.

[172] M. T. Frederick, N. A. VanderHorn, and A. K. Somani, Light trails: a sub-wavelength solution for optical networking. In *IEEE Workshop on High Performance Switching and Routing* (New York: IEEE Press, 2004), pp. 175–9.

[173] M. Gerla and L. Kleinrock, On the topological design of distributed computer networks. *IEEE Transactions on Communications*, **25** (1977), 48–60.

[174] M. Gerla, J. A. S. Monteiro, and R. Pazos, Topology design and bandwidth allocation in ATM nets. *IEEE Journal on Selected Areas in Communications*, **7**:8 (1989), 1253–62.

[175] M. Kodialam and T. V. Lakshman, Integrated dynamic IP and wavelength routing in IP over WDM networks. In *Proceedings of IEEE INFOCOM '01* (New York: IEEE Press, 2001).

[176] M. Kodialam and T. V. Lakshman, Dynamic routing of bandwidth guaranteed tunnels with restoration. In *Proceedings of IEEE INFOCOM '00* (New York: IEEE Press, 2000), pp. 902–11.

[177] M. Kovacevic and M. Gerla, A new optical signal routing scheme for linear lightwave networks. *IEEE Transactions on Communications*, **43**:12 (1995).

[178] M. Kovacevic and S. Acampora, Benefits of wavelength translation in all-optical clear-channel networks. *IEEE Journal of Selected Areas in Communications*, **14**:5 (1996), 868–80.

[179] M. Kovacevic and S. Acampora, On wavelength translation in all-optical networks. In *Proceedings of IEEE INFOCOM '95* (New York: IEEE Press, 1995), pp. 413–22.

[180] M. Mahony, D. Simeonidou, D. Hunter, and A. Tzanakaki, The application of optical packet switching in future communication networks. *IEEE Communication Magazine*, March (2001), 128–35.

[181] M. Schwartz, *Computer Communication Network Design and Analysis* (Englewood Cliffs, NJ: Prentice-Hall, 1977).

[182] M. Sridharan, *Design and Operation of Mesh-restorable WDM Networks*, Ph.D. dissertation. Iowa State University (2002).

[183] M. Sridharan and A. K. Somani, Design for upgradability in mesh-restorable optical networks. *Optical Networks Magazine*, May/June (2002).

[184] M. Sridharan and A. K. Somani, Revenue maximization in survivable WDM networks. In *OptiComm: Optical Networking and Communications* (2000), pp. 291–302.

[185] M. Sridharan, A. K. Somani, and M. Salapaka, Approaches for capacity and revenue optimization in survivable WDM network. *Journal of High Speed Network*, **10**:2 (2001), 109–25.

[186] M. Sridharan, R. Srinivasan, and A. K. Somani, Dynamic routing with partial information in mesh-restorable optical networks. *Sixth Working Conference on Optical Networks Design and Modelling* (2002).

[187] M. Sridharan, M. V. Salapaka, and A. K. Somani, A practical approach to operating survivable WDM networks. Special issue on WDM-based network architectures, *IEEE Journal of Selected Areas in Communications*, **20**:1 (2002), 34–46.

[188] M. Sridharan, M. V. Salapaka, and A. K. Somani, Operating mesh-survivable WDM transport networks. In *SPIE International Symposium on SPIE Terabit Optical Networking: Terabit Optical Networking* (Bellingham, WA: SPIE, 2000), pp. 113–23.

[189] M. Yoo, M. Jeong, and C. Qiao, A high speed protocol for bursty traffic in optical networks. *SPIE's All-Optical Communication Systems*, **3230** (1997), 79–90.

[190] N. G. Chattopadhyay, T. W. Morgan, and A. Raghuram, An innovative technique for backbone network design. *IEEE Transactions on Systems, Man and Cybernetics*, **19**:5 (1989), 1122–32

[191] N. Ghani, S. Dixit, and T. S. Wang, On IP-over-WDM integration. *IEEE Communications Magazine*, **38**:3 (2000), 72–84.

[192] N. Ghani and S. Dixit, Channel provisioning for higher layer protocols in WDM networks. *SPIE International Symposium on SPIE Terabit Optical Networking: Terabit Optical Networking* (Bellingham, WA: SPIE, 1999).

[193] N. Jose and A. K. Somani, Connection rerouting/network reconfiguration. In *Proceedings of Design of Reliable Communication Networks (DRCN) Workshop*, Banff (2003).

[194] N. Shacham and J. S. Meditch, An algorithm for optimal multicast of multimedia streams over heterogeneous networks. *Proceedings of IEEE INFOCOM '94*, Toronto (New York: IEEE Press, 1994), pp. 856–64.

[195] N. Wauters and P. Demeester, Wavelength conversion in optical multi-wavelength multi-fiber transport networks. *International Journal of Optoelectronics*, **11**:1 (1997), 53–70.

[196] N. Wauters and P. Demeester, Is wavelength conversion required? *IEEE International Conference on Communications '96* (New York: IEEE Press, 1996).

[197] O. Crochat, J. Y. Boudec, and O. Gerstel, Protection interoperability for WDM optical networks. *IEEE/ACM Transactions on Networking*, **8**:3 (2000), 384–95.

[198] O. Crochat and J. Y. Boudec, Design protection for WDM optical networks. *IEEE Journal of Selected Areas in Communications*, **16**:7 (1998), 1158–65.

[199] O. Gerstel, P. Lin, and G. Sasaki, Combined WDM and SONET design. In *Proceedings of IEEE INFOCOM '99* (New York: IEEE Press, 1999), pp. 734–43.

[200] O. Gerstel, P. J. Lin, and G. H. Sasaki, Wavelength assignment in WDM rings to minimize system cost instead of number of wavelengths. *Proceedings of IEEE INFOCOM '98*, vol. 1 (New York: IEEE Press, 1998), pp. 94–101.

[201] O. Gerstel, R. Ramaswami, and G. H. Sasaki, Cost-effective traffic grooming in WDM rings. *IEEE Transactions on Networking*, **8**:5 (2000), 618–30.

[202] O. Gerstel and R. Ramaswami, Optical layer survivability: a services perspective. *IEEE Communications* **38**:3 (2000), 104–13.

[203] O. Gerstel, R. Ramaswami, and G. Sasaki, Cost effective traffic grooming in WDM rings. In *Proceedings of IEEE INFOCOM '98* (New York: IEEE Press, 1998), pp. 69–77.

[204] O. Gerstel and R. Ramaswami, Optical layer survivability – an implementation perspective. *IEEE Journal of Selected Areas in Communications*, **18**:10 (2000), 1885–99.

[205] O. Gerstel and S. Kutten, Dynamic wavelength allocation in all-optical ring networks. In *Proceedings of the IEEE International Conference on Communications (ICC 1997)* (New York: IEEE Press, 1997), pp. 432–6.

[206] P. Batchelor, B. Daino, P. Heinzmann *et al.*, Ultra high capacity optical transmission networks. *Final report of action COST 239* (1999).

[207] P. Bonenfant, A. Rodrigues-Moral, and J. S. Manchester, *IP over WDM: the Missing Link.* Technical Report, white paper, Lucent Technologies (1999).

[208] P. Datta, M. T. Frederick, and A. K. Somani, Sub-graph routing: a novel fault-tolerant architecture for shared-risk link group failures in WDM optical networks. In *Proceedings of Design of Reliable Communication Networks (DRCN) Workshop*, Banff (2003).

[209] P. Datta, M. Sridharan, and A. K. Somani, A simulated annealing approach for topology planning and evolution of mesh-restorable optical networks. *Eighth IFIP Working Conference on Optical Networks Design and Modeling* (2003).

[210] P. F. Fonseca, Pan-European multi-wavelength transport networks: network design, architecture, survivability and SDH networking. *Proceedings of the 1st International Workshop on Reliable Communication Networks*, Paper P3 (1998).

[211] P. Gambini, M. Renaud, C. Guillemot *et al.*, Transparent optical packet switching: network architecture and demonstrators in the KEOPS project. *IEEE Journal on Selected Areas in Communications*, **16**:7 (1998), 1245–59.

[212] P. E. Green, Jr., *Fiber Optic Networks* (Englewood Cliffs, NJ: Prentice Hall, 1992).

[213] P. Ho and H. T. Mouftah, A framework for service-guaranteed shared protection in WDM mesh networks. *IEEE Communications Magazine*, **40**:2 (2002), 97–103.

[214] P. Sebos, J. Yates, D. Rubenstein, and A. Greenberg, Effectiveness of shared risk link group auto-discovery in optical networks. In *OFC'02* (Washington, DC: Optical Society of America, 2002), pp. 493–5.

[215] P. J. Wan, L. Liu, and O. Frieder, Grooming of arbitrary traffic in SONET/WDM rings. *Proceedings of IEEE Globecom '99* (New York: IEEE Press, 1999), pp. 1012–6.

[216] P. J. Wan, G. Calinescu, L. Liu, and O. Frieder, Grooming of arbitrary traffic in SONET/WDM BLSRs. In *Journal on Selected Areas in Communications*, **18** (2000), 1995–2003.

[217] R. K. Ahuja, T. L. Magnanti, and J. B. Orlin, *Network Flows: Theory, Algorithms, and Applications* (Englewood Cliffs, NJ: Prentice Hall, 1993).

[218] R. A. Barry and P. A. Humblet, Models of blocking probability in all-optical networks with and without wavelength changers. *IEEE Journal of Selected Areas in Communications*, **14**:5 (1996), 858–67.

[219] R. Battiti and M. Brunato, Reactive search for traffic grooming in WDM networks. In *Evolutionary Trends of the Internet. Proceedings of 2001 Tyrrhenian International Workshop on Digital Communications* (Berlin: Springer-Verlag, 2001), pp. 56–66.

[220] R. Berry and E. Modiano, Reducing electronic multiplexing costs in SONET/WDM rings with dynamically changing traffic. *Journal on Selected Areas in Communications*, **18** (2000), 1961–71.

[221] R. Berry and E. Modiano, Minimizing electronic multiplexing costs for dynamic traffic in unidirectional SONET ring networks. In *Proceedings of the IEEE International Conference on Communications (ICC 1999)* (New York: IEEE Press, 1999).

[222] R. Bhandari, *Survivable Networks: Algorithms for Diverse Routing* (Dordrecht: Kluwer Academic Publishers, 1999).

[223] R. Bellman, On a routing problem. *Quarterly Journal of Applied Mathematics*, **16** (1958), 87–90.

[224] R. Doverspike and B. Wilson, Comparison of capacity efficiency of DCS network restoration routing techniques. *Journal of Network and System Management*, **2**:2 (1994), 95–123.

[225] R. Dutta and G. N. Rouskas, Traffic grooming in WDM networks: past and future. *IEEE Network*, **16**:6 (2002), 46–56.

[226] R. Dutta and G. N. Rouskas, On optical traffic grooming in WDM rings. *IEEE Journal of Selected Areas in Communications*, **20**:1 (2002), 110–21.

[227] R. Dutta and G. N. Rouskas, Topology design in WDM rings to minimize electronic routing: efficient computation of tight bounds. In *Proceedings of the 2000 Allerton Conference on Communication, Control, and Computing* (2000), pp. 1284–93.

[228] R. Jain and S. Dharanikota, Internet protocol over DWDM – recent developments, trends and issues. In *Global Optical Communications – Business Briefing*, London, World Market Research Centre Ltd (www.wmrc.com) (2001).

[229] R. Lingampalli and P. Vengalam, Effect of wavelength and waveband grooming on all-optical networks with single layer photonic switching. In *Optical Fiber Communications Conference* (Washington, DC: Optical Society of America, 2002), pp. 501–2.

[230] R. Mahalati and R. Dutta, Reconfiguration of traffic grooming optical networks. In *BroadNets 2004: First International Conference on Broadband Networks*, San José (New York: IEEE Press, 2004).

[231] R. Parthiban, R. S. Tucker, and C. Leckie, Waveband grooming and IP aggregation in optical networks. *Journal of Lightwave Technology*, **21**:11 (2003), 2476–88.

[232] R. Perlman, *Interconnections: Bridges, Routers, Switches, and Internetworking Protocols*, 2nd edn (Reading, MA: Addison-Wesley, 1999).

[233] R. Ramaswami and A. Segall, Distributed network control for wavelength routed optical networks. *IEEE Transactions on Networking*, **5**:6 (1996), 936–43.

[234] R. Ramaswami and G. Sasaki, Multiwavelength optical networks with limited wavelength conversion. *IEEE/ACM Transactions on Networking*, **6**:6 (1998), 744–54.

[235] R. Ramaswami and K. N. Sivarajan, *Optical Networks: a Practical Perspective*, (San Francisco, CA: Morgan Kaufmann, 1998).

[236] R. Ramaswamy and K. N. Sivarajan, Routing and wavelength assignment in all-optical networks. *IEEE/ACM Transactions on Networking*, **3**:5 (1995), 489–500.

[237] R. Ranganathan, L. Blair, and J. Berthold, Benefits of grooming capable cross-connects in a paneuropean optical network. In *Proceedings 27th European Conference on Optical Communication* (2001), pp. 38–9.

[238] R. Ramamurthy, Z. Bogdanowicz, S. Samieian *et al.*, Capacity performance of dynamic provisioning in optical networks. *Journal of Lightwave Technology*, **19**:1 (2001), 40–8.

[239] R. Ramaswami, Multiwavelength lightwave networks for computer communication. *IEEE Communications Magazine*, **31**:2 (1993), 78–88.

[240] R. C. Prim, Shortest connection networks and some generalizations. *Bell System Technical Journal*, **36** (1957), 1389–401.

[241] R. Srinivasan, MICRON: a framework for connection establishment in optical networks. In *Proceedings of OPTICOMM*, Dallas (2003), pp. 139–50.

[242] R. Srinivasan and A. K. Somani, On achieving fairness and efficiency in high-speed shared-medium access. *IEEE Transactions on Networking*, **11**:1 (2003), 111–24.

[243] R. Srinivasan and A. K. Somani, Dynamic routing in WDM grooming networks. *Photonic Network Communications*, **5**:2 (2003), 123–35.

[244] R. Srinivasan and A. K. Somani, Analysis of multi-rate traffic in WDM grooming networks. In *Proceedings of International Conference on Computer Communications and Networks (ICCCN'02)*, Miami (2002), pp. 296–301.

[245] R. Srinivasan and A. K. Somani, Request-specific routing in WDM grooming networks. In *Proceedings of the IEEE International Conference on Communications (ICC 2002)*, vol. 5 (New York: IEEE Press, 2002), pp. 2876–80.

[246] R. Srinivasan and A. K. Somani, *Dynamic Routing in WDM Grooming Networks.* Technical Report (DCNL-ON-2001-001), Dependable Computing and Networking Laboratory, Department of Electrical and Computer Engineering, Iowa State University (2001).

[247] R. Srinivasan and A. K. Somani, A generalized framework for analyzing time-space switched optical networks. In *Proceedings of IEEE INFOCOM '01* (New York: IEEE Press, 2001), pp. 179–88.

[248] R. E. Tarjan, *Data Structures and Network Algorithms* (Philadelphia, PA: Society for Industrial and Applied Mathematics, 1983).

[249] S. Amstutz, Burst switching – an introduction. *IEEE Communications*, November (1983), 36–42.

[250] S. B. Choi and A. K. Somani, Design and performance analysis of load-distributing fault-tolerant network. *IEEE Transactions on Computers*, **45**:5 (1996), 540–51.

[251] S. B. Choi and A. K. Somani, Rearrangeable hypercube architecture for routing permutations. *Journal of Parallel Distributed Computing*, **19** (1993), 125–33.

[252] S. B. Choi and A. K. Somani, The generalized hyper-cube. In *Proceedings of ICPP-90* (1990), pp. I/372–5.

[253] S. Diez, Why should we have four-wave mixing? *Photonic Devices for Telecommunications*, ed. G. Guekos (Berlin: Springer-Verlag, 1999), pp. 273–9.

[254] S. E. Dreyfus, An appraisal of some shortest-path algorithms. *Operations Research*, **17** (1969), 395–412.

[255] S. Han and K. G. Shin, Efficient spare-resource allocation for fast restoration of real-time channels from network component failures. In *Proceedings of 18th IEEE Real-Time Systems Symposium* (New York: IEEE Press, 1997), pp. 99–108.

[256] S. Huang and R. Dutta, Research problems in dynamic traffic grooming in optical networks. In *Proceedings of First International Workshop on Traffic Grooming*, San José (2004).

[257] S. V. Kartalopoulos, *Introduction to DWDM Technology* (New York: IEEE Press, Bellingham, WA: SPIE Optical Engineering, 2000).

[258] S. Kim and S. S. Lumetta, Evaluation of protection reconfiguration for multiple failures in optical networks. In *Proceedings of the Optical Fiber Communication Conference*, Atlanta, GA (2003).

[259] S. Kini, M. Kodialam, T. V. Lakshman, S. Sengupta, and C. Villamizar, *Shared Backup Label Switched Path Restoration*. IETF, Internet draft (2001).

[260] S. Ramamurthy and B. Mukherjee, Survivable WDM mesh networks, part i – protection. *Proceedings of IEEE INFOCOM '99*, vol. 2 (New York: IEEE Press, 1999), pp. 744–51.

[261] S.-S. Roh, W.-H. So, and Y.-C. Kim, Design and performance evaluation of traffic grooming algorithms in WDM multi-ring networks. *Photonic Network Communications*, **3**:4 (2001), 335–48.

[262] S. Ramamurthy and B. Mukherjee, Fixed alternate routing and wavelength conversion in wavelength-routed optical networks. In *Proceedings of the Global Telecommunications Conference, GLOBECOM'98* (1998), pp. 2295–303.

[263] S. Ramesh, G. N. Rouskas, and H. G. Perros, Computing blocking probabilities in multi-class wavelength routing networks with multicast calls. *IEEE Journal on Selected Areas in Communications*, **20**:1 (2002), 89–96.

[264] S. Subramaniam, Role of wavelength converters in all-optical networks. Iowa State University (1997).

[265] S. Subramaniam, H. Choi, and H. A. Choi, On double-link failure recovery in WDM optical networks. *Proceedings of IEEE INFOCOM '02* (New York: IEEE Press, 2002), pp. 808–16.

[266] S. Subramaniam, A. K. Somani, M. Azizoğlu, and R. A. Barry, A performance model for wavelength conversion with non-Poisson traffic. In *Proceedings of IEEE INFOCOM '97* (New York: IEEE Press, 1997).

[267] S. Subramaniam, M. Azizoğlu, and A. K. Somani, On the optimal converter placement in wavelength-routed networks. *IEEE/ACM Transactions on Networking*, **7**:5 (1999), 754–66.

[268] S. Subramaniam, M. Azizoğlu, and A. K. Somani, On the optimal placement of wavelength converters in wavelength-routed networks. In *Proceedings of IEEE INFOCOM '98* (New York: IEEE Press, 1998), pp. 902–9.

[269] S. Subramaniam, M. Azizoğlu, and A. K. Somani, All-optical networks with sparse-wavelength conversion. *IEEE/ACM Transactions on Networking*, **4**:4 (1996), 544–57.

[270] S. Subramaniam, M. Azizoğlu, and A. K. Somani, Connectivity and sparse wavelength conversion in wavelength-routing networks. In *Proceedings of IEEE INFOCOM '96* (New York: IEEE Press, 1996), pp. 148–55.

[271] S. Subramaniam, M. Azizoğlu, and A. K. Somani, Effect of wavelength converter density on the blocking performance of all-optical networks. In *Proceedings of LEOS* (1995), pp. 210–11.

[272] S. Subramaniam and R. A. Barry, Wavelength assignment in fixed routing WDM networks. In *IEEE International Conference on Communications* (New York: IEEE Press, 1997), pp. 406–10.

[273] S. Tridandapani, J. S. Meditch, and A. K. Somani, The MaTPi protocol: masking tuning times through pipelining in WDM optical networks. In *Proceedings of IEEE INFOCOM '94*, Toronto (New York: IEEE Press, 1994), pp. 1528–35.

[274] S. Thiagarajan and A. K. Somani, Traffic grooming for survivable WDM mesh networks. *OptiComm: Optical Networking and Communications* (2001).

[275] S. Thiagarajan and A. K. Somani, Performance analysis of WDM networks with grooming capabilities. In *Proceedings of the SPIE Technical Conference on Terabit*

Optical Networking: Architecture, Control, and Management Issues (Bellingham, WA: SPIE, 2000), pp. 253–62.

[276] S. Thiagarajan and A. K. Somani, A capacity correlation model for WDM networks with constrained grooming capabilities. In *Proceedings of the IEEE International Conference on Communications (ICC 2001)*, vol. 5 (New York: IEEE Press, 2001), pp. 1592–6.

[277] S. Thiagarajan and A. K. Somani, Capacity fairness of WDM networks with grooming capabilities. *Optical Network Magazine*, **2**:3 (2001), 24–31.

[278] S. Thiagarajan and A. K. Somani, An efficient algorithm for optimal wavelength converter placement on wavelength-routed networks with arbitrary topologies. In *Proceedings of IEEE INFOCOM '99* (New York: IEEE Press, 1999).

[279] S. Thiagarajan, *Traffic Grooming and Wavelength Conversion in Optical Networks*. Ph.D. dissertation, Iowa State University (2001).

[280] S. Verma, H. Chaskar, and R. Ravikanth, Optical burst switching: a viable solution for terabit IP backbone. *IEEE Network*, November/December (2000), 48–53.

[281] S. J. B. Yoo, Wavelength conversion technologies for WDM network applications. *Journal of Lightwave Technology*, **14**:6 (1996), 955–66.

[282] T. Cinkler, D. Marx, C. P. Larsen, and D. Fogaras, Heuristic algorithms for joint configuration of the optical and electrical layer in multi-hop wavelength routing networks. In *Proceedings of IEEE INFOCOM '00*, Tel Aviv (New York: IEEE Press, 2000).

[283] T. Cinkler, R. Castro, and S. Johansson, Configuration and re-configuration of WDM networks. In *European Conference on Networks and Optical Communications*, Manchester (1998).

[284] T.-Y. Feng, A survey of interconnection networks. *Computer*, **14** (1981), 12–27.

[285] T. Nishida and H. Miyahara, Fault tolerant packet switched network design using capacity augmentation. In *Proceedings of IEEE INFOCOM '88* (New York: IEEE Press, 1988).

[286] T. D. Ndousse and N. Golmie, Differentiated optical services: quality of optical service model for WDM networks. *SPIE All Optical Networking: Architecture, Control and Management Issue* (Bellingham, WA: SPIE, 1999), pp. 79–88.

[287] T. Tripathi and K. N. Sivarajan, Computing approximate blocking probabilities in wavelength-routed all-optical networks with limited-range wavelength conversion. **1** (1999), 329–36.

[288] T. H. Wu, *Fiber Network Service Survivability* (Norwood, MA: Artech House, 1992).

[289] T.-H. Wu, D. T. Kong, and R. C. Lau, An economic feasibility study for a broadband virtual path SONET/ATM self-healing ring architecture. *IEEE Journal on Selected Areas in Communications*, **10**:9 (1992), 1459–73.

[290] T. M. Zhang and A. K. Somani, DIRSMIN: a dilated fault-tolerant switch design for B-ISDN applications. In *Proceedings of IEEE INFOCOM '95*, Boston, MA (New York: IEEE Press, 1995).

[291] V. Anand and C. Qiao, Dynamic establishment of protection paths in WDM networks. *Proceedings of International Conference on Computer Communications and Networks (ICCCN'00)* (2000), 198–204.

[292] V. E. Benes, *Mathematical Theory of Connecting Networks and Telephone Traffic* (New York: Academic, 1965).

[293] V. R. Konda and T. Y. Chow, Algorithm for traffic grooming in optical networks to minimize the number of transceivers. In *IEEE Workshop on High Performance Switching and Routing 2001* (New York: IEEE Press, 2001), pp. 218–21.

[294] V. Sharma and E. A. Varvarigos, Limited wavelength translation in all-optical WDM mesh networks. In *Proceedings of IEEE INFOCOM '98* (New York: IEEE Press, 1998), pp. 893–901.

[295] W. D. Grover, The self-healing network: a fast distributed restoration technique for networks using digital cross-connect machines. In *Proceedings of IEEE Globecom* (New York: IEEE Press, 1987), pp. 28.2.1–6.

[296] W. D. Grover, The protected working capacity envelope concept: an alternative paradigm for automated service provisioning. *IEEE Communications Magazine*, January (2004), 62–9.

[297] W. D. Grover, *Mesh-based Survivable Networks* (Englewood Cliffs, NJ: Prentice-Hall, 2003).

[298] W. D. Grover and D. Stamatelakis, Cycle-oriented distributed preconfiguration: ring-like speed and mesh-like capacity for self-planning network restoration. In *Proceedings of the IEEE International Conference on Communications (ICC 1998)*, vol. 1 (New York: IEEE Press, 1998), pp. 737–43.

[299] W. D. Grover and D. Stamatelakis, Bridging the ring-mesh dichotomy with p-cycles. In *Proceedings of DRCN workshop* (2000), pp. 92–104.

[300] W. D. Grover and J. Doucette, Advances in optical network design with p-cycles: joint optimization and pre-selection of candidate p-cycles. In *Proceedings of IEEE/LEOS Summer Topicals*, Mont Tremblant, PQ, Canada (2002), pp. 49–50 (paper WA2).

[301] W. D. Grover and D. Stamatelakis, Bridging the ring-mesh dichotomy with p-cycles. In *Proceedings of DRCN Workshop* (2000).

[302] W. He and A. K. Somani, Path-based protection for surviving double link failures. In *Proceedings of Globecom 2002* (New York: IEEE Press, 2002).

[303] W. He, M. Sridharan, and A. K. Somani, Capacity optimization for tolerating double link failures in WDM mesh optical networks. *Proceedings of SPIE*, **4874** (2002), 13–25.

[304] W. Qiangmin, R. Mengtian, and Z. Hongwen, Traffic grooming in an optical network based on minimizing number of transceivers. *Acta Photonica Sinica*, **32**:8 (2003), 936–9.

[305] X.-Y. Li, L. Liu, P.-J. Wan, and O. Frieder, Practical traffic grooming scheme for single-hub SONET/WDM rings. In *Proceedings of IEEE Conference on Local Computer Networks* (New York: IEEE Press, 2000), pp. 556–63.

[306] Y. Rekhter, B. Davie, D. Katz *et al.*, *Cisco Systems Tag Switching Architecture Overview*. Network Working Group Request for Comments: 2105 (1997).

[307] Y. Xin and G. N. Rouskas, A study of path protection in large-scale optical networks. *Photonic Network Communications*, **7**:4 (2004).

[308] X. Zhang and C. Qiao, An effective and comprehensive approach for traffic grooming and wavelength assignment in SONET/WDM rings. *IEEE/ACM Transactions on Networking*, **8** (2000).

[309] X. Zhang and C. Qiao, Wavelength assignment for dynamic traffic in multi-fiber WDM ring networks. In *Proceedings of International Conference on Computer Communications and Networks (ICCCN'98)* (1998), pp. 479–85.

[310] Y. Katsube, K. Nagami, and H. Esaki, *Toshiba's Router Architecture Extensions for ATM Overview*. Network Design Group Request for Comments: 2098 (1997).

[311] Y. Miyao and H. Saito, Optimal design and evaluation of survivable WDM transport networks. *IEEE Journal of Selected Areas in Communications*, **16**:7 (1998), 1190–8.

[312] Y. Yamada, K. Sasayama, K. Habara *et al.*, Optical output buffered ATM switch prototype based on FRONTIERNET architecture. *IEEE Journal on Selected Areas in Communications*, **16**:7 (1998), 1298–307.

[313] Z. Zhang and A. S. Acampora, A heuristic wavelength assignment algorithm for multihop WDM networks with wavelength routing and wavelength re-use. *IEEE/ACM Transactions on Networking*, **3**:3 (1995), 281–8.

[314] Z. Zhang, W.-D. Zhong, and B. Mukherjee, A heuristic method for design of survivable WDM networks with p-cycles. *IEEE Communications Letters*, **8**:7 (2004), 467–9.

Index

Printed in the United States
By Bookmasters